Frontiers in Nanomedicine

(Volume 2)

(Nanomedicine and Neurosciences: Advantages, Limitations and Safety Aspects)

Editor

Giovanni Tosi

Department of Life Sciences, University of Modena and Reggio Emilia, Via Campi 103, 41124 Modena, Italy

Frontiers in Nanomedicine

Volume #2

Nanomedicine and Neurosciences: Advantages, Limitations and Safety Aspects

Editor: Giovanni Tosi (Italy)

ISSN (print): 2405-9129

ISSN (online): 2405-9137

ISBN (online): 978-1-68108-493-0

ISBN (print): 978-1-68108-494-7

General:

1. Any dispute or claim arising out of or in connection with this License Agreement or the Work (including non-contractual disputes or claims) will be governed by and construed in accordance with the laws of the U.A.E. as applied in the Emirate of Dubai. Each party agrees that the courts of the Emirate of Dubai shall have exclusive jurisdiction to settle any dispute or claim arising out of or in connection with this License Agreement or the Work (including non-contractual disputes or claims).

2. Your rights under this License Agreement will automatically terminate without notice and without the need for a court order if at any point you breach any terms of this License Agreement. In no event will any delay or failure by Bentham Science Publishers in enforcing your compliance with this License Agreement constitute a waiver of any of its rights.

3. You acknowledge that you have read this License Agreement, and agree to be bound by its terms and conditions. To the extent that any other terms and conditions presented on any website of Bentham Science Publishers conflict with, or are inconsistent with, the terms and conditions set out in this License Agreement, you acknowledge that the terms and conditions set out in this License Agreement shall prevail.

Bentham Science Publishers Ltd.
Executive Suite Y - 2
PO Box 7917, Saif Zone
Sharjah, U.A.E.
Email: subscriptions@benthamscience.org

**BENTHAM
SCIENCE**

CONTENTS

FOREWORD

Although neurosurgery is one of the youngest surgical specialties, it is also the one that has undergone the most dramatic progress in recent years. As neurosurgery interfaces with other surgical areas, in particular ear, nose, throat specialty (ENT), ophthalmology and orthopedic surgery, new subspecialties have aroused such as: otoneurosurgery, neuro-ophthalmology and neuro-orthopedic spinal surgery. More particularly, new surgical approaches have appeared thanks to the improvements made in the areas of medical devices, imaging and information technology.

Nonetheless, surgery of the nervous system and the spine still has to face therapeutic challenges, including the incurability of most cerebral tumors, low back pain and its socioeconomic impact, as well as the neurodisability associated with the evolution of a large number of afflictions of the nervous system. These challenges can only be addressed through a new technological revolution.

For instance, Huntington's disease (HD) is an incurable neurodegenerative genetic disorder manifesting in adulthood and causing motor, psychiatric and cognitive disturbances. It is caused by a mutation in the huntingtin gene (htt), which at first leads to the degeneration of striatal GABAergic neurons and then to other neuronal areas. This mutation (mhtt) is involved in repression of several neuronal genes, particularly brain-derived neurotrophic factor. The use of trophic factors, targeting particularly BDNF in a neuronal protection strategy, may be particularly relevant for the treatment of HD where genetic screening can identify individuals at risk, providing a unique opportunity to intervene early in the onset of striatal degeneration.

The "NBIC convergence" (convergence between Nanotechnology, Biotechnology, Information technology and Cognitive sciences) is a concept that appeared in 2002, in a report from the National Science Foundation. This concept appeared following a reflection on the potential impact of this convergence in the improvement of human capabilities, both at the individual and societal level. While this new concept, in particular its potential applications, has generated a philosophical and ethical debate, it has already been a source of progress in health technologies.

For the first time, this e-book aims to depict the state of the art using nanotechnologies as a promising tool for therapy and diagnosis of neurodegenerative diseases. It focuses on anatomy and pathology of the main related-diseases, and gives a clear overview of the last advances in the so-called nanomedicine as to target the blood brain barrier or to image the brain defects accurately. All main issues linked to the development of new nanomedicine platforms (liposomes, targeting molecules, nanoconjugates…) and their fate, *in vivo,* (biopharmaceutical performances, interaction with biological media, toxicity…) are clearly presented with a translational approach.

This e-book gives the reader a perfect overview of this very exciting field of medical research. It is intended to help scientists, technologists, and students who may use or need to use some aspects of nanomedicine in their work or who wish to be trained in this emerging and promising area of investigation.

Frank Boury & Philippe Menei
INSERM U1066,
University & University hospital, Angers, France

PREFACE

The era of nanomedicine is claimed to be effective now, in these years. But we experiment that is not true, in the field of medicine in particular. Obviously, there is a plethora of papers published in the major scientific and highly impacted journals, but it is not enough to claim clearly that medical application of nanotechnology is currently on the edge of technological approaches.

Considering the brain, there are several pathological changes affecting the Central Nervous System (CNS): **neurodegenerative** (Alzheimer's, Parkinson's, retinal degeneration), **neurological/neuropsychiatric** (epilepsy, amyotrophic lateral sclerosis, autism), **brain tumors** (gliomas, astrocytomas, etc.) and **rare neurometabolic disorders** (i.e. inherited Lysosomal Storage Diseases), all considered **major contributors to human death**. Neurological disease management deeply impacts on patients health, care providers activity and represents a substantial socio-economic burden. Due to the absence of targeted and cost-effective therapies and limited diagnostic tools, the costs to the national health systems are high.

For disease management, it is fundamental to achieve a **deeper understanding of basic neurobiology underlying each distinct disorder** and as an urgent unmet need, to develop **novel targeted therapeutic strategies**. *Nanomedicine* represents a powerful new approach providing novel carriers to deliver drugs to specific sites in the brain as well as to other organs (lung/liver/ breast/ tumor sites). Only a **joint multidisciplinary research coordination** effort can facilitate the full development of nanomedicine as valuable treatment and diagnosis strategy for these diseases.

This book provides for the first step in order to "fill in the blanks" with a part of aspects which should be considered in order to produce and propose a real and applicable *nanomedicine* for the cure of neurodegenerative disorders and neurological diseases.

With critical behavior and with deep knowledge, scientists of high experience and skills in their field of research or clinical settings analyzed different aspects of nanomedicine for brain delivery and targeting of drugs.

From the bases of neurodegenerative diseases as anatomy and pathology of brain disorders, this book opens to wide overview of the applications of nanosystem to brain disorders, addressed by means of application of nanomedicines to neuronopathic lysosomal storage disorders, or by the application of nanoparticles to target mitochondria in neurodegenerative diseases. Drug delivery to the brain by liposomal systems along with nanotopography sensing are also approached, together with the targeting of nanomedicine in mucopolysaccharidoses and brain compromission, along with the validation of drug nanoconjugates in vivo. Finally, safety aspects and benefit/risk focus is given by means of analysis of protein corona affecting the *in vivo* efficiency of polymeric NPs and neuroendocrine aspects of nanoparticles into neurodegeneration.

Giovanni Tosi
Associate Professor, Department of Life Sciences,
University of Modena and Reggio Emilia,
Via Campi 103, 41124 Modena,
Italy

List of Contributors

Tasnuva Sarowar Neurology Dept., Neurocenter of Ulm University, WG Molecular Analysis of Synaptopathies, Ulm, Germany

Andreas M. Grabrucker Neurology Dept., Neurocenter of Ulm University, WG Molecular Analysis of Synaptopathies, Ulm, Germany
Ulm University, Institute for Anatomy and Cell Biology, Ulm, Germany

Michal Cagalinec Estonia Centre of Biosciences (former Institute of Molecular Physiology and Genetics), Slovak Academy of Sciences, Department of Pharmacology, Faculty of Medicine, University of Tartu, Tartu, Estonia Centre of Biosciences , Bratislava, Slovakia

Tonazzini Ilaria Istituto Nanoscienze-CNR and Scuola Normale Superiore, Piazza San Silvestro 12, NEST, 56127 Pisa, Italy

Cecchini Marco Istituto Nanoscienze-CNR and Scuola Normale Superiore, Piazza San Silvestro 12, NEST, 56127 Pisa, Italy
Piazza Velasca 5, Fondazione Umberto Veronesi, 20122 Milano, Italy

Anne Mahringer Institute of Pharmacy and Molecular Biotechnology, Im Neuenheimer Feld 329, Ruprecht-Karls University, 69120 Heidelberg, Germany

Gert Fricker Institute of Pharmacy and Molecular Biotechnology, Im Neuenheimer Feld 329, Ruprecht-Karls University, 69120 Heidelberg, Germany

Cinzia M. Bellettato Department of Women and Children Health, Via Giustiniani 3, Brains for Brains Foundation, 35128 Padova, Italy

David J. Begley Department of Women and Children Health, Via Giustiniani 3, Brains for Brains Foundation, 35128 Padova, Italy
Institute of Pharmaceutical Science, Franklin-Wilkins Building, 150 Stamford Street, Kings College London, London, SE1 9NH, UK

Christina Lampe Institute of Pharmaceutical Science, Franklin-Wilkins Building, 150 Stamford Street, Kings College London, London, SE1 9NH, UK
Department of Child and Adolescent Medicine, Center for Rare Diseases Helios Horst Schmidt Kliniken, Ludwig-Erhard-Straße 100 65199 Wiesbaden, Germany

Maurizio Scarpa Institute of Pharmaceutical Science, Franklin-Wilkins Building, 150 Stamford Street, Kings College London, London, SE1 9NH, UK
Department of Child and Adolescent Medicine, Center for Rare Diseases Helios Horst Schmidt Kliniken, Ludwig-Erhard-Straße 100 65199 Wiesbaden, Germany
Department of Women and Children Health, Via Giustiniani 3, University of Padova, 35128 Padova, Italy

Marika Salvalaio Department of Women's and Children's Health, University of Padova, Laboratory of Diagnosis and Therapy of Lysosomal Disorders, Padova, Italy
Institute "Città della Speranza", Pediatric Research, Padova, Italy

Laura Rigon Department of Women's and Children's Health, University of Padova, Laboratory of Diagnosis and Therapy of Lysosomal Disorders, Padova, Italy

Francesca D'Avanzo Department of Women's and Children's Health, University of Padova, Laboratory of Diagnosis and Therapy of Lysosomal Disorders, Padova, Italy
Foundation Onlus, Brains for Brain, Padova, Italy

Elisa Legnini	Department of Women's and Children's Health, University of Padova, Laboratory of Diagnosis and Therapy of Lysosomal Disorders, Padova, Italy
Valeria Balmaceda Valdez	Department of Women's and Children's Health, University of Padova, Laboratory of Diagnosis and Therapy of Lysosomal Disorders, Padova, Italy
Alessandra Zanetti	Department of Women's and Children's Health, University of Padova, Laboratory of Diagnosis and Therapy of Lysosomal Disorders, Padova, Italy
Rosella Tomanin	Department of Women's and Children's Health, University of Padova, Laboratory of Diagnosis and Therapy of Lysosomal Disorders, Padova, Italy
Ibane Abasolo	Vall d'Hebron Institut de Recerca (VHIR). Networking Research Center on Bioengineering, Biomaterials and Nanomedicine (CIBER-BBN), Functional Validation & Preclinical Research (FVPR). Drug Delivery & Targeting. CIBBIM-Nanomedicine, Barcelona, Spain
Yolanda Fernández	Vall d'Hebron Institut de Recerca (VHIR). Networking Research Center on Bioengineering, Biomaterials and Nanomedicine (CIBER-BBN), Functional Validation & Preclinical Research (FVPR). Drug Delivery & Targeting. CIBBIM-Nanomedicine, Barcelona, Spain
Simó Schwartz Jr.	Vall d'Hebron Institut de Recerca (VHIR). Networking Research Center on Bioengineering, Biomaterials and Nanomedicine (CIBER-BBN), Functional Validation & Preclinical Research (FVPR). Drug Delivery & Targeting. CIBBIM-Nanomedicine, Barcelona, Spain
Francesca Pederzoli	Department of Life Sciences, University of Modena and Reggio Emilia, Via Campi 103, Te.Far.T.I, 41124 Modena, Italy
Marianna Galliani	Department of Life Sciences, University of Modena and Reggio Emilia, Via Campi 103, Te.Far.T.I, 41124 Modena, Italy
Flavio Forni	Department of Life Sciences, University of Modena and Reggio Emilia, Via Campi 103, Te.Far.T.I, 41124 Modena, Italy
Maria Angela Vandelli	Department of Life Sciences, University of Modena and Reggio Emilia, Via Campi 103, Te.Far.T.I, 41124 Modena, Italy
Daniela Belletti	Department of Life Sciences, University of Modena and Reggio Emilia, Via Campi 103, Te.Far.T.I, 41124 Modena, Italy
Giovanni Tosi	Department of Life Sciences, University of Modena and Reggio Emilia, Via Campi 103, Te.Far.T.I, 41124 Modena, Italy
Barbara Ruozi	Department of Life Sciences, University of Modena and Reggio Emilia, Via Campi 103, Te.Far.T.I, 41124 Modena, Italy
Eva Rollerova	Faculty of Public Health, Department of Toxicology and Faculty of Medicine, Laboratory of Immunotoxicology, Slovak Medical University, Bratislava, Slovak Republic
Alzbeta Bujnakova Mlynarcikova	Institute of Experimental Endocrinology, Biomedical Research Center Slovak Academy of Sciences, Bratislava, Slovak Republic
Jana Tulinska	Faculty of Public Health, Department of Toxicology and Faculty of Medicine, Laboratory of Immunotoxicology, Slovak Medical University, Bratislava, Slovak Republic

Jevgenij Kovriznych	Faculty of Public Health, Department of Toxicology and Faculty of Medicine, Laboratory of Immunotoxicology, Slovak Medical University, Bratislava, Slovak Republic
Alexander Kiss	Institute of Experimental Endocrinology, Biomedical Research Center Slovak Academy of Sciences, Bratislava, Slovak Republic
Sona Scsukova	Institute of Experimental Endocrinology, Biomedical Research Center Slovak Academy of Sciences, Bratislava, Slovak Republic

CHAPTER 1

Nanomedicine and Neurodegenerative Diseases: An Introduction to Pathology and Drug Targets

Tasnuva Sarowar[1] and **Andreas M. Grabrucker**[2,*]

[1] *Institute for Anatomy and Cell Biology, Ulm University, Ulm, Germany*

[2] *Department of Biological Sciences, University of Limerick, Limerick, Ireland*

Abstract: Neurodegenerative diseases are debilitating conditions that result in progressive degeneration and death of neuronal cells. One of the hallmarks of neurodegenerative diseases is the formation of protein aggregates. Progressive accumulation of similar protein aggregates is recognized as a characteristic feature of many neurodegenerative diseases. Particularly in Parkinson's Disease (PD), aggregated forms of the protein α-synuclein (α-syn); and in Alzheimer's Disease (AD) and cerebral amyloid angiopathy (CAA), aggregated Aβ amyloid fibrils form the basis of parenchymal plaques and of perivascular amyloid deposits, respectively. In Amyotrophic Lateral Sclerosis (ALS), the RNA-binding protein TDP-43 is prone to aggregation. The focal aggregates at early disease stages later on result in the spreading of deposits into other brain areas and many neurodegenerative diseases display a characteristic spreading pattern. Here, we will summarize the anatomy and pathology of the predominant neurodegenerative diseases focusing on AD and PD and review their clinical manifestation to highlight the urge of novel therapeutic strategies. Additionally, given that development of treatments requires suitable animal models, the most commonly used model systems are introduced and their pathology compared to the human situation is mentioned briefly. Finally, possible drug targets in neurodegenerative diseases are discussed.

Keywords: Alzheimer's Disease, Amyotrophic Lateral Sclerosis, Animal models, Drug targets, Dementia, Lewy Bodies, Neurodegeneration, Parkinson's Disease, Synuclein, TDP-43 Tau pathology, β-Amyloid.

INTRODUCTION

The foundation for the definition of modern neurological disease entities was laid in the middle of the 19th century when Jean-Martin Charcot tried to relate - at this time mysterious - clinical phenotypes to neuro-anatomical findings. In *post*

---

* **Corresponding author Andreas M. Grabrucker:** Department of Biological Sciences, University of Limerick, Limerick, Ireland; Tel: +353 61 233240; E-mail: andreas.grabrucker@ul.ie

Giovanni Tosi (Ed)

mortem studies, he demonstrated such a relation for Amyotrophic Lateral Sclerosis (ALS) and Multiple Sclerosis (MS). Subsequently, the increasing interest in therapeutic approaches, including disease modification and prevention, fueled the interest in longitudinally studies that formally assess disease pathology. To that end, the use of molecular markers for a specific pathology such as synuclein for Parkinson´s Disease (PD) and tau for Alzheimer´s Disease (AD) became a useful tool to describe the pre-symptomatic and symptomatic stages of a disorder. Findings from these studies led to the current understanding of the pathology of neurodegenerative diseases, which is characterized by an initiation- and propagation phase of the disease process.

Today, the term "Neurodegenerative disease" is used for a wide range of conditions primarily affecting neurons in the brain and spinal cord. Given the inability of neurons to perform cell division and to replace themselves, progressive neuronal cell death is an irrevocable and, over time, cumulative process. The most prominent examples of neurodegenerative diseases include Parkinson's, Alzheimer's, Huntington's Disease (HD) and Amyotrophic Lateral Sclerosis (ALS). Neurodegenerative diseases may be hereditary or sporadic conditions.

Ongoing neuronal loss ultimately leads to problems with movement (called ataxias), or mental functioning (called dementias). With approximately 60-70% of cases, AD represents the greatest burden within the group of dementias. Other neurodegenerative diseases are Prion Disease, Multiple Sclerosis, Spino-cerebellarataxia (SCA) or Spinal Muscular Atrophy (SMA). However, hundreds of different disorders fulfill the criteria for a neurodegenerative disease.

Currently, the life expectancies of the general populations in both developed and developing countries are increasing, which affect the prevalence of neurodegenerative disorders (Table. **1**). This creates an enormous socio-economic burden with a total cost of hundreds of billion Euro per year in Europe alone [1].

Table 1. Age and gender specific prevalence rates (%) of dementia and PD in Europe [2].

	Dementia		Parkinson	
Age group (years)	♂	♀	♂	♀
65-69	1.8	1.4	0.7	0.6
70-74	3.2	3.8	1	1
75-79	7	7.6	2.7	2.8
80-84	14.5	16.4	4.3	3.1
85-89	20.9	28.5	3.8	3.4

(Table 1) contd.....

	Dementia		Parkinson	
Age group (years)	♂	♀	♂	♀
>90	32.4	48.8	2.2	2.6

Thus, research in the field of neurodegenerative disorders and the translation of the findings in this area to novel treatment strategies are an urgent and important goal. Fortunately, in recent years, our understanding of the anatomy and pathology of neurodegenerative diseases have made good progress.

CLINICAL REPRESENTATIONS

Alzheimer's Disease

AD is a progressive neurodegenerative disorder, which is described as the most common form of dementia nowadays. It was first described in 1907 by the German psychiatrist and neuropathologist Dr. Alois Alzheimer after observing a 55 years old patient named Auguste Deter. In general, AD patients suffer from disturbances in cognitive function or information processing like reasoning, planning, language & perception; which lead to a significant decrease in the quality of life. Besides other factors, age is the main contributing factor (Table. **1**) where 30% of individuals aged more than 85, develop the disease. A new case of AD is diagnosed worldwide every 7 seconds [3] and it is estimated that at least 34 million people will be suffering from AD by 2025, in both industrialized and developing countries [4].

Core Features of AD

AD can be divided into two groups based on the onset of the disease- early onset AD and late onset AD. In early onset, the disease occurs before the age of 65 and in late onset, the disease occurs after 65. Most of the patients are usually late onset as early onset accounts for only around 5% of the total disease occurrence. However, studies show that early onset AD is associated with high mortality and morbidity whereas late onset is much more common with less morbidity and mortality [5]. The disease progression of early onset AD is often predictable and it is possible to express the stage numerically using scales like Global Deterioration Scale [6] or Clinical Dementia Rating Scale [7]. The symptoms usually start around the age of 70. Patients show impairment in memory, problem solving, planning, judgment, language and visual perception. Some also suffer from hallucination and delusion. Eventually, the condition worsens and the patients are unable to carry out normal day-to-day functions and become bed-ridden. They need extensive palliative care and often die of other medical conditions [8 - 10].

Symptoms & Diagnosis of AD

The accuracy of diagnosis has increased many-fold since 1970. Today, it is possible to diagnose AD with more than 90% accuracy. However, except for brain autopsy, there is currently no definitive way to diagnose AD. To differentiate AD accurately from other forms of dementia, physicians rely on neuroimaging, psychiatric assessments, laboratory assessments and various other diagnostic tools.

Since 1984, AD has been diagnosed according to the National Institute of Neurological and Communicative Disorders and Stroke and the Alzheimer's Disease and Related Disorders Association (NINCDS-ADRDA) [11]. Such diagnosis can state the patient as probable, possible and definite case of AD. Probable AD is diagnosed based on clinical syndromes like progressive decline in memory for at least six months. Possible AD is diagnosed if memory disturbances are increased and most often, if a second disease contributes to the dementia syndrome. However, memory disturbances can be accompanied by other cognitive disturbances resulting in diverse types of clinical symptoms like aphasia, visuospatial disturbances, or posterior cortical atrophy. Often AD patients are diagnosed via reduced verbal fluency or difficulty in generative naming tasks like producing as many flower names as possible in one minute. They might have difficulty in performing normal daily or familiar tasks and often confuse time or passage. Sometimes patients show spatial disorientation, or do not recognize familiar places. Patients might also have trouble in understanding color and contrast, judging distances, making decisions, placing the right things in the right places etc. Such symptoms are often accompanied by alteration in the metabolism in different brain regions, which can be diagnosed by neuroimaging techniques like Positron Emission Tomography (PET) [12 - 15].

Additionally, Computerized Tomography (CT) and Magnetic Resonance Imaging (MRI) are performed to show any abnormality in the structure of the brain. Structural MRI can be used to detect brain atrophy, whereas functional MRI can be used to detect normal neuronal activity or any task-induced activity [16]. Nowadays, PET is often performed to identify a change in the blood flow to the brain. Both PET and Single-Photon Emission Computed Tomography (SPECT) can be used to investigate neurotransmitters and metabolic changes in different brain regions [13].

There are some other neuropsychiatric symptoms that accompany AD while the disease progresses like depression, disinhibition, delusion, hallucination, agitation, anxiety and aggression [13]. Such psychiatric symptoms may or may not correlate with cognitive impairments. However, studies have shown that the cognitive disturbances worsen rapidly in AD patients with psychosis compared to AD patients without psychosis [17 - 21]. Patients suffering from both AD and depression exhibit greater impairment in terms of memory, language, attention, and other functions than patients suffering from AD only. *Post mortem* brain tissue analysis often shows that AD patients suffering from mood changes have reduced norepinephirne and serotonin levels in cortical regions [22, 23].

Currently, there are no reliable biomarkers available that can be used to diagnose the disease or predict the progression of the disease. However, there are a few molecules whose measurement can be useful to increase the specificity of the diagnosis. For example, several molecules such as amyloid β (Aβ), cholesterol, and homocysteine can be detected in plasma, although such detection might not be consistent between patients. Due to the relatively free transport between cerebrospinal fluid (CSF) and brain, biomarkers extracted from CSF can be assessed as a measurement of their brain levels. Therefore, Aβ, total tau and phosphorylated tau extracted from CSF can provide a valuable insight on pathological changes in the brain.

In *post mortem* tissues, Aβ senile plaques and neurofibrillary tangles are observed in AD brain. Aβ plaques are historically named after the repetitive β-sheet pattern of the filamentous peptides. Upon staining with congo red or thioflavin, the Aβ plaques are doubly refractive and can be viewed under polarized light. However, plaques and tangles can also be found in normal individuals who have not been diagnosed with dementia. Typically, AD is diagnosed once the abundance of plaques and tangles has reached a certain threshold. An intermediate state between normal cognition and dementia has been termed Mild Cognitive Impairment (MCI). People suffering from MCI have a higher chance of developing AD in later stage of life.

Parkinson's Disease

Parkinson's disease is one of the multi-factorial neurodegenerative disorders. James Parkinson first described the symptoms with involuntary tremor in his essay entitled "An Essay on the Shaking Palsy" in 1817 and later the disease was named after him by Jean Martin Charcot, the father of modern neurology [24]. The prevalence of PD depends on the age. Around 1% of the population is affected at the age of 65, which increases to 4-5% at the age of 85 years. PD prevalence is around 15-19 in per 100,000 people [25, 26]. The median onset of disease is 60 years and the mean duration from disease diagnosis to death is 15 years. The striatum, which is at the center of the regulation of movement control, is depleted of the neurotransmitter dopamine accompanied by progressive neuronal death in PD patients. Therefore, the most prominent and common features of PD are involuntary movement while resting, rigidity, bradykinesia, and postural instability. All these symptoms are collectively termed "parkinsonism". Besides PD, the symptoms of parkinsonism have been associated with some other neurodegenerative disorders like dementia with Lewy bodies, Multiple System Atrophy, Frontotemporal Dementia with parkinsonism [27]. The hallmark of PD is the loss of dopaminergic neurons in the substantia nigra of the brain and the presence of Lewy bodies in the remaining intact neurons. However, recently it has been observed that degeneration of the substantia nigra can occur even without Lewy bodies.

Symptoms & Core Features of PD

Usually the first disease symptoms are impairment in dexterity and slowness in movement. Often such early motor symptoms are overlooked and the diagnosis occurs several years after the first appearance of such symptoms. Sometimes patients suffer from insomnia and hyposmia (the reduced ability to smell and to detect odors). In the later stages of disease progression, the face becomes expressionless, the speech becomes monotonous, and the posture becomes flexed. Freezing of the gait can occur occasionally. Patients might require assistance in dressing, eating, feeding or getting out of the bed. Constipation, chewing and swallowing, drooling of saliva and incontinence in urine are common complaints [24, 28 - 31].

Besides these symptoms, there are many non-motor symptoms in patients suffering from PD such as cognitive decline, psychiatric disturbances, autonomic failure, sleep disturbances and pain syndrome [32]. Indeed, the non-motor symptoms are more associated with the decrease in quality of life and life expectancy as well as the desire for treatment and the cost of caretaker compared to the motor symptoms. Anxiety, depression, apathy and schizophrenia are the most common psychiatric symptoms of PD. Patients suffering from depressive disorders are more likely to develop PD than the healthy control group [33, 34]. In the late stages of PD, dementia is very common [35].

Diagnosis of PD

Around 80% of the patients suffering from parkinsonism develop PD later in life. However, a definite diagnosis requires *post mortem* tissue analysis. The brains from the patients suffering from PD show depigmentation of the substantia nigra as well as the presence of Lewy bodies [36]. When investigating a patient, certain movements are assessed such as the time required to undress, whether the face is expressionless, or whether it takes time to show any expression, the speech lacks rhythm and melody, or whether repetitive finger or foot tapping occurs. The most popular test is rapid repetitive finger tapping of the index finger on the thumb for

20 seconds. Furthermore, bradykinesia can be confirmed by reduction of speed and amplitude on sequential motor tasks. Another symptom of PD is slow, pill rolling tremor of the hands in resting state (4-6 cycles per second) [24]. The Queen Square brain bank (QSBB) has outlined the symptoms and diagnosis of PD (Table **2**). These are the most commonly used symptoms for the diagnosis of PD.

Table 2. QSBB clinical diagnosis criteria for PD.

Step 1. Diagnosis of Parkinsonian Syndrome 　• Bradykinesia 　• At least one of the following 　　▪ Muscular rigidity 　　▪ 4-6 Hz rest tremor 　　　▪ postural instability not caused by primary visual, vestibular, cerebellar, or proprioceptive dysfunction
Step 2. Exclusion criteria for Parkinson's disease 　• history of repeated strokes with stepwise progression of parkinsonian features 　• history of repeated head injury 　• history of definite encephalitis 　• oculogyric crises 　• neuroleptic treatment at onset of symptoms 　• more than one affected relative 　• sustained remission 　• strictly unilateral features after 3 years 　• supranuclear gaze palsy 　• cerebellar signs 　• early severe autonomic involvement 　• early severe dementia with disturbances of memory, language, and praxis 　• Babinski sign 　• presence of cerebral tumor or communication hydrocephalus on imaging study 　• negative response to large doses of levodopa in absence of malabsorption 　• MPTP exposure
Step 3. Supportive prospective positive criteria for Parkinson's disease Three or more required for diagnosis of definite PD in combination with step one 　• Unilateral onset 　• Rest tremor present 　• Progressive disorder 　• Persistent asymmetry affecting side of onset most 　• Excellent response (70-100%) to levodopa 　• Severe levodopa-induced chorea 　• Levodopa response for 5 years or more 　• Clinical course of ten years or more

When any individual is diagnosed with PD, already a significant amount of neurons in the substantia nigra has degenerated. Therefore, there is an urge to develop biomarkers that can detect PD in early stages. For example, the levels of α-synuclein in CSF, blood, urine, saliva, and the gastrointestinal tract are increased in PD patients [37 - 39]. Several neuroimaging techniques can provide valuable information of the brain regions like Single Photon Emission Tomography (SPECT), Positron Emission Tomography (PET), Magnetic Resonance Imaging (MRI) and Transcranial Sonography (TCS) [40]. However, often, the information provided are not conclusive enough.

Amytropohic Lateral Sclerosis

Amylotropic Lateral Sclerosis (ALS) is the most common adult onset motor neuron disease. In USA, this disease is often referred to as Lou Gehrig's Disease, named after a famous baseball player suffering from the disease. In Europe, 2-3 individuals in every 100,000 persons are diagnosed with ALS. The typical phenotype is progressive paralysis with loss of bulbar and limb function, and eventually death by respiratory failure. Usually 60% of patients die within 2-3 years of disease onset [41]. Unlike other neurodegenerative disorders i.e. Alzheimer's Disease or Parkinson's Disease, the risk of developing ALS is high between the age of 50 and 75; after this age, the chances of developing ALS are low [42 - 44].

Core Features of ALS

The patients can be suffering from an array of motor symptoms like limb symptoms (majority of the patients manifest this), bulbar dysfunction or respiratory-onset disease [42]. Weight loss and impairment in emotional ability have also been reported. Features with the combination of upper and lower motor neurons are detected in neurological examinations like spasticity combined with wasting and fasciculations [45]. Usually the sensory abilities remain normal.

Cognitive impairment and change in behavior are also very common in patients suffering from ALS. However, such changes also occur in other neurodegenerative disorders like Frontotemporal Dementia. Therefore, the identification of such changes is often inconclusive in terms of the diagnosis of ALS. Behavioral changes are sometimes observed by spouses or close relatives, which, however, is usually not reflected in formal diagnostic tests. Verbal fluency may be impaired and a simple 2 minute word generation test can identify a probable patient. These features should be studied in details in order to identify the disease conclusively [46 - 48].

Symptoms & Diagnosis of ALS

Some features of ALS overlap with those of other disorders like Frontotemporal Dementia (FTD), Kennedy Disease (Spinobulbar Muscular Atrophy), or Multifocal Motor Neuropathy. It is often hypothesized that a shared environmental and genetic susceptibility might be responsible for such neurodegenerative disorders. Diagnosis of ALS is made based on internationally recognized consensus criteria, which diagnose ALS after excluding conditions that can mimic ALS. An ALS Functional Rating Scale can predict the severity and estimate survival time of the patient. The phenotypes of ALS can be classified based on bulbar-onset and spinal-onset disease. These phenotypes can be further classified based on upper and lower motor neuron involvement such as Primary Lateral Sclerosis (PLS), Progressive Muscular Atrophy (PMA) and Progressive Bulbar Palsy. The diagnostic phenotypes for ALS include both upper and lower motor neuron degeneration whereas PLS patients only suffer from upper motor neurons symptoms, PMA patients suffers from lower motor neurons phenotypes, and progressive bulbar palsy is diagnosed as a lower motor neuron (more specifically cranial nerves) phenotype.

The El Escorial criteria were developed to diagnose ALS for research and clinical trial purposes in 1994 by the World Federation of Neurology. Later, some modifications were done in order to be more specific [49, 50]. A definite diagnosis of ALS is based on the identification of lower motor neuron degeneration validated by clinical, electrophysiological or neuropathological examination; evidence of upper motor neuron degeneration by clinical examination and progression of motor syndrome within a regions or to other regions. However, these criteria are often criticized to be over-restrictive and inappropriate for the regular diagnosis.

Thus, currently there is no definite test to diagnose ALS. Routine investigations include the measurement of erythrocyte sedimentation, liver functions, thyroid functions, serum and urine protein electrophoresis, and electrolytes. In some specific ethnic groups, some other tests are done i.e. beta hexoaminidase activity in Ashkenazi Jewish people [50]. Besides these tests, there are electrodiagnostic studies. Electromyography can be used to identify the loss of lower motor neurons, and unaffected brain regions [51]. It is possible to identify fasciculation, spontaneous denervation discharges (hallmark of ongoing motor neuron loss) and polyphasic units (indicative of re-innervation) [52]. At least 5% of the familiar cases of ALS follow a Mendelian pattern. Therefore, it is possible to identify a mutation in the responsible gene using genetic testing. The most useful neuroimaging technique for ALS is MRI. MRI of the brain and spinal cord can distinguish similar diseases from ALS [53]. Diffusion tensor imaging, voxel-based morphometry and resting functional MRI can be used for cross-sectional and longitudinal studies of ALS.

ANATOMY AND PATHOLOGY

Initiation and Progression

The pathology of neurodegenerative diseases is characterized by an initiation- and propagation phase of the disease process. In line with the early ideas of Oskar and Cecilie Vogt on selective vulnerability ("Pathoklise"), initiation might be localized to a specific cell population and pathology propagated following higher order connectivity.

Progressions of Aβ and tau deposition in Alzheimer's Disease as well as α-synuclein (α-syn) inclusion aggregation in Parkinson's Disease are reported to follow the so-called Braak stages. Using these molecular markers, symptomatic and pre-symptomatic stages can be described [54, 55], showing that neurodegenerative diseases are characterized by a continuous, systematic spreading of the pathology, which initiates at a disease-specific focus. In particular, the spreading pathology is characterized by the tracking of the specific pathological marker protein deposits and aggregates throughout the brain (Fig. 1). Several factors may contribute to the mechanisms underlying initiation and propagation of abnormal proteins that will be discussed below. However, clearly, processes regulating protein turnover and degradation of modified proteins are essential in preventing accumulation of toxic products.

Once neurodegenerative processes are triggered, or even contributing to this process, activated microglia and astrocytes, together with other cells, cause a neuroinflammatory response at sites of protein-aggregate insults. Inflammation is a common feature in neurodegenerative diseases. It is the primary response of the innate immune system to damage and invasion.

Alzheimer's Disease

The pathology of AD is characterized on histological level by the presence of cerebral Aβ deposits, neuritic plaques (NP), neurofibrillary tangles (NFT), and neuropil threads (NT) [56]. Considering the distribution of neurons presenting intracellular NFTs and the severity of neurofibrillary pathology, six stages (NFT stages) can be distinguished [54]. Additionally, the deposition of Aβ plaques seems to occur throughout the brain in a hierarchical pattern. This makes it possible to stage plaque expansion in five phases of disease propagation.

In phase 1, Aβ deposits are found exclusively in the neocortex, while in phase 2, additional involvement of all cortical brain regions can be seen. In phase 3, Aβ accumulations can be found additionally in diencephalic nuclei, the striatum, and the cholinergic nuclei of the basal forebrain. In phase 4, several brainstem nuclei become involved. Finally, in phase 5, cerebellar Aβ deposit occur. Progression through the stages correlates with clinical features such as gradually developing intellectual decline [57].

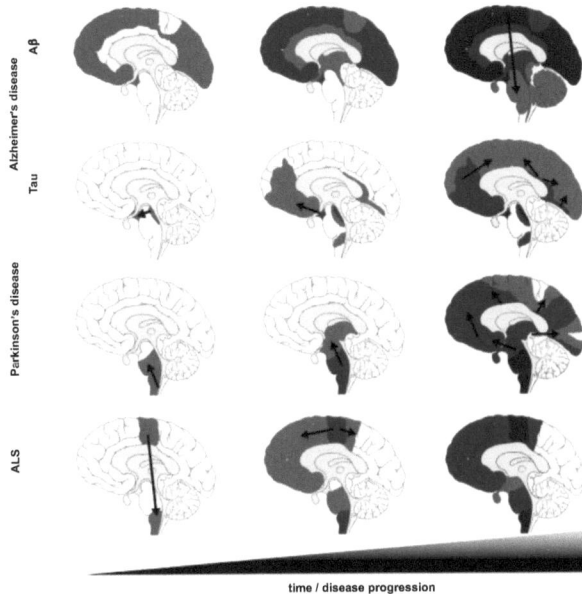

Fig. (1). Spreading of pathology detectable by molecular markers for AD (Tau, Aβ), PD (α-syn), and ALS (TDP-43) over time. Aβ aggregates (senile plaques) in patients with AD are found and their occurrence can be used to distinguish five stages of disease progression (see text). NFT pathology is similarly spreading in a specific pattern and thus categorized in six Braak stages (see text). α-syn inclusions in brains of patients with PD invade the vagal nerve dorsal nucleus and spread through the brain stem nuclei into the basal ganglia to finally affect the neocortex. TDP-43 inclusions in brains of patients with ALS initially appear in agranular motor cortex and might spread to the bulbar motor nuclei and anterior horn cells of the spinal cord. Later, precerebellar, striatal and hippocampal nuclei are affected. Arrows indicate the putative spread of pathological proteins (data based on [73]).

Since many years it is known from studies in animal models that Aβ propagates within the brain and it has been shown that the formation of Aβ aggregates can be induced in Aβ-producing mice upon intracerebral injection of Aβ plaque-containing AD brain tissue [58 - 61]. This also holds true for the second hallmark of AD, the tau pathology. For example, it was shown that after injections of exogenous tau aggregates, a tauopathy can be induced in wild type animals [62]. Moreover, propagation of tau was observed *in vivo* after intracerebral injection of human aggregated tau protein into tau transgenic animals [63].

The six stages of progression of NFTs (so-called Braak stages) can be distinguished by first NFTs appearing in the transentorhinal (perirhinal) region (stage 1) and subsequently in the entorhinal cortex proper, followed by the CA1 region of the hippocampus (stage 2). In stage 3, NFT pathology is found in limbic structures such as the subiculum of the hippocampal formation, progressing to the amygdala, thalamus, and claustrum (stage 4). Finally, NFTs develop in all isocortical areas, where associative areas are affected first (stage 5), followed by the primary sensory, motor, and visual areas (stage 6). In addition, NFTs may accumulate in the striatum and substantia nigra during the later stages [54, 64, 65].

Parkinson's Disease

In PD, aggregates of α-synuclein (α-syn), the major component of Lewy bodies and Lewy neurites, are the basis of spreading patterns in the brain described during disease progression, although the pathology may start outside the central nervous system (CNS). A recent hypothesis points to the gastrointestinal (GI) tract and the enteric nervous system (ENS) as site of „initiation". Spreading of the pathology then follows the structure of the vagal nerve, invades its dorsal nucleus, and propagates through the brain stem nuclei into the basal ganglia to finally affect the neocortex [66]. In parallel, α-synpathology also shows a pattern that hints towards an additional descend from the nucleus of the vagus into the spinal cord [66, 67].

The hierarchical pattern of progression suggests a cell-to-cell transmission of α-syn as the basis for spreading. Thus, interconnected neuronal pathways may transmit pathological α-syn species. However, release of α-syn by extracellular vesicles has been additionally reported. Extracellular vesicles released from many cell types are capable of carrying mRNAs, miRNAs, noncoding RNAs, and proteins [68].

Similar to AD, in PD patients, the intracerebral formation of Lewy bodies and Lewy neurites follows a specific pattern in the brain and thus specific stages, also termed Braak stages, have been proposed [55]. In PD, 6 stages are defined based on the presence of Lewy bodies and Lewy neurites. Stages 1, 2 and 3 are usually present in the pre-symptomatic phase of PD, while the symptomatic phase is characterized by stage 3, 4, 5 and 6. In particular, in stage 1, lesions occur in the dorsal glossopharyngeal/vagal motor nucleus and in the anterior olfactory nucleus. Along with this, a "Lewy pathology" can be found in the enteric nervous system. In stage 2, the pathology found in the anterior olfactory nucleus progresses to areas of the brain stem including the caudal raphe nuclei, gigantocellular reticular nucleus, and the coeruleus–subcoeruleus complex that might also already be affected in stage 1.

In stage 3, additionally midbrain lesions, mostly in the substantia nigra pars compacta, can occur. In stage 4, the prosencephalon and the temporal mesocortex (transentorhinal region) and allocortex (CA2-plexus) are affected in addition but not the neocortex, which, however, is involved in stage 5 with pathology in sensory association areas and prefrontal neocortex. Finally, in stage 6, lesions in further sensory association areas, and pre-motor areas are found.

The progressive accumulation of α-syn aggregates in individuals with PD correlates well with the decline in motor and/or cognitive function [69].

Other Neurodegenerative Disorders

The sequential and continuously progression of motor deficits and pareses over the body as clinical symptom of ALS patients hints already to a possible underlying pathology similarly spreading throughout the CNS. Although in ALS, the molecular marker that can be selected to describe this process is less clear compared to AD and PD, the continuous spreading of the protein TDP-43 in the CNS may mirror some of the stages observed in disease progression [70, 71]. Although it is currently not known where TDP-43 aggregation begins in very early stages of disease; the initiation of the CNS pathology might occur in the agranular motor cortex and rapid spreading via monosynaptic pathways into the bulbar motor nuclei and the anterior horn cells of the spinal cord can be seen. In later stages, again via monosynaptic pathways, precerebellar, striatal and hippocampal nuclei are affected. In parallel, the molecular neuropathology continuously spreads via cortical association fibers into the frontal cortex and the post-centralgyrus [72].

Brain Morphology, Cellular Pathology and Neurobiology of the Disease

The progressive involvement of different and distant brain structures in the progression of neurodegenerative diseases has been recently linked to the spreading of protein deposits within the CNS which is based on transport of a pathogenic protein from site to site and progressive invasion of specific neuronal networks. Aβ, α-syn and TDP-43 pathology in AD, PD and ALS, respectively, affect neuronal structures in the brain with a stereotyped pattern of progression.

The regional focality of lesions might account for the distinctive characteristics in patients with neurodegenerative disorders. Although the attributes of focal distribution of lesions will initially lead to specific pathologies, a "pseudo-diffuse" pattern of lesions in the end stages of the diseases might lead to the characteristic global symptomatology.

A number of cellular mechanisms underlying the observations of initiation and progression of the pathology in neurodegenerative disorders have been proposed [74, 75]. For example, an interaction of neuronal dysfunction and molecular pathology such as neurofibrillary tangles and senile plaques will ultimately lead to a loss of neurons and synapses.

In addition to intrinsic properties of the characteristic proteins for the disorder aggregating (Fig. **2**) and spreading throughout the CNS, disease progression (aggregation and propagation of the marker protein) is likely modulated by excitotoxicity, inflammation, alterations in trace metals or other mechanisms.

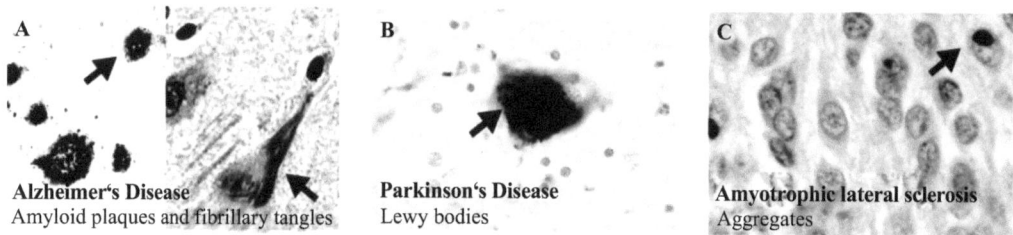

Fig. (2). Examples for pathological protein aggregates found in neurodegenerative disorders. A) Amyloid beta immunostaining showing senile plaques (left) and fibrillary tangles (hyperphsophorylated tau) visible in Hematoxylin and Eosin staining (right). B) α-synuclein immunostaining counterstained with Mayer's Hematoxylin showing Lewy bodies. C) Immunostaining of spinal cord sections for total TDP-43 shows immunoreactivity of cytoplasmic aggregates additional to nuclear TDP-43.

Alzheimer's Disease

The pathogenesis of AD is characterized by Aβ deposition in the form of senile plaques and by changes in the content of modified forms of Aβ (N-termina--truncated and pyroglutamate modified ($A\beta_{N3pE}$) and serine-8-phosphorylated (pAβ)) within these aggregates. Additionally, the presence of intracellular neurofibrillary tangles made up of hyperphosphorylated tau protein is a hallmark of AD (Table **3**). Cerebral amyloid angiopathy is also a frequent feature in AD. Furthermore, so-called Hirano bodies can be found in the hippocampal formation. Regarding the gross anatomical features of AD, cortical atrophy affecting the medial temporal lobes but mostly sparing the primary motor, sensory, and visual cortices, can be observed. This thinning of cortical structures may lead to a dilated appearance of the lateral ventricles.

A prominent feature in AD is neuronal loss and synapse loss that largely parallel tangle formation, although it is currently under debate whether tangles are causative factors in these processes. In AD, the characteristic loss of neurons, neuropil, and synaptic elements is the best molecular correlate for the observed cognitive decline of patients. The underlying pathological features, such as amyloid plaques can be found throughout the cortical mantle, while fibrillary tangles are found primarily in limbic and association cortices [76].

The neurobiology of AD is not fully understood. Of the two kinds of abnormal, pathological structures observed in the post mortem brain from AD patients, one is intracellular (microtubule-associated hyper-phosphorylated tau protein) and the other one is mainly extracellular (amyloid based neuritic plaques). Amyloid plaques are composed of Aβ peptide, which is a product of cleaved Amyloid Precursor Protein (APP).

Table 3. Pathological hallmarks of AD.

Pathology	Description
Amyloid plaques	Extracellular deposits of Aβ. Amyloid plaques are commonly found in the cortex of AD patients. Amyloid plaques can be subdivided into dense-core plaques and diffuse plaques. Dense-core plaques are typically surrounded by dystrophic neurites, reactive astrocytes and activated microglial cells. Diffuse plaques are usually nonneuritic.
Cerebral amyloid angiopathy	Aβ deposits in leptomeningeal arteries and cortical capillaries, as well as small arterioles and arteries. Severe can cerebral amyloid angiopathy can weaken the wall of blood vessels and lead to hemorrhages.
Neurofibrillary tangles	Intracellular aggregates of hyperphosphorylated tau. Extracellular tangles can occur as remainder of dead neurons.
Neuropil threads	Segments of neurons (axonal and dendritic) bearing hyperphosphorylated tau resulting from neurofibrillary tangles.

Amyloid Precursor Protein (APP) Gene

Aβ has been established at the center of the pathology in AD. However, the physiological role of APP and its association with AD is not conclusive. The coding region of APP consist of 18 exon spanning in 290 kbp region. Exon 16 and 17 code for Aβ [77]. Human APP is mainly expressed in the brain. However, there are other splice variants available with different protein length [78]. APP695, the most common form with 695 amino acids is mainly expressed in hippocampal neurons, whereas APP770 (with 770 amino acids) is mainly expressed in microglia, astrocytes and oligodendrocytes [79].

Structurally, APP is a type 1 transmembrane protein, which can be structurally and functionally sub-divided into several domains (Fig. **3**). The ectodomain can be subdivided into ectodomain 1 (E1) & ectodomain 2 (E2) which are linked via an acidic amino acid rich domain. E1 can be further subdivided into several sub-domains, including a growth factor domain that can bind heparin; as well as a tightly interacting copper binding domain and a zinc - binding domain. In the E2 domain, there is a growth-promoting factor binding domain, heparin binding domain as well as other glycosylation and carbohydrate attachment sites. In the vicinity of transmembrane region, the cleavage sites for three secretase - enzymes is found [80, 81].

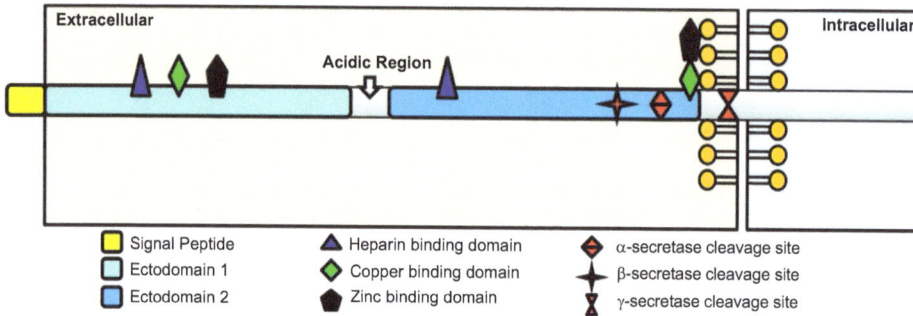

Fig. (3). Schematic representation of the domain architecture of APP. APP is a transmembrane protein. The extracellular part can be subdivided into ectodomain 1 and 2, linked by an acidic region. In the vicinity of the transmembrane domain the three cleavage sites of alpha, beta and gamma secretase are found. In addition, there are binding sites for copper and zinc in the ectodomain as well as in the Aβ peptide. There is also a heparin - binding domain in ectodomain 1 and other carbohydrate attachment sites in ectodomain 2 (not shown).

APP Processing and Aβ Generation

APP is synthesized in the endoplasmic reticulum, later glycosylated at the Golgi apparatus and exocytosed at the cell membrane. The processing of APP is mediated and controlled by several enzyme complexes and signaling molecules. When APP is cleaved by gamma and beta secretase, the resulting fragments are APP intracellular domain (AICD), soluble APP beta (sAPPβ) and Aβ with 39-42 amino acids. The AICD fragment can interact with transcription factors like Fe65 or Tip60 and thereby participate in gene regulation. The longer Aβ peptide has a higher tendency to form aggregates. On the other hand, when APP is processed via the non-amyloidogenic pathway, then gamma and alpha secretases cleave APP, resulting in AICD, soluble APP alpha (sAPPα) and P3 fragment for which the function is still unknown. Both sAPPα and sAPPβ play role in cell survival, neuronal excitability, synaptogenesis and synaptic plasticity, neurite outgrowth, and neuronal excitability (Fig. **4**) [82].

Under normal physiological condition, $A\beta_{40}$ is generated which can be degraded easily. However, in the brain of the AD patient, most of the produced Aβ is $A\beta_{42}$ which has a higher tendency to become fibrilized and insoluble, and therefore, resistant to degradation.

The size of neurite plaques ranges between 10 and 120 μm in cross sectional diameter [83]. As a peptide, Aβ has an amphipathic nature. Amino acids 12-14 at the C-terminal are hydrophobic, therefore the cleavage site at the C-terminus is within the vicinity of the transmembrane region. In aqueous solution, $A\beta_{40}$ can remain in random coil for longer periods than $A\beta_{42}$. $A\beta_{42}$ has a tendency to form β-sheets rapidly [84]. There are several levels of structural hierarchy in Aβ plaque formation starting from dimers, trimers, to small oligomers, protofibrils, fibrils, and finally plaques. Each intermediated structure is distinct in terms of neurotoxicity [85]. The minimal concentration for $A\beta_{40}$ to form such a fibril is 10-40 μM, which is five times higher compared to the minimal concentration for $A\beta_{42}$.

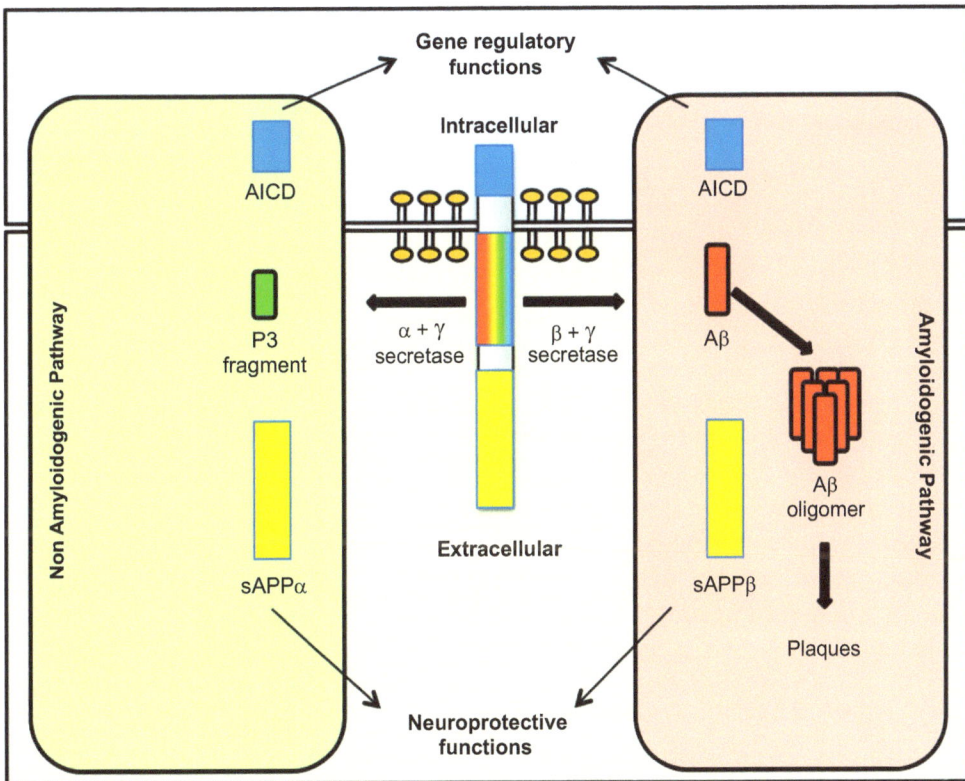

Fig. (4). Generation of Aβ by APP processing. In the non-amyloidogenic pathway, APP is cleaved by alpha and gamma secretases resulting in P3 fragment, APP intracellular domain (AICD), and soluble APP alpha (sAPPα). In amyloidogenic pathway, APP is cleaved by beta and gamma secretases, thus yielding Aβ peptide with 39-42 amino acids, AICD and sAPPβ. AICD can interact with transcription factors, thus has a role in gene expression. Both sAPPα and sAPPβ have neuroprotective function. $A\beta_{40}$ can easily be degraded, while $A\beta_{42}$ is prone to form oligomers and fibrils.

The Aβ peptide is thought to exhibit neurotoxicity via oxidative stress, membrane and ion channel disruption, alteration in cellular process like inflammation, and apoptosis [85]. There is debate over the cytotoxicity of soluble and aggregated Aβ. Recent studies have favored the soluble form of Aβ as the source of synaptotoxicity in AD [86], which might be mediated by changes in the balance between long-term potentiation (LTP) and long-term depression (LTD) in dendritic spines. In contrast, Aβ aggregates can act as a site for neuro-inflammation and such immunological reactions may contribute to synapse loss [87]. Astrocytes proliferate and microglial cells become activated as part of an inflammatory response in areas surrounding senile plaques [88]. However, while there is strong evidence that inflammation contributes to neuronal death in AD, the precise role of microglia in AD pathology is currently not well understood.

Zinc Amyloid Hypothesis

Upon synaptic activation, zinc stored in presynaptic glutamatergic (zincergic) terminals is released into the synaptic cleft where the concentration may reach as much as 300 μM. Due to N-Methyl-D-Aspartate receptor (NMDAR) mediated activation, postsynaptic copper transporter ATP7a also releases copper into the synaptic cleft where the concentration reaches around 15μM [89]. In addition, Aβ peptide is also released constitutively via the cleavage of APP in response to neuronal activity. Thus, the synapse is the major site, where Aβ can interact with zinc and copper ions (Fig. **5**) [90].

Aβ can bind a maximum of 2.5 mole metal ions. Zinc loaded Aβ is resistant to degradation. The amyloid plaques act as metal sink and the metal ions reach abnormal high concentrations (Cu: ~400 μM, Zn: ~1 mM, and Fe: ~1 mM) [91]. The zinc mediated Aβ aggregation is reversible using zinc chelators [92]. However, both loss of zinc by chelators or sequestration in amyloid plaques lead to a reduction in the pool of physiologically active zinc. In line with this, ZnT3 knock-out mice exhibit age-dependent learning and memory deficits thought to represent synaptic and memory deficits of AD [93].

The zinc amyloid hypothesis can explain certain aspects of AD. Aβ peptides may aggregate only in cortical and neocortical regions in the brain despite being broadly expressed due to the fact that the zinc ion concentration is highest in these regions. Moreover, rats and mice have specific alterations in the zinc and copper binding amino acid sequence in Aβ, rendering the peptide insensitive to the metals. Therefore, they develop plaques only upon expression of human APP (see animal models below). Further, female mice have higher expression levels of ZnT3 which may be associated with increased zinc levels. Correlating with this, female transgenic mice expressing human APP show greater Aβ pathology. In addition, Lee *et al.* showed that preamyloid Aβ aggregates do not develop into congophilic plaques in the cerebellum of APP transgenic mice where synaptic vesicular zinc is not present [94].

Histochemically reactive zinc is present in olfactory bulb, cerebral cortex and the limbic areas that correlate with the regional distribution of plaques. Plaque load is diminished and the ratio of soluble/insoluble Aβ increases in hAPP$^+$:ZnT3$^{-/-}$ mice which further indicates that synaptic zinc has a pivotal role in shifting the equilibrium of plaque accumulation [95].

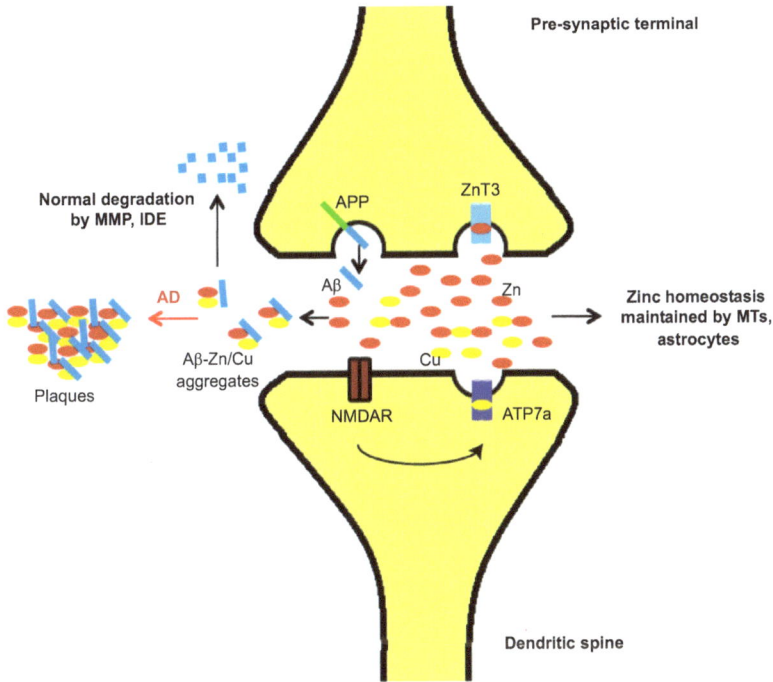

Fig. (5). Zinc amyloid hypothesis. Zn is released with glutamate in high concentrations from presynaptic glutamatergic terminals into the synaptic cleft upon neuronal activation. The zinc transporter protein ZnT3 is primarily responsible for loading Zn into the synaptic vesicles. Copper also plays a role in regulating synaptic function as it is released from the post-synaptic side of the cleft following NMDAR activation, facilitated by trafficking by ATP7a. Under normal physiological condition, the concentrations of the free metals are kept low due to buffering by metallothioneins and uptake by astrocytes. Aβ is released in the synaptic cleft via APP cleavage. Because of the metal binding sites, Aβ binds copper and zinc. In AD brains, the high concentration of Aβ-Zn/Cu species leads to the formation of senile plaques.

Tau

In the damaged brain region of AD patients, abnormal neuronal function and death are promoted by the loss of synaptic proteins via dysregulation of cytoskeletal components. The most common cytoskeletal component associated with this is the microtubule associated protein tau. Hyperphosphorylated tau protein is very poorly soluble. Therefore, filamentous inclusions are observed in the diseased brain. Interestingly, the severity of dementia in AD patients correlates with the amount of tau tangles, rather than Aβ plaques [96]. The amyloid cascade hypothesis, proposed in 1991, attempts to explain the correlation between Aβ and tau [97]. In this hypothesis it is assumed that tau acts a mediator of Aβ plaques with leads to neuronal loss. Later the hypothesis was modified and intracellular calcium was added as a central element [98]. Imbalance in calcium homeostasis leads to hyperphosphorylation of tau and later neurodegeneration. Due to hyperphosphorylation, tau aggregates are transformed to a paired helical form, which disrupts microtubule dynamics [99].

Tau as a protein plays important role in neuronal regulation via binding to microtubules which is essential for microtubule assembly and stability. The phosphorylation of tau protein is mediated by several kinases and phosphates such as glycogen synthase kinase 3β (GSK3β), cyclin dependent kinase (CDK5), and ERK [100, 101]. GSK3β is able to cause memory deficit via phosphorylating tau on multiple sites and inhibiting LTP [102]. Some other phosphatases can dephosphorylate hyperphosphorylated tau protein and thus can rescue the effect. One such phosphatase is PP2A [103, 104]. Thus activation of GSK3β and inhibition of PP2A create a vicious cycle that can rapidly induce the accumulation of hyperphosphorylated tau protein. Phosphorylation of tau can mislocalize it into dendritic spines and thus might cause synaptic deficit [105]. Moreover metal ions like Fe play a role in tau aggregation.

The accumulation of hyperphosphorylated tau protein can also be found in some other neurodegenerative diseases like corticobasal degeneration (CBD), Niemann-Pick type C disease, Down syndrome, frontotemporal dementia with parkinsonism linked to chromosome 17 (FTDP-17) [106]. Besides Aβ and tau, further protein aggregates associated with neurodegenerative diseases other than AD are found sometimes in the AD brain. One such protein is prion protein. Prion protein might be required for Aβ induced inhibition of LTP.

Parkinson's Disease

Aggregation of α-syn occurs as a nucleation-dependent process [107], meaning that the polymerization is greatly accelerated by the presence of even very small quantities of already aggregated or fibrillar α-syn that is able to serve as nucleation site. The pathological form of α-syn consists of oligomers and fibrils rich in β-sheets. The conversion of the original α-helical structure to the β-sheet rich fibril is a key step in the pathogenesis of PD. It has been speculated that α-syn might display prion-like activities, which might be involved in the conversion [108]. β-sheet rich α-syn was shown to be the major component of Lewy bodies and Lewy neurites both in sporadic PD, dementia with Lewy bodies. Under physiological conditions, α-syn is soluble but also forms filamentous aggregates that resemble the ultrastructural elements of Lewy bodies. However, α-syn mutations associated with familiar PD exhibit accelerated filament formation

[109]. Further factors have been proposed mediating the conversion of α-syn into a β-sheet conformation followed by fibrillogenesis and aggregation. For example, post-translational modifications and oxidative stress seem to be promoting factors. Indeed, oxidative stress induced by iron overload seems to result in preferential aggregation of β-sheet rich α-syn. Furthermore, the concentrations of fatty acids, phospholipids, and metal ions were shown to induce and/or modulate α-syn oligomerization [110]. Thus, it is likely that in addition to genetic abnormalities, environmental factors might trigger α-syn pathology (see below).

Abnormal degradation of α-syn can be caused by interference with the three mechanisms: chaperone-mediated autophagy, macroautophagy, and the ubiquitin-proteasome. Impairment of these pathways might lead to alternative clearance mechanism for α-syn that ultimately may result in a release of α-syn from the cell. Indeed, a number of studies suggest that a specific conformation of α-syn is released into the extracellular space [111, 112] and taken up and transported by neighboring cells. This increased release to the extracellular space could provide a basis for the spreading of the protein to different brain regions [113]. In addition, α-syn aggregates may display pathogenic actions also in the extracellular milieu.

α-synuclein

The α-syn protein plays important roles in cognitive function in the brain. It is expressed throughout the CNS and mostly localized to the presynapse. Although its exact function is currently not fully elucidated, it has been suggested that α-syn interacts with neurotransmitters, lipids, carbohydrates, membrane bound receptors, and other proteins in the brain in order to control synaptic membrane processes and to participate in the process of neurotransmitter release. The latter might be mediated by interactions with members of the SNARE family.

α-syn consists of 140 amino acids (14.5 kDa) and is highly conserved in vertebrates. Two further family members, β- and γ-synuclein, have been described. The α-syn protein contains three distinct regions. The amino-terminal sequence (residues 1–60) harbors a highly conserved hexamer motif similar to that found in the amphipathic helices of apolipoproteins. Mutations associated with familiar forms of PD such as A30P, A53T, and E46K are found in this part of the protein. The central NAC (non-amyloid-β component) region (residues 61–95) contains two additional motifs and possesses high hydrophobicity. It is this region that is able to change the conformation to a β-sheet structure. The third, C-terminal region (residue 96–140) plays a role in the regulation of aggregation and fibril formation. Inappropriate cleavage of the C-terminal part of α-syn might affect dopamine homeostasis in PD given that the C-term interacts with dopamine [114] (Fig. **6**).

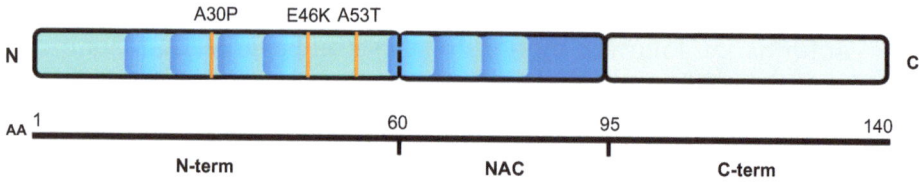

Fig. (6). Schematic representation of α-syn. The N-terminal region from amino acids (AA) 1–60 is responsible for α-syn interaction with membranes. Repeats of hexamer motifs can be found in this domain. Point mutations of α-syn associated with PD are indicted (A30P, E46K, and A53T). AA 61-95 represent the central region, termed NAC (non-β amyloid component). The hydrophobic portion of the protein is involved in aggregation processes by folding into a β-sheet secondary structure. The C-terminal domain (AA 96–140) contains acidic residues and several negative charges. Serine at position 129 in this region is phosphorylated in Lewy bodies.

Once β-structures are formed, the α-syn protein is harder to degrade compared to the protein forming α-helixes. Thus, removal of α-syn by the proteasome system is impaired which might lead to the cellular accumulation of misfolded proteins. Misfolded proteins usually are exported from the endoplasmic reticulum (ER) to the cytosol. An accumulation in the ER may lead to chronic ER stress and trigger cell death cascades [115].

In addition, α-syn might confer cellular toxicity through the permeabilization of membranes. Oligomers might be able to interfere with cellular membranes to form pore-like structures, resulting in abnormal ion influx with consequent neurodegeneration.

Mitochondria in PD

Besides ER stress, proteasome impairment, and/or the disruption of the plasma membrane, α-syn in its oligomeric state may cause mitochondrial dysfunction and oxidative stress [116, 117]. For example, wild-type as well as mutated α-syn (A53T) were shown to localize at mitochondria *in vitro*. There, overexpression of α-syn resulted in the release of cytochrome c. Indeed, the expression of α-syn (wild-type or A53T) leads to excess of mitochondrial Ca^{2+}, which might initiate apoptotic pathways via activation of caspases [117].

α-syn may regulate mitochondrial fission by directly interacting with the mitochondrial membranes. It has been speculated that this process might be accompanied by a decline in respiration. This decline may be underlying the selective vulnerability of neurons in the substantia nigra pars compacta given that they have been identified as very "energy-demanding". This demand is based on an extended axonal field and number of synapses per axon that are high compared to that of other neurons.

Additionally, the expression of mutated α-syn (A53T) increased the intracellular level of reactive oxygen species (ROS) and by that causes oxidative stress. Oxidative stress in general is caused by an imbalance of the levels of ROS produced in the cell and the ability of the system to detoxify the reactive intermediates. In neurons but also glial cells, ROS can be generated from several sources. For example, direct interactions between redox-active metals and oxygen species or indirect pathways involving the activation of enzymes such as nitric oxide synthase (NOS) or NADPH oxidases may generate ROS. The superoxide anion radical (O_2^{2-}), the hydroxyl radical (*OH), and hydrogen peroxide (H_2O_2) are ROS abundantly found in organisms [118].

The electron transport chain found at mitochondria is the major contributor. Thus mitochondrial dysfunction, but also the metabolism of dopamine, Neuromelanin, Glutathione, lipids, iron, neuroinflammation, and calcium are involved in mechanisms generating ROS. For example, dopamine itself is an unstable molecule that undergoes auto-oxidation forming free radicals. In addition, neuroinflammation leads to the activation of microglia. Microglia generate superoxide and nitric oxide, which in turn contribute to oxidative and nitrative stress in the brain.

Special attention has been given to the role of iron in oxidative stress in PD. Iron ions (Fe^{2+}/Fe^{3+}) can generate ROS by reacting directly with superoxide and hydrogen peroxide, generating the highly reactive hydroxyl free radical. Intriguingly, abnormal, progressive deposition of iron and increased concentration of free iron have been reported in specific brain regions in PD (Fig. 7).

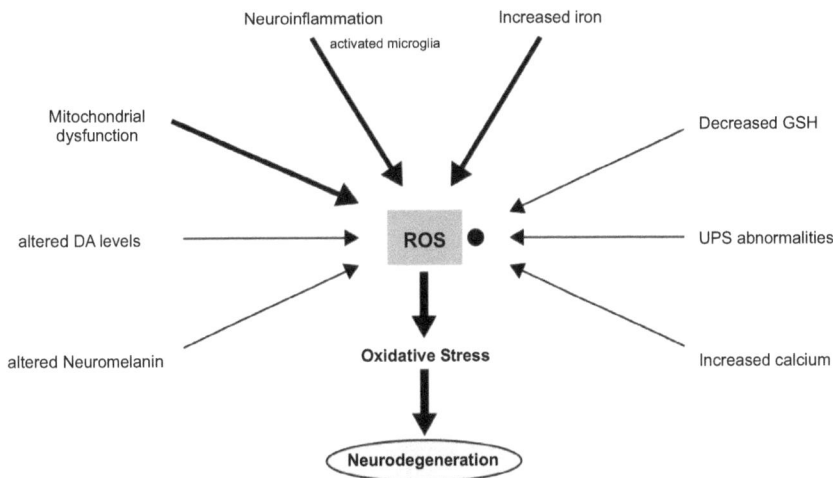

Fig. (7). Mechanisms generating oxidative stress in PD. DA = dopamine; ROS = reactive oxygen species; UPS = ubiquitin-proteasome system; GSH = glutathione.

However, oxidative stress is not specifically associated with PD only, but a common mechanism contributing to neuronal degeneration in several other neurodegenerative disorders such as AD, HD, and ALS.

Brain Morphology in PD

The loss of dopaminergic neurons in the substantia nigra pars compacta is one of the most prominent features of PD. This loss most likely is responsible for the characteristic motor symptoms seen in PD and results in the depletion of dopamine, which causes impairment in dopaminergic neurotransmission. The appearance of non-motor symptoms in turn is linked to the degeneration of non-dopaminergic systems following the propagation and accumulation of Lewy bodies in different brain regions.

The main anatomical changes found in PD can be divided into three different areas: mesencephalic (loss of dopaminergic neurons), basal ganglia (dopaminergic depletion) and cortical (functional reorganization) [119]. Abnormalities in the substantia nigra can be detected using 7 Tesla MRI and diffusion tensor MRI. In addition, magnetisation transfer can visualize a loss of melanin in the substantia nigra. In advanced stages of PD, cortical atrophy of the frontal or temporal lobe can be seen. The medial temporal lobe atrophy may be responsible for memory dysfunction seen in PD. However, the first regions to be affected by atrophy in PD seem to be the parietal cortex, the putamen, and the inferior-posterior midbrain [120] based on imaging studies.

Amyloid Lateral Sclerosis

ALS mainly affects lower motor neurons in the brainstem and ventral horn of the spinal cord and upper motor neurons in the cortex. These neurons establish the corticospinal tract, which descends through the lateral spinal cord. The defining pathology in the majority of ALS patients is the presence of neuronal aggregates of TDP-43 (43-kDa TAR DNA-binding protein). Mutations in TDP-43 can cause rare forms of familial ALS, but also most sporadic cases of ALS show abnormalities in TDP-43 location and function such as nucleo-cytoplasmic redistribution, post-translational modification and cytoplasmic aggregation of TDP-43. Additionally, a small subgroup of ALS patients is characterized by the aggregation of other proteins, such as SOD [121].

Mutations in the SOD1 gene are found in about 20% of the familiar cases of ALS. SOD1 is an ubiquitously expressed enzyme that catalyzes the conversion of superoxide free radicals to hydrogen peroxide. The latter is subsequently further detoxified to water and oxygen by other enzymes. Currently, the mechanism through which mutant SOD1 induces motor neuron degeneration is not well understood and several models have been proposed [122]. Given that SOD1 is an important enzyme in the defense against superoxide anions, increased protein damaged by oxidation that might lead to increased excitotoxicity is discussed as possible patho-mechanism. Alternatively, oxidative stress may be caused by mitochondrial dysfunction. An abnormal recruitment of mutant SOD1 to the mitochondrial compartment has been reported in a cell- and tissue-specific manner that may result in reduced energy production and increased free radical generation, interference with mitochondrial Ca^{2+} buffering, and/or initiation of apoptosis [122].

Another model for the patho-mechanism of SOD1 mutations focuses on the feature of mutant SOD1 to form aggregates. Mutations in SOD1 may affect the folding and assembly of SOD1. Misfolding in turn initiates the so-called unfolded protein response (UPR), which leads to refolding of the protein or, in case this cannot be achieved, export of the protein from the endoplasmic reticulum (ER) to the cytosol (see above: α-syn in PD). There, ubiquitinated proteins are degraded by the proteasome system. However, mutant SOD1 has been found to accumulate in the ER and thus, to escape the transport to the cytosol [123]. This induces ER stress, and may exhaust the UPR response and/or proteasome system. Cytoplasmic misfolded SOD1 may additionally undergo unwanted protein-protein interactions or attach to mitochondrial membranes (see above), and aggregate forming higher molecular species and intracellular inclusions.

Similarly, mutant TDP-43 proteins and misfolded TDP-43 were found to be mislocated, and abnormally processed and ubiquitinated in ALS. TDP-43 binds to various heterogenous nuclear nucleoproteins (hnRNPs), and under physiological conditions, is localized in the nucleus, where it plays a role in a variety of processes such as processing, stabilization and transport of RNA [124]. However, mutant and misfolded TDP-43are mainly found in the cytoplasm. Thus, mislocalization of TDP-43 may result in abnormal RNA metabolism. Intriguingly, another protein implicated in ALS, namely FUS (hnRNPP2), is an hnRNP. FUS plays a role in pre-mRNA splicing and the export of mRNA to, and transport within the cytoplasm [125]. In addition, FUS may be an inhibitor of DNA transcription especially in the context of DNA damage [126].

Finally, the identification of expanded GGGGCC hexanucleotide repeats in the first intron located between the first and second non-coding exons of C9orf72, a ubiquitously expressed but highly enriched protein in neurons with largely unknown function, further links aberrant RNA processing with ALS [127]. Sense and antisense transcripts of the expanded repeat interact with various RNA-binding proteins and form nuclear structures. These so-called RNA foci sequester RNA-binding proteins, which leads to an impairment in their function.

Furthermore, alternative modes of translation, occurring in the absence of an initiating codon, generate abnormal poly(GA), poly(GP), poly(GR), poly(PR), and poly(PA) - rich peptides (C9RAN proteins) that form neuronal inclusions. The inclusions may play a role as patho-mechanism in ALS [128].

In ALS, an intriguing hypothesis postulates a prion-like propagation of misfolded proteins as possible mechanism of aggregation and spreading. Indeed, TDP-43 has a prion-like domain [129].

Additionally, ALS candidate genes like TDP-43, Optineurin, FUS, SOD1, SQSTM1 and VCP may contribute to the initiation of inflammatory processes. Given that these proteins can all be associated with the capacity to increase NF-κB activity/activation [130, 131], NF-κB-mediated neuroinflammation caused by protein aggregates might be a common feature in ALS.

GENES AND ENVIRONMENT

Most neurodegenerative disorders are multi-factorial and therefore, both genetic predisposition and environmental exposure contribute to their pathology. With improved and fast technologies to identify genetic mutations, the association of novel genetic variants in diseased individuals with a neurodegenerative disorder was facilitated within the last years. Unfortunately, the majority of cases of neurodegenerative disorders are sporadic and often not mimicked in the animal models with the same genetic mutation. To investigate the genetics of the most common neurodegenerative disorders, hundreds of genetic variants have been comprehensively studied in order to identify a possible risk factor in sporadic cases, but the results are often inconclusive. This gives rise to a model of complex gene-to-gene and gene-to-environment interactions (Fig. **8**).

Genetics of Neurodegenerative Diseases

There are several ways to identify a possible genetic predisposition. One is a so–called candidate gene approach, where a certain candidate gene is studied to identify an association between the gene of interest and the disease. Another way is to perform a genome wide association study (GWAS), where many DNA samples are studied simultaneously in order to find a connection to a disease phenotype. Such studies also include single nucleotide polymorphisms (SNPs). For complex and common neurodegenerative disorders, databases have been established that include all known genetic risk factors associated with a disorder. In general, a mutation acts as risk factor either via a gain-of-function or loss-of-function mechanism.

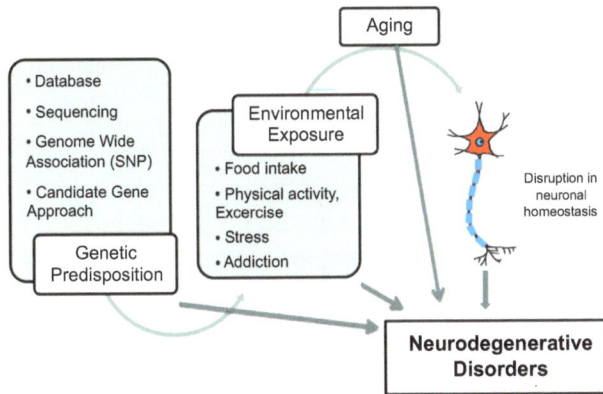

Fig. (8). Complex gene-environment interactions lead to neurodegenerative diseases. Genetic predisposition can be identified using candidate gene approaches or genome wide association studies. Genetic predisposition contributes to familial forms of the diseases, which account only for 5-10% of the total cases. Therefore, besides genetic mutations with high penetrance, environmental factors possibly linked to predisposing genetic variants may contribute to the pathology of neurodegenerative diseases. Environmental exposure includes many factors (see below). Along with these genetic predispositions and environmental factors, age is another contributing factor for the development of these diseases.

Candidate Genes for Alzheimer's Disease

AD is a genetically complex disorder. Besides age, genetic predisposition is the most common risk factor for AD. Familiar and twin studies have revealed a strong association between AD and genetic mutations. However, so far, mutations in only three genes have been associated with familial early onset (before 65 years) of the disease including amyloid precursor protein (*APP*), presinilin 1 (*PSEN1*) and presinilin 2 (*PSEN2*). The majority of AD cases is sporadic and has been additionally associated with several low penetrance genes (Table **4**). Among them, specific genetic variants of Apolipoprotein E (*APOE*) are commonly found in the patients suffering from late onset AD [132].

Table 4. Common candidate genes associated with AD.

Protein	Gene	Locus	Inheritance	Disease Onset	Common Mutations	Possible Pathology
Amyloid Precursor Protein	*APP*	21q21.2	Dominant	Early	Duplication, Pathologic mutations	Increased total $A\beta$ and $A\beta_{42}$
Presinilin 1	*PSEN1*	14q24.3	Dominant	Early	Point mutation	Increased ratio of $A\beta_{42}/A\beta_{40}$
Presinilin 2	*PSEN2*	1q31-q42	Dominant	Early	Point mutation	Increased ratio of $A\beta_{42}/A\beta_{40}$
Apolipoprotein	*APOE*	19q13	Dominant	Late	ε4 allele (Cys112, Arg158)	Impaired $A\beta$ clearance

(Table 4) contd.....

Protein	Gene	Locus	Inheritance	Disease Onset	Common Mutations	Possible Pathology
Microtubule Associated Protein Tau	*MAPT*	17q21.31	Recessive	Early	R406W	Unclear

Mutations in APP: In 1991, a linkage between AD and an APP missense mutation was first identified. Later, several further mutations have been found in APP. Most of these mutations are situated in the vicinity of the secretase cleavage sites. Therefore, it is possible that these "pathological" mutations are mostly associated with the production and aggregation of Aβ (Table **5**). However, besides these pathological mutations, duplication or overexpression of APP alone is sufficient to cause early onset AD [133].

Table 5. Pathological Mutations in APP(770) affecting production and aggregation of Aβ [80, 134, 135].

Notation of the mutation	Mutation	Pathology/Phenotype
Swedish	K670N, M671L	Increased Aβ production, increased beta secretase potentiation
Flemish	A692G	Make Aβ resistant to proteolytic clearance (mainly by neprilysin), thus increasing the half-life of Aβ and facilitating fibril formation.
Dutch	E693G	
Arctic	E693Q	
Austrian	T714I	Increased intraneuronal accumulation of Aβ
French	V715M	Unclear
Florida	I716V	Unclear
London	V717F	Impairment in long term memory
Indiana	V717L	Increased fibril production, increased Aβ production
Australian	L723P	Unclear

Mutations in Presinilins: By the early 1990's, it was clear that the mutations in the APP gene account only for a very small percentage of the early onset AD cases. In 1995, two other mutations were identified on chromosome 14 and 1, namely in the presinilin 1 and 2 (*PSEN1* and *PSEN2*) genes. Till now, more than 150 mutations have been listed in the Frontotemporal Dementia Mutation Database for *PSEN1* and more than 10 mutations have been listed for *PSEN2* [136 - 138].

Though presinilins were initially identified for their role in AD pathology, later their biological functions were characterized. Generation of Aβ from APP requires the cleavage by gamma secretase. The gamma secretase complex is made of integral membrane proteins including presinilin, nicastrin, Aph1 and Pen-2.

Presinilin1 lies at the catalytic core of this secretase complex and both presinilins facilitate the cleavage of APP by gamma secretase by acting as aspartyl proteases. The mutations in the presinilins cause a subtle shift in the cleavage of the transmembrane domain of APP thus altering the $A\beta_{40}/A\beta_{42}$ ratio, which is key to the development of senile plaques [139, 140].

Variants of Apolipoprotein E: One allelic variant of the apolipoprotein E (*APOE*) gene has been associated with late onset AD, which is the ε4 allele. There are three major alleles of the APOE gene, based on the amino acid in the positions 112 and 158. The ε2 allele has cysteine in both the positions, whereas ε4 has arginine. The ε3 allele has cysteine at position 112 and arginine at position 158. The most common allele is ε3. The presence of the ε4 allele increases the risk for AD several folds, whereas the ε2 allele decreases the risk. The ε4 allele also acts in a dose dependent manner. This allele has been associated with high plasma cholesterol levels. Physiologically APOE has a role on lipid metabolism and transport. In the AD brain, it is postulated to have a role in Aβ clearance [141, 142].

Other Early Onset Candidate Genes: From genome wide association studies and systemic meta analysis, several hundred genes and single nucleotide polymorphism have been identified. For more information, please refer to the publicly available database AlzGene. Among these genes, the most promising candidates are discussed briefly.

Angiotensin converting enzyme (ACE): ACE1 codes for a zinc metalloprotease which is expressed ubiquitously. The main function of this protein is to regulate blood pressure. Three different polymorphisms have been associated with AD. Besides these, the transposon *Alu* insertion in ACE1 is a risk factor variant for AD. ACE has also been associated with some cardio and cerebro-vascular diseases [143 - 145].

Cholesterol 25 hydroxylase (CH25H): CH25H codes for cholesterol 25 hydroxylase, a transcriptional regulator involved in lipid and cholesterol metabolism. The genetic locus of this gene is on the long arm of chromosome 10, within a broad AD linkage region. Some haplotypes of CH25H have been associated with high lathosterol (a precursor of cholesterol) levels in brain leading to increased accumulation of Aβ [146, 147].

Microtubule associated protein tau (MAPT): Accumulation of microtubule associated hyperphosphorylated tau protein is another hallmark of AD. Some SNP variants in MAPT have been associated with frontotemporal lobe degeneration and Parkinson's related dementia [148, 149].

Parkinson's Disease

PD has long been considered as a sporadic disease. The first indication that this disease can be inherited came from the study of Leroux in 1880 where it was suggested that there could be a possible genetic background for PD. So far, around 7 genes have been definitely proven to be associated with the pathogenecity of PD (Table **6**) and several further genes increasing the susceptibility for PD have been identified using genome wide association studies. However, these genes need to be verified to be included in the routine identification process. However, despite having the same mutation, the age of onset, penetrance of the mutation, and severity of symptoms may show high variability.

Candidate Genes for Parkinson's Disease

Alpha synuclein (SNCA): α-syn is the major structural component of the Lewy body. The very first genetic study for PD was the study of the pathogenic mutations in the SNCA gene. SNCA codes for the protein α-syn, which is expressed throughout the brain and has functions on dopamine synthesis, synaptic plasticity, and vesicle dynamics [150, 151]. Several point mutations have been identified in the coding region and such mutations are dominantly inherited. The N-terminus of α-syn forms an amphipathic alpha helical structure and thereby is able to bind with lipids. It is a potent inhibitor of phospholipase D2 and competitive inhibitor of tyrosine hydroxylase. The conversion of tyrosine to L-DOPA is the rate-limiting step in the L-DOPA biosynthesis and tyrosine hydroxylase catalyzes this step. Therefore, the equilibrium between cytoplasmic and lipid associated α-syn is crucial for neurotransmitter generation and vesicle dynamics. Point mutations in α-syn increase the cytoplasmic portion of the protein, which favors the formation of a beta sheet conformation and ultimately fibrils [152]. Overexpression of α-syn can induce similar effects.

Leucine-rich Repeat Kinase 2 (LRRK 2): The leucine-rich repeat kinase 2 codes for a protein with 2527 amino acids length with 5 conserved domains [153, 154]. The protein is expressed in many brain regions including the substantia nigra. LRRK2 is found in Lewy bodies, but the co-localization is not exclusive. One domain of the LRRK2 encoded protein is called a leucine-rich region because it contains a large amount of leucine and mediates interactions with other proteins. Furthermore, the protein has an enzymatic function acting as kinase, thus phos-phorylating target proteins. However, the protein may have a second enzymatic function referred to as a GTPase activity. More than 100 *LRRK2* gene mutations have been identified in families with late-onset Parkinson disease. However, how the point mutations cause neuropathy is currently not clearly understood.

Parkin: The Parkin protein has an ubiquitin like domain at the N-terminus and two ring finger domains. It is postulated that Parkin works as E3 ubiquitin ligase, marking inappropriately folded protein for degradation [155]. Besides point mutations, duplication, deletion and rearrangements have been identified within the genomic region. These mutations are relatively rare and cause pathology mainly via a loss of function mechanism. However, intracellular aggregation of ubiquitinylated proteins is common in many neurodegenerative diseases. Prior to the use of α-syn as marker in immunohistochemistry, ubiquitin staining was used to identify Lewy bodies. Thus, disruption of the ubiquitination system is an important part of the PD pathology.

PTEN Induced Kinase 1 (PINK1): PINK1 is expressed ubiquitously and has a highly conserved kinase domain. It also functions as a mitochondrial targeting motif. Mutations in this protein affect protein stability, localization and kinase activity [156, 157].

Oncogene DJ1: Recessively inherited deletion and missense mutations are found in oncogene DJ1. Such mutations can cause early onset of the disease, but are very rare. The protein belongs to a group of molecular chaperones and is induced after oxidative stress. A complete loss of function is required for this protein to cause the pathology of PD [158, 159].

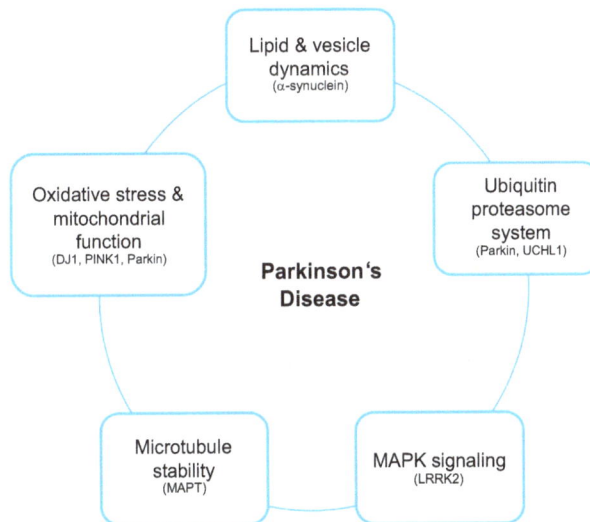

Fig. (9). Role of the mutations of different genes in PD pathology. Mutation in α-syn can modify lipid and vesicle dynamics, whereas DJ1, PINK1 and Parkin can be associated with oxidative stress and improper mitochondrial function. Parkin also has a role in ubiquitin proteosome systems, besides UCHL1. Mutation in LRKK2 can alter MAPK signaling and mutation in MAPT can destabilize microtubule. This figure also points towards the pathway that can be targeted as therapeutics for treating PD.

Besides these genes, microtubule associated protein tau (MAPT) and carboxy terminal esterase L1 (UCHL1) might contribute to the genetics of PD. These genes and their possible role in PD pathology have been summarized in Table **6** and (Fig. **9**).

Table 6. Risk factor genes associated with PD (AA: amino acids).

	Protein	Gene	Locus	Common Mutations	Physiological Function	Pathology
Dominantly Inherited	Alpha Synuclein (SNCA) (140 AA)	*PARK1* & *PARK4*	4q21	Dominant Substitute: 209G>A (Ala53Thr) 88G>C(Ala30Pro) 188G>AGlu46Lys Copy number variation	Synaptic Plasticity, Vesicle Dynamics, Dopamine Synthesis, Inhibitor of Phospholipase D2 and tyrosine hydroxylase	Aggregation and Lewy body formation
	Leucine-rich Repeat Kinase 2 (LRRK 2) (2527 AA)	*PARK8*	12p11.2	Dominant Substitute: Gly2019Ser, Ile2012Thr, Arg1441Cys/Gly/His, Tyr1699Cys, Ile2020Thr	Protein phosphorylation, Protein-protein interaction, Substrate binding	Mainly Lewy body disease
Recessively Inherited	Parkin (465 AA)	*PARK2*	6q25.2-q27	Point and deletion mutation	E3 ubiquitin ligase	Nigral neuronal loss
	pTEN Induced Kinase 1 (PINK1) (581 AA)	*PARK6*	1p35-p36	Haploinsufficiency, Missense and deletion mutation	Mitochondrial targeting motif	Unclear
	Oncogene DJ1	*PARK7*	1p36	Missense and deletion mutation	25CM telomeric of PINK, Chaperone	Unclear
	UCHL1	*PARK5*	4p14	Substitution (Ile93Met), Polymorphism	Ubiquitin C terminal hydrolase	Prominent component of Lewy body

Candidate Genes for ALS

The earliest identification of ALS was done in the 1840s. In 1848, Aran reported several cases of muscle weakness. One of these patients was suffering from clamp and upper limb paralysis and he died two years later. Several of the family members of another patient were also suffering from similar symptoms, which gave a hint that this disease might be hereditary [160]. In 1873, however, Charcot postulated that ALS was never hereditary and used this criterium to distinguish ALS from spinal muscular atrophy [161]. In contrast, indeed, there are both familiar and sporadic forms of ALS (termed as FALS and SALS respectively). It is very hard to distinguish between FALS and SALS biologically. So far, more that 10 genes have been associated with FALS. Mutations in these genes can be dominantly inherited, recessively inherited and the penetrance also varies.

Superoxide dismutase 1 (SOD1): The cellular function of superoxide dismutase 1 is to catalyze the reduction of superoxide anions to O_2 and H_2O_2. This enzyme is often also called Cu-Zn SOD1 as there is a Cu ion in its catalytic core and a Zn ion stabilizes the structure. It is expressed in all cell types [162]. First in 1991, 11 missense mutations were identified in the SOD1 sequence using linkage analysis [163]. Today, around 166 mutations have been identified. Most of them are missense mutations, but there are also nonsense mutations and deletions. Such mutations are able to generate truncations and affect the enzymatic activity of the protein [164 - 167].

TAR DNA-binding protein (TDP-43): TDP-43 is a transcription factor. Many patients suffering from ALS or frontotemporal dementia (FTD) have ubiquitin positive neuronal cytoplasmic inclusions. Around 44 mutations in the TDP-43 coding sequence have been identified so far and almost all of them are missense mutations. Due to such mutations, TDP-43 is redistributed from nucleus to neuronal cytoplasm and in glia cells in the spinal cord [168 - 173].

Fused in sarcoma (FUS): FUS is a protein similar to TDP-43. The main function of this protein is postulated to be in transcription factor regulation and genomic maintenance. Several missense mutations have been identified in the coding sequence, most affecting the C-terminus. The resulting phenotype can be ALS, FTD and/or parkinsonism [174 - 178].

Some rare mutations in the genes alsin (GTPase regulator), senataxin (involved in RNA processing), VAMP associated protein type B (involved in intracellular membrane transportation), angiotensin (involved in vascular endothelial growth factor metabolic pathway), optineurin (role in vesicle trafficking), ataxin 2, ubiquillin 2 (regulates proteosomal degradation of ubiquitinated proteins), and hexanucleotide repeat expansions in C9ORF72 have also been associated with ALS [162, 131].

Environmental Risk Factors

Multiple environmental factors have been reported to contribute to the etiology of neurodegenerative disorders such as aging, abnormalities in trace metal status [179], head injury [180], and exposure to certain toxins and chemical compounds [181] (Fig. **10**. However, it remains unknown whether a single agent or a combination of risk factors, or a combination of a risk factor and a certain genetic predisposition, contribute to disease onset and progression.

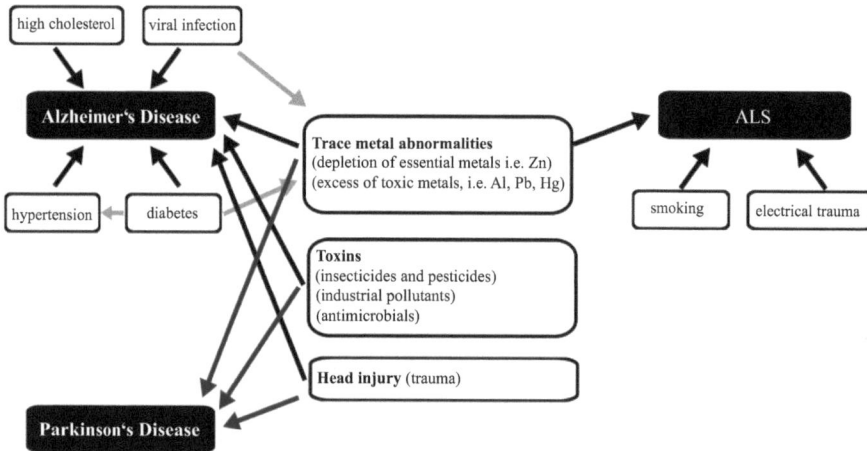

Fig. (10). Environmental risk factors of neurodegenerative disorders. Several risk factors have been associated with neurodegenerative disorders (black arrows) but are also themselves linked (light grey arrows).

Trace Metal Alterations

Several lines of evidence hint towards an influence of biometal concentrations in the CNS on the initiation and progression of neurodegenerative disorders. On the one hand, candidate genes associated with AD, PD or ALS and other neurodegenerative diseases encode for proteins that bind trace metals or that are involved in trace metal homeostasis. In AD for example, presenilins are important for cellular Cu and Zn turnover [182] and metals have been implicated to specifically interact with Aβ and tau. Especially Cu and Zn are known to associate with senile plaques [183]. In PD, α-syn has been shown to bind Cu and Fe, which enhances α-syn fibrillation rate [184]. In ALS, TDP-43 and ANG interact with metal ions and the loss of Zn from SOD1 results in the remaining Cu in SOD1 to become toxic to motor neurons [185, 186]. In Huntigton's disease, Huntingtin was found to interact with Cu [187]. Moreover, the prion protein (PrP) is known to be a metal binding protein [188, 189] and might be involved in neuronal Zn homeostasis due to an evolutionary link to the Zrt/ Irt-like protein (ZIP) zinc ion transporter family [190, 191].

On the other hand, specific alterations in biometal profiles in patients can be observed [179]. For example, a decrease in Zn in serum and blood of AD patients, while an increase of Zn in CSF has been shown [179]. In PD, Ag, Cd, Co, Fe, Se and Zn were found decreased in serum while an increase in Al, Ca, Cr, Hg, Mg, Mn, and Pb was noticed [192]. Differences for Zn and Cu concentrations were also found in serum and/or CSF of older patients with ALS [193] and an increase in plasma l-Ferritin, a Fe binding protein has been reported, which has been correlated with survival time of patients [194].

However, changes in trace metals in neurodegenerative disorders are complex given that the capacity of protein aggregates to sequester some metal ions will lead to mislocalized trace metals rather than simply a decrease or increase on systemic level. In contrast, altered trace metal levels preceding neuropathological changes might indeed influence the initiation of aggregation and progression of proteins. For example, high arsenic levels can induce hyperphosphorylation of tau [195] and APP transcription [196]. Recently, also a particular role of Al in AD was proposed [197 - 199]. Al overload is able to induce neurofibrillary degeneration, oxidative stress and inflammatory responses. Additionally, Al may act as a cross-linker for amyloid-β oligomerization as shown by *in vitro* studies [200].

Since the distribution of metals in the brain is inhomogeneous, local differences in metal ions may provide an interesting model for the brain region specific distribution of aggregates and selective vulnerability of some neuronal sub-populations. For example, multiple studies have shown an increase in Fe levels in brains of PD patients. There, dopaminergic cells seem to be especially vulnerable to damage by excessive Fe. A decrease in Neuromelanin that is able to sequester free Fe might mediate this cellular toxicity [201] with too high free Fe in turn promoting α-syn aggregation.

Given that trace metal absorption takes place in the gastro-intestinal (GI) system, altered trace metal levels such as zinc deficiency due to GI problems, will ultimately affect the brain. An influence of the so-called gut-brain axis in neurodegeneration is supported by the finding that specific dietary patterns may modify the pathology of neurodegenerative disorders. Furthermore, alterations in the GI system i.e. causing inflammation may be part of a relationship between the GI system, lifestyle factors such as diet, and cognitive decline. Additionally, a link between several vascular and metabolic diseases that are highly associated with dietary risk factors, and dementia has been shown.

Other Risk Factors

Genetic predisposition is believed to account for the majority of the overall risk to develop a neurodegenerative disorder with limited contribution to the remaining risk by factors such as trace metal imbalances, obesity, smoking, lack of exercise, mid-life hypertension, diabetes, head injury, and exposure during life to environmental agents [202, 203]. Among them, alterations in biometals or exposure to toxic heavy metals seem to show the strongest association with the development of neurodegenerative disorders. Thus, other risk factors will only be discussed briefly here.

Environmental agents discussed as risk factors include pesticides (organochlorine and organophosphate insecticides), industrial chemicals (flame retardants) and air pollutants (particulate matter). It is speculated that the long-term exposure to these contaminants in combination with their accumulation within the body over a life-span may trigger the development of neurodegenerative disorders mainly by induction of neuroinflammation.

For example, in AD, an association between occupational exposure to pesticides and AD was observed and a statistically significantly increased risk to develop AD was observed for fumigants and defoliants. The concentration of pesticides measured in the brain of individuals was additionally reported to be associated with AD. However, not all studies could verify these findings [204].

Some epidemiologic studies have reported an association between traumatic brain injury (TBI) and increased risk of AD and other neurodegenerative disorders even years after the injury. Additionally, AD appears to be closely tied to cardiovascular disease, supporting also an association of high cholesterol levels and hypertension with AD [205]. Also, the association of the ε4 variant of apolipoprotein E (APOE), a major lipid transporter, suggests an altered role in both normal and aberrant lipid metabolism in the pathogenesis of AD [206].

Increasing numbers of recent observations have detected *Herpes simplex* virus type-1 (HSV-1) and other viruses and bacteria in the brain of individuals with AD [207] and several studies showed that HSV-1 is a significant risk factor in carriers of the APOE ε4 variant. This again might be linked to an increase in neuroinflammation.

In PD, multiple epidemiologic studies have found a positive association between PD and the exposure to pesticides, herbicides, and insecticides. Increased levels of pesticides were found in the brains of PD cases compared to healthy controls . In line with this, farming occupation and farm residence that are related to pesticide exposure have been reported to be associated with PD. However, these findings have not been replicated by all studies investigating the relationship between pesticide exposure and PD [204].

Similar to AD, meta-analysis indicates that a history of head trauma is associated with an increased higher risk to develop PD. Additionally, neuronal loss in the substantia nigra, altered dopaminergic metabolism, or altered synuclein pathology was reported in animal models of TBI [208, 209].

Far fewer studies have examined the association of pesticides and ALS than for both AD and PD with inconsistent results. However, smoking was reported as risk factor for ALS with a gender bias towards women, especially post-menopausal women, being at higher risk [210]. Furthermore, meta-analysis suggests a slight but statistically significant increase in ALS risk in individuals associated with "electrical" occupations and/or exposed to electromagnetic fields [210].

ANIMAL MODELS

Research in the past decades has led to the development of various animal models for neurodegenerative diseases. Especially genetic studies on hereditary forms of AD, PD, and ALS provided useful targets for genetic manipulation in mice and other species. Many of these newly developed animal models show phenotypic similarities to human patients and have provided invaluable information on the pathogenesis and pathophysiology of the disorders.

However, most animal models display only partial aspects of the specific disorders, which created an obstacle especially for translational approaches, diagnostics and prevention. For example, since age-related neurodegenerative diseases are largely human-specific, the animals might not develop the full neuropathology or all clinical phenotypes observed in humans due to a limited life-span. Nevertheless, animal models are an important tool and can be divided into genetic and environmental models. Unfortunately, most models take only one of these factors into account, which might be contributing to the only partial overlapping phenotype compared to the human situation. Some of the most important mouse models will be described below.

Animal Models for AD

Animal models for AD can be divided into models with amyloid and models with tau pathology. Unfortunately, almost no mouse model for AD so far displays both hallmarks for AD together. Most amyloid mouse models overexpress human APP (sometimes including mutations found in AD) under control of a brain-specific promoter. These mice display an age-dependent accumulation of Aβ including amyloid plaques, neuritic dystrophy, microgliosis and astrocytosis. On behavioral level, mild cognitive impairment in animals with severe Aβ deposition is observed. Additionally to APP, overexpression or knock-in of mutant presenilin (PS1 or PS2) has been performed (APPxPS mice). PS mice alone show elevated $A\beta_{42}$ levels, but few other phenotypes. In combination with APP overexpression, the phenotype of APP mice is strengthened. For example, earlier age-of- onset of

Aβ accumulation is observed.

Mouse models with tau pathology include the mice that overexpress human tau that usually contains described mutations known to results in tauopathy. These mice display tau inclusions that are identical to those of AD patients in terms of morphology, structure and biochemical composition. On behavioral level, motor disturbance and cognitive impairments occur.

Other AD mouse models include models with human ApoE ε2 /ε3 / ε4 knock-in. However, these mice do not develop amyloid or tau pathology but show alterations in cholesterol metabolism. Again, crossings with APP mice have been performed. Furthermore, the Ts65Dn mouse, a model of Down syndrome, shows synaptic and cognitive phenotypes that depend on APP gene dosage.

Apart from genetic models, seeding models have been established. In this case, infusion of brain extracts with Aβ or tau and/or synthetic Aβ peptides into young APP or tau mice leads to accelerated progression of the pathology.

Animal Models for PD

Animal models for PD have been established through genetic overexpression of full length or fragments of wild type or mutated α-syn. Although these animals show only slight behavioral abnormalities, α-syn deposits are widespread in the brain of these mice. Similarly, mice overexpressing mutant or wild type LRRK2 do not show behavioral deficits such as motor defects. Other mouse models include mice with deletion of parkin, Pink1, Pitx3, and DJ1. Along with altered dopamine transmission, some effects on locomotor activity were observed in these animals but no α-syn deposits. Similarly, mice with heterozygous knock-out of the transcription factor Nurr1 show no deposits but an age-dependent nigral cell loss and decrease in striatal dopamine leading to a decreased motor coordination and locomotor activity. In contrast, α-syn deposits can be observed in mice carrying a null mutation in c-Rel. These mice additionally show age-dependent locomotor deficits.

Besides seeding models that also have been reported for α-syn, environmentally triggered PD mouse models are available. For example, application of 6-Hydroxydopamine (6-OHDA), 1-methyl-4-phenyl-1,2,3,6-tetrahydropyridine (MPTP), Rotenone, Paraquat, Annonacin, or l-trans-pyrrolidine-2,4-dicarboxylate, which in some cases can be systemic or has to be local, all lead to some aspects of PD pathology.

Animal Models for ALS

Animal models for ALS are based on the introduction of ALS candidate genes carrying mutations in the genome of mice. For example, several SOD1 mice have been generated that show SOD1 aggregates, motor neuron degeneration and paralysis causing premature death. Furthermore, mice with knock-out or overexpression of TDP-43 have been established. Some of these show nuclear and cytosolic aggregates of TDP-43 and motor neuron degeneration. The overexpression of human PrP-FUS similarly causes loss of motor neurons. Currently, several mouse lines targeting C9ORF72 are under development.

Animal Models for HD

Given that HD is a monogenic disorder, many mouse models with genetic manipulation using various variants of the HTT gene have been established. Most of these models show aggregates of Htt such as nuclear inclusions. Additionally, characteristic effect on grip strength, gait, motor coordination and locomotor activity can be observed along with changes in metabolism.

DRUG TARGETS IN NEURODEGENERATIVE DISORDERS

Currently, there are no therapies available to cure neurodegenerative diseases, although for some of the diseases, medication that alleviates symptoms is available. For example, Memantine and Donepezil are used in AD and reported to slow the progression of dementia symptoms in some affected individuals. In PD, Levodopa is used to increase dopamine level in the brain, which is able to relieve some core-symptoms in patients. Thus, new treatment strategies are highly desired. To that end, in the past, several interesting drug targets have been in the focus of research.

Drug Targets in AD

Currently there are only five drugs approved by FDA to treat AD- Rivastigmine, Galantamine, Tacrine, Donepezil and Memantine [211]. These drugs target the acetylcholine esterase, and NMDA receptors and provide only symptomatic relief. Therefore, efforts are put into generating molecules that target the neurobiology of AD directly.

Since Aβ lies at the center of AD pathology and is a cleaved product from APP, the proteases that cleave APP to generate Aβ are an interesting target. One such target is the beta secretase or beta-site APP-cleaving enzyme (BACE). BACE belongs to the group of aspartic proteases. Several drugs have been developed for influencing such proteases for other diseases like HIV. Moreover, the structure of

BACE has been crystalized. BACE knock-out mice show decreased Aβ generation. Some peptidomimetic inhibitors of BACE have been already generated like GRL-8234. However, one problem arises due to the larger and structurally open enzymatic catalytic core of BACE [212 - 216].

Like beta secretases, gamma secretases have been a therapeutic target for treatment of AD. In 2004, Eli Lilly reported in a phase 1 trial with gamma secretase inhibitor LY450139, reduced Aβ level in brain. However, the gamma secretase is also an important element in the Notch signaling pathway. Therefore, inhibiting gamma secretase raises the concern about toxic side effects [217 - 220]. In addition, several non-steroid anti-inflammatory drugs (NSAIDs) have been tried to lower the $A\beta_{42}$ level in the brain like R-flurbiprofen. These NSAIDs work in a cyclooxygenase independent manner, but most likely involve gamma secretase pathways [221, 222].

Perhaps the most exciting therapeutic target in AD is Aβ vaccine. Transgenic mice have been immunized both with Aβ peptides as active immunization and with monoclonal antibody generated against Aβ peptide as passive immunization. Such immunization may result in active soluble Aβ clearance from the periphery, microglial clearance of Aβ oligomers via antibody-mediated pathways or even clearance of brain Aβ aggregates. However, in one case, patients treated with $A\beta_{42}$ antibody developed meningoencephalitis, which hampered further development of Aβ vaccination in humans. Still, several clinical and preclinical studies are ongoing showing success in slowing the cognitive decline seen in AD [223 - 229]. Another strategy to target Aβ aggregation is via inhibiting fibril formation. To that end, glycosaaminoglycans (GAG) mimetics that compete with binding of GAGs to amyloid fibrils can be useful. The biotechnology company Neurochem is investigating such a GAG molecule in treating AD in a phase 3 clinical trial [230, 231].

Biometal status - The zinc amyloid hypothesis suggests the sequestration of zinc ions within Aβ aggregates. This sequestration has effects on multiple processes. First of all, it reduces the zinc concentration in the extracellular space, but also affects intracellular levels. Zinc cannot exert its normal metabolic function in neurons [232 - 234]. For example, one pathway for Aβ degradation is the degradation by metalloproteinases, MMPs that bind zinc ions for activation and proper function. Thus, a re-distribution of zinc from Aβ aggregates towards their physiological binding partners is highly desired. Several attempts have been made in the past to reach this goal with promising results. *In vitro* studies and experiments using mouse models have shown that metal protein attenuating compounds (MPACs) can promote the solubilization and clearance of Aβ. For example, a specific effect of clioquinol (CQ, PBT1) and PBT2 as novel therapeutic approach for the treatment of AD was proposed. CQ and PBT2 sequester extracellular metal ions. CQ/PBT2-metal complexes are cell permeable and thus increase intracellular levels of the metals after intracellular localization. The activity of these MPACs thus seems to be mediated by prevention of toxic interactions between Aβ and metal ions outside the cell, and the redistribution of the metal ions into the cell to enable their physiological function. However, in clinical trials, no positive benefit of CQ for patients with AD has been found. PBT2 in turn improved cognition in a phase 2 clinical trial with AD patients and appeared to be safe and well tolerated [235]. Both CQ and PBT2 also inhibit the GSK3 enzyme which phosphorylates tau protein. Therefore, such therapeutic approach not only targets the interaction between metal ions and Aβ, but also the hyperphosphorylation of tau protein-the two hallmarks of AD [236].

In general, degradation of Aβ aggregates or decrease of Aβ production are prime goals in AD treatment. Here, ACE1 has been shown to degrade naturally occurring Aβ *in vitro*. In some studies, it has been observed that individuals suffering from high blood pressure in mid-life, have a higher tendency to develop AD later in life. Due to the role of ACE1 in blood pressure regulation, the anti-hypertensive medication for ACE1 regulation has shown some success in treating the symptoms of AD. ACE1 concentration is high in some brain regions like perivascular areas.

Cholesterol might also have a role in AD. Some variants of cholesterol 25 hydroxylase increase the lathosterol level in the brain, which makes the brain more susceptible to Aβ aggregation.

Proteins associated with internalization of APP or Aβ are also possible drug targets. One of such protein is sortilin related receptor *(SORL1)*. SORL1 is highly expressed in brain. It has a vacuolar protein-sorting motif and therefore, can bind both APP and APOE. SORL1 expression is reduced AD brain, which is associated with reduced Aβ internalization *in vitro* [237, 238].

Since aggregates of hyperphosphorylated tau protein are also abundant in AD patient brains, there are attempts to target the filament formation of tau protein and the enzymes that phosphorylates tau (i.e. GSK3β, drug in phase 3- tideglusib). However, such approaches need further *in vivo* validation [239 - 243]. Because of the tau pathology, the structure of microtubules is destabilized. Several drugs like paclitaxel have been tried in lieu of this to stabilize microtubules in mouse models. However, no such trial has been carried out in humans [244, 245].

Drug Targets in PD

Several promising drugs and drug targets have been identified in the last years.

However, still, Dopamine-replacement therapy remains the prominent treatment of PD and addresses mostly motor symptoms. Although synthetic dopamine agonists are available, Levodopa (L-DOPA) still is the major drug used for dopamine replacement, but the search for novel approaches continues. In line with this, dopamine receptors in the brain are potential targets. However, so far, most currently used drugs only activate D_2 and D_3 dopamine receptors but not D_1, which might be a major target for anti-Parkinsonian agents [246].

In addition, cholinergic drugs, namely anticholinergics, were among the first drugs used in PD. Given that many cholinergic systems are affected in PD, a modification of the balance between dopamine and acetylcholine levels in PD seems a desired goal. Some cholinesterase inhibitors, such as Rivastigmine and Donepezil were shown to improve cognition and in some cases motor function. However, anticholinergics and anticholinesterases produce unwanted side effects. Serotoninergic drugs on the other hand, such as the $5HT_{1A}$ receptor agonists sarizotan, 8-hydroxy-2-di-n-propylamino-tetralin (8-OH-DPAT), and Buspirone also showed some beneficial effects such as an extended action of L-DOPA and thus might be a promising class of drugs for PD. However, the diversity of 5-HT receptors raises the issue of a possible wanted or unwanted selectivity [246].

Furthermore, drugs targeting glutamatergic neurotransmission might reduce motor complications associated with L-DOPA therapy and AMPA receptor antagonists, such as E-2007, GYKI-47261 and the non-competitive inhibitor Talampanel, have entered into clinical trials in PD.

Another interesting class of drugs in PD are Opioid drugs. In particular, targeting opioid transmission in the basal ganglia might be a promising goal. However, so far, the use of non-selective opioid receptor antagonists such as Naloxone and Naltrexone has provided inconclusive results in PD patients and the use of selective opioid-receptor antagonists might be more interesting.

Finally, Adenosine A_{2A} receptor antagonists have been studied as PD drug targets. A_{2A} antagonists consistently reverse motor deficits in preclinical models of PD and initial studies in patients using an A_{2A} antagonist have shown slight symptomatic improvements [246].

Several other approaches have been taken to find novel treatment strategies in PD of which only a few can be mentioned here. For example, multiple attempts have been made to regenerate or deliver dopaminergic neurons. Here, delivery of stem cells might be a future approach. However, so far, infusions of glial-derived nerve growth factor (GDNF) seemed to increase the number of tyrosine hydroxylase-positive nigral neurons and dopamine levels in animal models. However, in early studies on human patients, mixed results were obtained depending on the site of

GDNF application (intraventricular injections vs. putaminal infusion) and a subsequent larger clinical trial generated negative results and was terminated prematurely because of side effects [246].

Furthermore, given that mutants of LRRK2 associated with PD hinted towards a role of enhanced kinase activity as possible patho-mechanism, inhibitors of LRRK2 kinase activity were proposed as treatment option. Two of these inhibitors, GW5074 and Sorafenib, showed promising results in model organisms [247].

TREATMENT OF NEURODEGENERATIVE DISORDERS USING NANOMEDICINE

Despite much efforts put into the development of treatments, the translation of findings into clinical applications remained rather low in the recent years. Thus, further research is needed to provide more sophisticated diagnoses, and to better understand the neurobiology of neurodegenerative disorders and the mechanisms involved (pathways/cell types, genetic- and more so environmental factors). Since the brain is one of the most vital organs in the body, substances that can enter the brain are strictly controlled via physical barriers like the blood-brain barrier (BBB). This makes the treatment of neurodegenerative diseases extremely complex as drugs need to cross the BBB and reach the brain in a sufficient dose dependent manner.

There are some pharmaceutical agents that can successfully treat many symptoms when they are administered using the proper route and in case they can reach the brain in a proper dose. Nanoparticles can be very efficient and useful in this respect. There are many formulations available to encapsulate in and coat pharmaceutical substance to nanocarriers. Such coupling of drugs and nanoparticles has many-fold advantages over traditional therapeutics. Using nanoparticles, it is possible to deliver both hydrophilic and hydrophobic substances across the BBB. Moreover, the degradation of drugs in systemic circulation is high when administered via a traditional route. However, using proper nanotechnology, it is possible to keep the drug intact until it reaches the target site, thus preventing it from enzymatic degradation. Therefore, a nanomedicine can ensure sustainable drug release, higher penetration, reduced side effects and targeted therapeutic effects [248, 249].

CONCLUSIONS

Most neurodegenerative disorders are chronic and incurable conditions lasting for years or even decades. Thus, the economic costs of neurodegenerative brain disorders are correspondingly large. Here, not only the cost of treatment, but also the lost productivity of patients and their caregivers has to be considered. Therefore, the development of new strategies to treat neurodegenerative diseases is one of the key priorities to decrease the socioeconomic burden caused by these disorders.

Currently, the emergence of nano-neuroscience provides novel approaches in treatment of neurodegenerative disorders given the potential of nano-carriers to bypass the BBB and other cellular barriers, and to specifically interact with biological systems at molecular level. Thus, in the upcoming years, with the advent of nanomedicine, new opportunities will merge that hopefully will lead to effective treatment of neurodegenerative disorders in the future.

CONFLICT OF INTEREST

The authors confirm that they have no conflict of interest to declare for this publication.

ACKNOWLEDGEMENT

TS is supported under the Postgraduate Scholarships Act of the Land of Baden-Wuerttemberg (LGFG), Germany.

REFERENCES

[1]   Gustavsson A, Svensson M, Jacobi F, *et al.* Cost of disorders of the brain in Europe 2010. Eur Neuropsychopharmacol 2011; 21(10): 718-79.
        [http://dx.doi.org/10.1016/j.euroneuro.2011.08.008] [PMID: 21924589]

[2]   Domínguez A, Álvarez A, Hilario E, Suarez-Merino B, Goñi-de-Cerio F. Central nervous system diseases and the role of the blood-brain barrier in their treatment. Neurosci Discov 2013; 1: 3.
        [http://dx.doi.org/10.7243/2052-6946-1-3]

[3]   Ferri CP, Prince M, Brayne C, *et al.* Global prevalence of dementia: a Delphi consensus study. Lancet 2005; 366(9503): 2112-7.
        [http://dx.doi.org/10.1016/S0140-6736(05)67889-0] [PMID: 16360788]

[4]   St George-Hyslop PH. Piecing together Alzheimers. Sci Am 2000; 283(6): 76-83.
        [http://dx.doi.org/10.1038/scientificamerican1200-76] [PMID: 11103462]

[5]   Duckett L. Alzheimers dementia: morbidity and mortality. J Insur Med 2001; 33(3): 227-34.
        [PMID: 11558402]

[6]   Reisberg B, Ferris SH, de Leon MJ, Crook T. The Global Deterioration Scale for assessment of primary degenerative dementia. Am J Psychiatry 1982; 139(9): 1136-9.
        [http://dx.doi.org/10.1176/ajp.139.9.1136] [PMID: 7114305]

[7]   Hughes CP, Berg L, Danziger WL, Coben LA, Martin RL. A new clinical scale for the staging of dementia. Br J Psychiatry 1982; 140: 566-72.
        [http://dx.doi.org/10.1192/bjp.140.6.566] [PMID: 7104545]

[8]   Bekris LM, Yu CE, Bird TD, Tsuang DW. Genetics of Alzheimer disease. J Geriatr Psychiatry Neurol 2010; 23(4): 213-27.
        [http://dx.doi.org/10.1177/0891988710383571] [PMID: 21045163]

[9]   Campion D, Dumanchin C, Hannequin D, *et al.* Early-onset autosomal dominant Alzheimer disease: prevalence, genetic heterogeneity, and mutation spectrum. Am J Hum Genet 1999; 65(3): 664-70. [http://dx.doi.org/10.1086/302553] [PMID: 10441572]

[10]   Brickell KL, Steinbart EJ, Rumbaugh M, *et al.* Early-onset Alzheimer disease in families with late-onset Alzheimer disease: a potential important subtype of familial Alzheimer disease. Arch Neurol 2006; 63(9): 1307-11. [http://dx.doi.org/10.1001/archneur.63.9.1307] [PMID: 16966510]

[11]   McKhann GM, Knopman DS, Chertkow H, *et al.* The diagnosis of dementia due to Alzheimers disease: recommendations from the National Institute on Aging-Alzheimers Association workgroups on diagnostic guidelines for Alzheimers disease. Alzheimers Dement 2011; 7(3): 263-9. [http://dx.doi.org/10.1016/j.jalz.2011.03.005] [PMID: 21514250]

[12]   Cummings JL. Alzheimers disease management. J Clin Psychiatry 1998; 59 (Suppl. 13): 4-5. [PMID: 9771824]

[13]   Cummings JL. Cognitive and behavioral heterogeneity in Alzheimers disease: seeking the neurobiological basis. Neurobiol Aging 2000; 21(6): 845-61. [http://dx.doi.org/10.1016/S0197-4580(00)00183-4] [PMID: 11124429]

[14]   Bandera L, Della Sala S, Laiacona M, Luzzatti C, Spinnler H. Generative associative naming in dementia of Alzheimers type. Neuropsychologia 1991; 29(4): 291-304. [http://dx.doi.org/10.1016/0028-3932(91)90043-8] [PMID: 1857501]

[15]   Henderson AS, Jorm AF. Some contributions to the epidemiology of dementia and depression. Int J Geriatr Psychiatry 1997; 12(2): 145-54. [http://dx.doi.org/10.1002/(SICI)1099-1166(199702)12:2<145::AID-GPS579>3.0.CO;2-3] [PMID: 9097207]

[16]   Reitz C, Mayeux R. Alzheimer disease: epidemiology, diagnostic criteria, risk factors and biomarkers. Biochem Pharmacol 2014; 88(4): 640-51. [http://dx.doi.org/10.1016/j.bcp.2013.12.024] [PMID: 24398425]

[17]   Chui HC, Lyness SA, Sobel E, Schneider LS. Extrapyramidal signs and psychiatric symptoms predict faster cognitive decline in Alzheimers disease. Arch Neurol 1994; 51(7): 676-81. [http://dx.doi.org/10.1001/archneur.1994.00540190056015] [PMID: 8018040]

[18]   Drevets WC, Rubin EH. Psychotic symptoms and the longitudinal course of senile dementia of the Alzheimer type. Biol Psychiatry 1989; 25(1): 39-48. [http://dx.doi.org/10.1016/0006-3223(89)90145-5] [PMID: 2912509]

[19]   Förstl H, Burns A, Levy R, Cairns N, Luthert P, Lantos P. Neuropathological correlates of behavioural disturbance in confirmed Alzheimers disease. Br J Psychiatry 1993; 163: 364-8. [http://dx.doi.org/10.1192/bjp.163.3.364] [PMID: 8401967]

[20]   Mortimer JA, Ebbitt B, Jun SP, Finch MD. Predictors of cognitive and functional progression in patients with probable Alzheimers disease. Neurology 1992; 42(9): 1689-96. [http://dx.doi.org/10.1212/WNL.42.9.1689] [PMID: 1513455]

[21]   Rosen J, Zubenko GS. Emergence of psychosis and depression in the longitudinal evaluation of Alzheimers disease. Biol Psychiatry 1991; 29(3): 224-32. [http://dx.doi.org/10.1016/0006-3223(91)91284-X] [PMID: 2015329]

[22]   Zubenko GS, Moossy J, Kopp U. Neurochemical correlates of major depression in primary dementia. Arch Neurol 1990; 47(2): 209-14. [http://dx.doi.org/10.1001/archneur.1990.00530020117023] [PMID: 1689144]

[23]   Chen CP, Alder JT, Bowen DM, *et al.* Presynaptic serotonergic markers in community-acquired cases of Alzheimers disease: correlations with depression and neuroleptic medication. J Neurochem 1996; 66(4): 1592-8. [http://dx.doi.org/10.1046/j.1471-4159.1996.66041592.x] [PMID: 8627315]

[24] Lees AJ, Hardy J, Revesz T. Parkinsons disease. Lancet 2009; 373(9680): 2055-66.
[http://dx.doi.org/10.1016/S0140-6736(09)60492-X] [PMID: 19524782]

[25] Fahn S. Description of Parkinsons disease as a clinical syndrome. Ann N Y Acad Sci 2003; 991: 1-14.
[http://dx.doi.org/10.1111/j.1749-6632.2003.tb07458.x] [PMID: 12846969]

[26] Van Den Eeden SK, Tanner CM, Bernstein AL, *et al.* Incidence of Parkinsons disease: variation by age, gender, and race/ethnicity. Am J Epidemiol 2003; 157(11): 1015-22.
[http://dx.doi.org/10.1093/aje/kwg068] [PMID: 12777365]

[27] Farrer MJ. Genetics of Parkinson disease: paradigm shifts and future prospects. Nat Rev Genet 2006; 7(4): 306-18.
[http://dx.doi.org/10.1038/nrg1831] [PMID: 16543934]

[28] Doty RL, Bromley SM, Stern MB. Olfactory testing as an aid in the diagnosis of Parkinsons disease: development of optimal discrimination criteria. Neurodegeneration 1995; 4(1): 93-7.
[http://dx.doi.org/10.1006/neur.1995.0011] [PMID: 7600189]

[29] Iranzo A, Santamaría J, Rye DB, *et al.* Characteristics of idiopathic REM sleep behavior disorder and that associated with MSA and PD. Neurology 2005; 65(2): 247-52.
[http://dx.doi.org/10.1212/01.wnl.0000168864.97813.e0] [PMID: 16043794]

[30] Meseguer E, Taboada R, Sánchez V, Mena MA, Campos V, García De Yébenes J. Life-threatening parkinsonism induced by kava-kava. Mov Disord 2002; 17(1): 195-6.
[http://dx.doi.org/10.1002/mds.1268] [PMID: 11835463]

[31] Cosentino C, Torres L, Scorticati MC, Micheli F. Movement disorders secondary to adulterated medication. Neurology 2000; 55(4): 598-9.
[http://dx.doi.org/10.1212/WNL.55.4.598] [PMID: 10953205]

[32] Modugno N, Lena F, Di Biasio F, Cerrone G, Ruggieri S, Fornai F. A clinical overview of non-motor symptoms in Parkinsons Disease. Arch Ital Biol 2013; 151(4): 148-68.
[PMID: 24873924]

[33] Leentjens AF, Van den Akker M, Metsemakers JF, Lousberg R, Verhey FR. Higher incidence of depression preceding the onset of Parkinsons disease: a register study. Mov Disord 2003; 18(4): 414-8.
[http://dx.doi.org/10.1002/mds.10387] [PMID: 12671948]

[34] Schuurman AG, van den Akker M, Ensinck KT, *et al.* Increased risk of Parkinsons disease after depression: a retrospective cohort study. Neurology 2002; 58(10): 1501-4.
[http://dx.doi.org/10.1212/WNL.58.10.1501] [PMID: 12034786]

[35] Aarsland D, Kurz MW. The epidemiology of dementia associated with Parkinson disease. J Neurol Sci 2010; 289(1-2): 18-22.
[http://dx.doi.org/10.1016/j.jns.2009.08.034] [PMID: 19733364]

[36] Hughes AJ, Daniel SE, Ben-Shlomo Y, Lees AJ. The accuracy of diagnosis of parkinsonian syndromes in a specialist movement disorder service. Brain 2002; 125(Pt 4): 861-70.
[http://dx.doi.org/10.1093/brain/awf080] [PMID: 11912118]

[37] Parnetti L, Castrioto A, Chiasserini D, *et al.* Cerebrospinal fluid biomarkers in Parkinson disease. Nat Rev Neurol 2013; 9(3): 131-40.
[http://dx.doi.org/10.1038/nrneurol.2013.10] [PMID: 23419373]

[38] Gelpi E, Navarro-Otano J, Tolosa E, *et al.* Multiple organ involvement by alpha-synuclein pathology in Lewy body disorders. Mov Disord 2014; 29(8): 1010-8.
[http://dx.doi.org/10.1002/mds.25776] [PMID: 24395122]

[39] Visanji NP, Marras C, Hazrati LN, Liu LW, Lang AE, Lang AE. Alimentary, my dear Watson? The challenges of enteric α-synuclein as a Parkinsons disease biomarker. Mov Disord 2014; 29(4): 444-50.
[http://dx.doi.org/10.1002/mds.25789] [PMID: 24375496]

[40]   Miller DB, OCallaghan JP. Biomarkers of Parkinsons disease: present and future. Metabolism 2015; 64(3) (Suppl. 1): S40-6.
       [http://dx.doi.org/10.1016/j.metabol.2014.10.030] [PMID: 25510818]

[41]   Rowland LP, Shneider NA. Amyotrophic lateral sclerosis. N Engl J Med 2001; 344(22): 1688-700.
       [http://dx.doi.org/10.1056/NEJM200105313442207] [PMID: 11386269]

[42]   Logroscino G, Traynor BJ, Hardiman O, *et al.* Incidence of amyotrophic lateral sclerosis in Europe. J Neurol Neurosurg Psychiatry 2010; 81(4): 385-90.
       [http://dx.doi.org/10.1136/jnnp.2009.183525] [PMID: 19710046]

[43]   Chiò A, Mora G, Calvo A, Mazzini L, Bottacchi E, Mutani R. Epidemiology of ALS in Italy: a 10-year prospective population-based study. Neurology 2009; 72(8): 725-31.
       [http://dx.doi.org/10.1212/01.wnl.0000343008.26874.d1] [PMID: 19237701]

[44]   OToole O, Traynor BJ, Brennan P, *et al.* Epidemiology and clinical features of amyotrophic lateral sclerosis in Ireland between 1995 and 2004. J Neurol Neurosurg Psychiatry 2008; 79(1): 30-2.
       [http://dx.doi.org/10.1136/jnnp.2007.117788] [PMID: 17634215]

[45]   Kiernan MC, Vucic S, Cheah BC, *et al.* Amyotrophic lateral sclerosis. Lancet 2011; 377(9769): 942-55.
       [http://dx.doi.org/10.1016/S0140-6736(10)61156-7] [PMID: 21296405]

[46]   Phukan J, Pender NP, Hardiman O. Cognitive impairment in amyotrophic lateral sclerosis. Lancet Neurol 2007; 6(11): 994-1003.
       [http://dx.doi.org/10.1016/S1474-4422(07)70265-X] [PMID: 17945153]

[47]   Raaphorst J, de Visser M, Linssen WH, de Haan RJ, Schmand B. The cognitive profile of amyotrophic lateral sclerosis: A meta-analysis. Amyotroph Lateral Scler 2010; 11(1-2): 27-37.
       [http://dx.doi.org/10.3109/17482960802645008] [PMID: 19180349]

[48]   Abrahams S, Leigh PN, Harvey A, Vythelingum GN, Grisé D, Goldstein LH. Verbal fluency and executive dysfunction in amyotrophic lateral sclerosis (ALS). Neuropsychologia 2000; 38(6): 734-47.
       [http://dx.doi.org/10.1016/S0028-3932(99)00146-3] [PMID: 10689049]

[49]   Brooks BR. El Escorial World Federation of Neurology criteria for the diagnosis of amyotrophic lateral sclerosis. Subcommittee on Motor Neuron Diseases/Amyotrophic Lateral Sclerosis of the World Federation of Neurology Research Group on Neuromuscular Diseases and the El Escorial Clinical limits of amyotrophic lateral sclerosis workshop contributors. J Neurol Sci 1994; 124 (Suppl.): 96-107.
       [http://dx.doi.org/10.1016/0022-510X(94)90191-0] [PMID: 7807156]

[50]   Miller RG, Munsat TL, Swash M, Brooks BR. Consensus guidelines for the design and implementation of clinical trials in ALS. World Federation of Neurology committee on Research. J Neurol Sci 1999; 169(1-2): 2-12.
       [http://dx.doi.org/10.1016/S0022-510X(99)00209-9] [PMID: 10540001]

[51]   Daube JR. Electrodiagnostic studies in amyotrophic lateral sclerosis and other motor neuron disorders. Muscle Nerve 2000; 23(10): 1488-502.
       [http://dx.doi.org/10.1002/1097-4598(200010)23:10<1488::AID-MUS4>3.0.CO;2-E]   [PMID: 11003783]

[52]   Eisen A, Swash M. Clinical neurophysiology of ALS. Clin Neurophysiol 2001; 112(12): 2190-201.
       [http://dx.doi.org/10.1016/S1388-2457(01)00692-7] [PMID: 11738189]

[53]   Turner MR, Kiernan MC, Leigh PN, Talbot K. Biomarkers in amyotrophic lateral sclerosis. Lancet Neurol 2009; 8(1): 94-109.
       [http://dx.doi.org/10.1016/S1474-4422(08)70293-X] [PMID: 19081518]

[54]   Braak H, Braak E. Neuropathological staging of Alzheimer-related changes. Acta Neuropathol 1991; 82(4): 239-59.
       [http://dx.doi.org/10.1007/BF00308809] [PMID: 1759558]

[55]   Braak H, Del Tredici K, Rüb U, de Vos RA, Jansen Steur EN, Braak E. Staging of brain pathology related to sporadic Parkinsons disease. Neurobiol Aging 2003; 24(2): 197-211.
[http://dx.doi.org/10.1016/S0197-4580(02)00065-9] [PMID: 12498954]

[56]   Braak H, Alafuzoff I, Arzberger T, Kretzschmar H, Del Tredici K. Staging of Alzheimer disease-associated neurofibrillary pathology using paraffin sections and immunocytochemistry. Acta Neuropathol 2006; 112(4): 389-404.
[http://dx.doi.org/10.1007/s00401-006-0127-z] [PMID: 16906426]

[57]   Thal DR, Rüb U, Orantes M, Braak H. Phases of A beta-deposition in the human brain and its relevance for the development of AD. Neurology 2002; 58(12): 1791-800.
[http://dx.doi.org/10.1212/WNL.58.12.1791] [PMID: 12084879]

[58]   Baker HF, Ridley RM, Duchen LW, Crow TJ, Bruton CJ. Evidence for the experimental transmission of cerebral beta-amyloidosis to primates. Int J Exp Pathol 1993; 74(5): 441-54.
[PMID: 8217779]

[59]   Baker HF, Ridley RM, Duchen LW, Crow TJ, Bruton CJ. Induction of beta (A4)-amyloid in primates by injection of Alzheimers disease brain homogenate. Comparison with transmission of spongiform encephalopathy. Mol Neurobiol 1994; 8(1): 25-39.
[http://dx.doi.org/10.1007/BF02778005] [PMID: 8086126]

[60]   Kane MD, Lipinski WJ, Callahan MJ, *et al.* Evidence for seeding of beta -amyloid by intracerebral infusion of Alzheimer brain extracts in beta -amyloid precursor protein-transgenic mice. J Neurosci 2000; 20(10): 3606-11.
[PMID: 10804202]

[61]   Rosen RF, Fritz JJ, Dooyema J, *et al.* Exogenous seeding of cerebral β-amyloid deposition in βAPP-transgenic rats. J Neurochem 2012; 120(5): 660-6.
[http://dx.doi.org/10.1111/j.1471-4159.2011.07551.x] [PMID: 22017494]

[62]   Clavaguera F, Akatsu H, Fraser G, *et al.* Brain homogenates from human tauopathies induce tau inclusions in mouse brain. Proc Natl Acad Sci USA 2013; 110(23): 9535-40.
[http://dx.doi.org/10.1073/pnas.1301175110] [PMID: 23690619]

[63]   Clavaguera F, Bolmont T, Crowther RA, *et al.* Transmission and spreading of tauopathy in transgenic mouse brain. Nat Cell Biol 2009; 11(7): 909-13.
[http://dx.doi.org/10.1038/ncb1901] [PMID: 19503072]

[64]   Hyman BT, Van Hoesen GW, Damasio AR, Barnes CL. Alzheimers disease: cell-specific pathology isolates the hippocampal formation. Science 1984; 225(4667): 1168-70.
[http://dx.doi.org/10.1126/science.6474172] [PMID: 6474172]

[65]   Arnold SE, Hyman BT, Flory J, Damasio AR, Van Hoesen GW. The topographical and neuroanatomical distribution of neurofibrillary tangles and neuritic plaques in the cerebral cortex of patients with Alzheimers disease. Cereb Cortex 1991; 1(1): 103-16.
[http://dx.doi.org/10.1093/cercor/1.1.103] [PMID: 1822725]

[66]   Del Tredici K, Duda JE. Peripheral Lewy body pathology in Parkinsons disease and incidental Lewy body disease: four cases. J Neurol Sci 2011; 310(1-2): 100-6.
[http://dx.doi.org/10.1016/j.jns.2011.06.003] [PMID: 21689832]

[67]   Brettschneider J, Del Tredici K, Lee VM, Trojanowski JQ. Spreading of pathology in neurodegenerative diseases: a focus on human studies. Nat Rev Neurosci 2015; 16(2): 109-20.
[http://dx.doi.org/10.1038/nrn3887] [PMID: 25588378]

[68]   Candelario KM, Steindler DA. The role of extracellular vesicles in the progression of neurodegenerative disease and cancer. Trends Mol Med 2014; 20(7): 368-74.
[http://dx.doi.org/10.1016/j.molmed.2014.04.003] [PMID: 24835084]

[69]   Klucken J, Ingelsson M, Shin Y, *et al.* Clinical and biochemical correlates of insoluble alpha-synuclein in dementia with Lewy bodies. Acta Neuropathol 2006; 111(2): 101-8.
[http://dx.doi.org/10.1007/s00401-005-0027-7] [PMID: 16482476]

[70]   Brettschneider J, Del Tredici K, Irwin DJ, *et al.* Sequential distribution of pTDP-43 pathology in behavioral variant frontotemporal dementia (bvFTD). Acta Neuropathol 2014; 127(3): 423-39.
[http://dx.doi.org/10.1007/s00401-013-1238-y] [PMID: 24407427]

[71]   Brettschneider J, Del Tredici K, Toledo JB, *et al.* Stages of pTDP-43 pathology in amyotrophic lateral sclerosis. Ann Neurol 2013; 74(1): 20-38.
[http://dx.doi.org/10.1002/ana.23937] [PMID: 23686809]

[72]   Kassubek J, Müller HP, Del Tredici K, *et al.* Diffusion tensor imaging analysis of sequential spreading of disease in amyotrophic lateral sclerosis confirms patterns of TDP-43 pathology. Brain 2014; 137(Pt 6): 1733-40.
[http://dx.doi.org/10.1093/brain/awu090] [PMID: 24736303]

[73]   Jucker M, Walker LC. Self-propagation of pathogenic protein aggregates in neurodegenerative diseases. Nature 2013; 501(7465): 45-51.
[http://dx.doi.org/10.1038/nature12481] [PMID: 24005412]

[74]   Braak H, Brettschneider J, Ludolph AC, Lee VM, Trojanowski JQ, Del Tredici K. Amyotrophic lateral sclerosis a model of corticofugal axonal spread. Nat Rev Neurol 2013; 9(12): 708-14.
[http://dx.doi.org/10.1038/nrneurol.2013.221] [PMID: 24217521]

[75]   Braak H, Zetterberg H, Del Tredici K, Blennow K. Intraneuronal tau aggregation precedes diffuse plaque deposition, but amyloid-β changes occur before increases of tau in cerebrospinal fluid. Acta Neuropathol 2013; 126(5): 631-41.
[http://dx.doi.org/10.1007/s00401-013-1139-0] [PMID: 23756600]

[76]   Serrano-Pozo A, Frosch MP, Masliah E, Hyman BT. Neuropathological alterations in Alzheimer disease. Cold Spring Harb Perspect Med 2011; 1(1): a006189.
[http://dx.doi.org/10.1101/cshperspect.a006189] [PMID: 22229116]

[77]   Yoshikai S, Sasaki H, Doh-ura K, Furuya H, Sakaki Y. Genomic organization of the human amyloid beta-protein precursor gene. Gene 1990; 87(2): 257-63.
[http://dx.doi.org/10.1016/0378-1119(90)90310-N] [PMID: 2110105]

[78]   Kong GK, Adams JJ, Harris HH, *et al.* Structural studies of the Alzheimers amyloid precursor protein copper-binding domain reveal how it binds copper ions. J Mol Biol 2007; 367(1): 148-61.
[http://dx.doi.org/10.1016/j.jmb.2006.12.041] [PMID: 17239395]

[79]   Mattson MP. Cellular actions of beta-amyloid precursor protein and its soluble and fibrillogenic derivatives. Physiol Rev 1997; 77(4): 1081-132.
[PMID: 9354812]

[80]   Evin G, Weidemann A. Biogenesis and metabolism of Alzheimers disease Abeta amyloid peptides. Peptides 2002; 23(7): 1285-97.
[http://dx.doi.org/10.1016/S0196-9781(02)00063-3] [PMID: 12128085]

[81]   Dahms SO, Könnig I, Roeser D, *et al.* Metal binding dictates conformation and function of the amyloid precursor protein (APP) E2 domain. J Mol Biol 2012; 416(3): 438-52.
[http://dx.doi.org/10.1016/j.jmb.2011.12.057] [PMID: 22245578]

[82]   Hicks DA, Nalivaeva NN, Turner AJ. Lipid rafts and Alzheimers disease: protein-lipid interactions and perturbation of signaling. Front Physiol 2012; 3: 189-207.
[http://dx.doi.org/10.3389/fphys.2012.00189] [PMID: 22737128]

[83]   Thomas P, Fenech M. A review of genome mutation and Alzheimers disease. Mutagenesis 2007; 22(1): 15-33.
[http://dx.doi.org/10.1093/mutage/gel055] [PMID: 17158517]

[84]   Barrow CJ, Zagorski MG. Solution structures of beta peptide and its constituent fragments: relation to amyloid deposition. Science 1991; 253(5016): 179-82.
[http://dx.doi.org/10.1126/science.1853202] [PMID: 1853202]

[85]   Finder VH, Glockshuber R. Amyloid-beta aggregation. Neurodegener Dis 2007; 4(1): 13-27.
[http://dx.doi.org/10.1159/000100355] [PMID: 17429215]

[86]   Selkoe DJ. Alzheimers disease is a synaptic failure. Science 2002; 298(5594): 789-91.
[http://dx.doi.org/10.1126/science.1074069] [PMID: 12399581]

[87]   Rosen AM, Stevens B. The role of the classical complement cascade in synapse loss during development and glaucoma. Adv Exp Med Biol 2010; 703: 75-93.
[http://dx.doi.org/10.1007/978-1-4419-5635-4_6] [PMID: 20711708]

[88]   Glass CK, Saijo K, Winner B, Marchetto MC, Gage FH. Mechanisms underlying inflammation in neurodegeneration. Cell 2010; 140(6): 918-34.
[http://dx.doi.org/10.1016/j.cell.2010.02.016] [PMID: 20303880]

[89]   Duce JA, Tsatsanis A, Cater MA, *et al.* Iron-export ferroxidase activity of β-amyloid precursor protein is inhibited by zinc in Alzheimers disease. Cell 2010; 142(6): 857-67.
[http://dx.doi.org/10.1016/j.cell.2010.08.014] [PMID: 20817278]

[90]   Kenche VB, Barnham KJ. Alzheimers disease & metals: therapeutic opportunities. Br J Pharmacol 2011; 163(2): 211-9.
[http://dx.doi.org/10.1111/j.1476-5381.2011.01221.x] [PMID: 21232050]

[91]   Lovell MA, Robertson JD, Teesdale WJ, Campbell JL, Markesbery WR. Copper, iron and zinc in Alzheimers disease senile plaques. J Neurol Sci 1998; 158(1): 47-52.
[http://dx.doi.org/10.1016/S0022-510X(98)00092-6] [PMID: 9667777]

[92]   Huang X, Atwood CS, Moir RD, *et al.* Zinc-induced Alzheimers Abeta140 aggregation is mediated by conformational factors. J Biol Chem 1997; 272(42): 26464-70.
[http://dx.doi.org/10.1074/jbc.272.42.26464] [PMID: 9334223]

[93]   Adlard PA, Parncutt JM, Finkelstein DI, Bush AI. Cognitive loss in zinc transporter-3 knock-out mice: a phenocopy for the synaptic and memory deficits of Alzheimers disease? J Neurosci 2010; 30(5): 1631-6.
[http://dx.doi.org/10.1523/JNEUROSCI.5255-09.2010] [PMID: 20130173]

[94]   Lee JY, Mook-Jung I, Koh JY. Histochemically reactive zinc in plaques of the Swedish mutant beta-amyloid precursor protein transgenic mice. J Neurosci 1999; 19(11): RC10.
[PMID: 10341271]

[95]   Lee JY, Cole TB, Palmiter RD, Suh SW, Koh JY. Contribution by synaptic zinc to the gender-disparate plaque formation in human Swedish mutant APP transgenic mice. Proc Natl Acad Sci USA 2002; 99(11): 7705-10.
[http://dx.doi.org/10.1073/pnas.092034699] [PMID: 12032347]

[96]   Arriagada PV, Growdon JH, Hedley-Whyte ET, Hyman BT. Neurofibrillary tangles but not senile plaques parallel duration and severity of Alzheimers disease. Neurology 1992; 42(3 Pt 1): 631-9.
[http://dx.doi.org/10.1212/WNL.42.3.631] [PMID: 1549228]

[97]   Hardy J, Allsop D. Amyloid deposition as the central event in the aetiology of Alzheimers disease. Trends Pharmacol Sci 1991; 12(10): 383-8.
[http://dx.doi.org/10.1016/0165-6147(91)90609-V] [PMID: 1763432]

[98]   Hardy JA, Higgins GA. Alzheimers disease: the amyloid cascade hypothesis. Science 1992; 256(5054): 184-5.
[http://dx.doi.org/10.1126/science.1566067] [PMID: 1566067]

[99] Alonso AC, Zaidi T, Grundke-Iqbal I, Iqbal K. Role of abnormally phosphorylated tau in the breakdown of microtubules in Alzheimer disease. Proc Natl Acad Sci USA 1994; 91(12): 5562-6.
[http://dx.doi.org/10.1073/pnas.91.12.5562] [PMID: 8202528]

[100] Buée L, Bussière T, Buée-Scherrer V, Delacourte A, Hof PR. Tau protein isoforms, phosphorylation and role in neurodegenerative disorders. Brain Res Brain Res Rev 2000; 33(1): 95-130.
[http://dx.doi.org/10.1016/S0165-0173(00)00019-9] [PMID: 10967355]

[101] Avila J, Lucas JJ, Perez M, Hernandez F. Role of tau protein in both physiological and pathological conditions. Physiol Rev 2004; 84(2): 361-84.
[http://dx.doi.org/10.1152/physrev.00024.2003] [PMID: 15044677]

[102] Zhu LQ, Wang SH, Liu D, *et al.* Activation of glycogen synthase kinase-3 inhibits long-term potentiation with synapse-associated impairments. J Neurosci 2007; 27(45): 12211-20.
[http://dx.doi.org/10.1523/JNEUROSCI.3321-07.2007] [PMID: 17989287]

[103] Wang JZ, Gong CX, Zaidi T, Grundke-Iqbal I, Iqbal K. Dephosphorylation of Alzheimer paired helical filaments by protein phosphatase-2A and -2B. J Biol Chem 1995; 270(9): 4854-60.
[http://dx.doi.org/10.1074/jbc.270.9.4854] [PMID: 7876258]

[104] Wang JZ, Grundke-Iqbal I, Iqbal K. Restoration of biological activity of Alzheimer abnormally phosphorylated tau by dephosphorylation with protein phosphatase-2A, -2B and -1. Brain Res Mol Brain Res 1996; 38(2): 200-8.
[http://dx.doi.org/10.1016/0169-328X(95)00316-K] [PMID: 8793108]

[105] Hoover BR, Reed MN, Su J, *et al.* Tau mislocalization to dendritic spines mediates synaptic dysfunction independently of neurodegeneration. Neuron 2010; 68(6): 1067-81.
[http://dx.doi.org/10.1016/j.neuron.2010.11.030] [PMID: 21172610]

[106] Wang JZ, Wang ZH, Tian Q. Tau hyperphosphorylation induces apoptotic escape and triggers neurodegeneration in Alzheimers disease. Neurosci Bull 2014; 30(2): 359-66.
[http://dx.doi.org/10.1007/s12264-013-1415-y] [PMID: 24627329]

[107] Wood SJ, Wypych J, Steavenson S, Louis JC, Citron M, Biere AL. alpha-synuclein fibrillogenesis is nucleation-dependent. Implications for the pathogenesis of Parkinsons disease. J Biol Chem 1999; 274(28): 19509-12.
[http://dx.doi.org/10.1074/jbc.274.28.19509] [PMID: 10391881]

[108] Chu Y, Kordower JH. The prion hypothesis of Parkinsons disease. Curr Neurol Neurosci Rep 2015; 15(5): 28.
[http://dx.doi.org/10.1007/s11910-015-0549-x] [PMID: 25868519]

[109] Duda JE, Lee VM, Trojanowski JQ. Neuropathology of synuclein aggregates. J Neurosci Res 2000; 61(2): 121-7.
[http://dx.doi.org/10.1002/1097-4547(20000715)61:2<121::AID-JNR1>3.0.CO;2-4] [PMID: 10878583]

[110] Lashuel HA, Overk CR, Oueslati A, Masliah E. The many faces of α-synuclein: from structure and toxicity to therapeutic target. Nat Rev Neurosci 2013; 14(1): 38-48.
[http://dx.doi.org/10.1038/nrn3406] [PMID: 23254192]

[111] Luk KC, Kehm V, Carroll J, *et al.* Pathological α-synuclein transmission initiates Parkinson-like neurodegeneration in nontransgenic mice. Science 2012; 338(6109): 949-53.
[http://dx.doi.org/10.1126/science.1227157] [PMID: 23161999]

[112] Luk KC, Song C, OBrien P, *et al.* Exogenous alpha-synuclein fibrils seed the formation of Lewy body-like intracellular inclusions in cultured cells. Proc Natl Acad Sci USA 2009; 106(47): 20051-6.
[http://dx.doi.org/10.1073/pnas.0908005106] [PMID: 19892735]

[113] Lopes da Fonseca T, Villar-Piqué A, Outeiro TF. The Interplay between Alpha-Synuclein Clearance and Spreading. Biomolecules 2015; 5(2): 435-71.
[http://dx.doi.org/10.3390/biom5020435] [PMID: 25874605]

[114] Gallegos S, Pacheco C, Peters C, Opazo CM, Aguayo LG. Features of alpha-synuclein that could explain the progression and irreversibility of Parkinsons disease. Front Neurosci 2015; 9: 59.
[http://dx.doi.org/10.3389/fnins.2015.00059] [PMID: 25805964]

[115] Saleh H, Saleh A, Yao H, Cui J, Shen Y, Li R. Mini review: linkage between α-Synuclein protein and cognition. Transl Neurodegener 2015; 4: 5.
[http://dx.doi.org/10.1186/s40035-015-0026-0] [PMID: 25834729]

[116] Zaltieri M, Longhena F, Pizzi M, Missale C, Spano P, Bellucci A. Mitochondrial Dysfunction and α-Synuclein Synaptic Pathology in Parkinson's Disease: Who's on First? Parkinsons Dis 2015; 108029.

[117] Parihar MS, Parihar A, Fujita M, Hashimoto M, Ghafourifar P. Mitochondrial association of alpha-synuclein causes oxidative stress. Cell Mol Life Sci 2008; 65(7-8): 1272-84.
[http://dx.doi.org/10.1007/s00018-008-7589-1] [PMID: 18322646]

[118] Dias V, Junn E, Mouradian MM. The role of oxidative stress in Parkinsons disease. J Parkinsons Dis 2013; 3(4): 461-91.
[PMID: 24252804]

[119] Meijer FJ, Goraj B. Brain MRI in Parkinsons disease. Front Biosci (Elite Ed) 2014; 6: 360-9.
[http://dx.doi.org/10.2741/e711] [PMID: 24896211]

[120] Fioravanti V, Benuzzi F, Codeluppi L, et al. MRI correlates of Parkinson's disease progression: a voxel based morphometry study 2015.
[http://dx.doi.org/10.1155/2015/378032]

[121] Al-Chalabi A, Jones A, Troakes C, King A, Al-Sarraj S, van den Berg LH. The genetics and neuropathology of amyotrophic lateral sclerosis. Acta Neuropathol 2012; 124(3): 339-52.
[http://dx.doi.org/10.1007/s00401-012-1022-4] [PMID: 22903397]

[122] Bento-Abreu A, Van Damme P, Van Den Bosch L, Robberecht W. The neurobiology of amyotrophic lateral sclerosis. Eur J Neurosci 2010; 31(12): 2247-65.
[http://dx.doi.org/10.1111/j.1460-9568.2010.07260.x] [PMID: 20529130]

[123] Nishitoh H, Kadowaki H, Nagai A, et al. ALS-linked mutant SOD1 induces ER stress- and ASK1-dependent motor neuron death by targeting Derlin-1. Genes Dev 2008; 22(11): 1451-64.
[http://dx.doi.org/10.1101/gad.1640108] [PMID: 18519638]

[124] Geser F, Lee VM, Trojanowski JQ. Amyotrophic lateral sclerosis and frontotemporal lobar degeneration: a spectrum of TDP-43 proteinopathies. Neuropathology 2010; 30(2): 103-12.
[http://dx.doi.org/10.1111/j.1440-1789.2009.01091.x] [PMID: 20102519]

[125] Zinszner H, Sok J, Immanuel D, Yin Y, Ron D. TLS (FUS) binds RNA in vivo and engages in nucleo-cytoplasmic shuttling. J Cell Sci 1997; 110(Pt 15): 1741-50.
[PMID: 9264461]

[126] Wang X, Arai S, Song X, et al. Induced ncRNAs allosterically modify RNA-binding proteins in cis to inhibit transcription. Nature 2008; 454(7200): 126-30.
[http://dx.doi.org/10.1038/nature06992] [PMID: 18509338]

[127] Barmada SJ. Linking RNA dysfunction and neurodegeneration in amyotrophic lateral sclerosis. Neurotherapeutics 2015; 12(2): 340-51.
[http://dx.doi.org/10.1007/s13311-015-0340-3] [PMID: 25689976]

[128] Gendron TF, Belzil VV, Zhang YJ, Petrucelli L. Mechanisms of toxicity in C9FTLD/ALS. Acta Neuropathol 2014; 127(3): 359-76.
[http://dx.doi.org/10.1007/s00401-013-1237-z] [PMID: 24394885]

[129] Nonaka T, Masuda-Suzukake M, Arai T, et al. Prion-like properties of pathological TDP-43 aggregates from diseased brains. Cell Reports 2013; 4(1): 124-34.
[http://dx.doi.org/10.1016/j.celrep.2013.06.007] [PMID: 23831027]

[130] Swarup V, Phaneuf D, Dupré N, *et al.* Deregulation of TDP-43 in amyotrophic lateral sclerosis triggers nuclear factor κB-mediated pathogenic pathways. J Exp Med 2011; 208(12): 2429-47.
[http://dx.doi.org/10.1084/jem.20111313] [PMID: 22084410]

[131] Renton AE, Chiò A, Traynor BJ. State of play in amyotrophic lateral sclerosis genetics. Nat Neurosci 2014; 17(1): 17-23.
[http://dx.doi.org/10.1038/nn.3584] [PMID: 24369373]

[132] Tanzi RE. The genetics of Alzheimer disease. Cold Spring Harb Perspect Med 2012; 2(10): a006296.
[http://dx.doi.org/10.1101/cshperspect.a006296] [PMID: 23028126]

[133] St George-Hyslop PH, Tanzi RE, Polinsky RJ, *et al.* The genetic defect causing familial Alzheimers disease maps on chromosome 21. Science 1987; 235(4791): 885-90.
[http://dx.doi.org/10.1126/science.2880399] [PMID: 2880399]

[134] Mullan M, Crawford F, Axelman K, *et al.* A pathogenic mutation for probable Alzheimers disease in the APP gene at the N-terminus of beta-amyloid. Nat Genet 1992; 1(5): 345-7.
[http://dx.doi.org/10.1038/ng0892-345] [PMID: 1302033]

[135] Murrell J, Farlow M, Ghetti B, Benson MD. A mutation in the amyloid precursor protein associated with hereditary Alzheimers disease. Science 1991; 254(5028): 97-9.
[http://dx.doi.org/10.1126/science.1925564] [PMID: 1925564]

[136] Sherrington R, Rogaev EI, Liang Y, *et al.* Cloning of a gene bearing missense mutations in early-onset familial Alzheimers disease. Nature 1995; 375(6534): 754-60.
[http://dx.doi.org/10.1038/375754a0] [PMID: 7596406]

[137] Levy-Lahad E, Wasco W, Poorkaj P, *et al.* Candidate gene for the chromosome 1 familial Alzheimers disease locus. Science 1995; 269(5226): 973-7.
[http://dx.doi.org/10.1126/science.7638622] [PMID: 7638622]

[138] Rogaev EI, Sherrington R, Rogaeva EA, *et al.* Familial Alzheimers disease in kindreds with missense mutations in a gene on chromosome 1 related to the Alzheimers disease type 3 gene. Nature 1995; 376(6543): 775-8.
[http://dx.doi.org/10.1038/376775a0] [PMID: 7651536]

[139] De Strooper B, Saftig P, Craessaerts K, *et al.* Deficiency of presenilin-1 inhibits the normal cleavage of amyloid precursor protein. Nature 1998; 391(6665): 387-90.
[http://dx.doi.org/10.1038/34910] [PMID: 9450754]

[140] Wolfe MS, Xia W, Ostaszewski BL, Diehl TS, Kimberly WT, Selkoe DJ. Two transmembrane aspartates in presenilin-1 required for presenilin endoproteolysis and gamma-secretase activity. Nature 1999; 398(6727): 513-7.
[http://dx.doi.org/10.1038/19077] [PMID: 10206644]

[141] Corder EH, Saunders AM, Risch NJ, *et al.* Protective effect of apolipoprotein E type 2 allele for late onset Alzheimer disease. Nat Genet 1994; 7(2): 180-4.
[http://dx.doi.org/10.1038/ng0694-180] [PMID: 7920638]

[142] Strittmatter WJ, Weisgraber KH, Huang DY, *et al.* Binding of human apolipoprotein E to synthetic amyloid beta peptide: isoform-specific effects and implications for late-onset Alzheimer disease. Proc Natl Acad Sci USA 1993; 90(17): 8098-102.
[http://dx.doi.org/10.1073/pnas.90.17.8098] [PMID: 8367470]

[143] Kehoe PG, Russ C, McIlory S, *et al.* Variation in DCP1, encoding ACE, is associated with susceptibility to Alzheimer disease. Nat Genet 1999; 21(1): 71-2.
[http://dx.doi.org/10.1038/5009] [PMID: 9916793]

[144] Hu J, Igarashi A, Kamata M, Nakagawa H. Angiotensin-converting enzyme degrades Alzheimer amyloid beta-peptide (A beta); retards A beta aggregation, deposition, fibril formation; and inhibits cytotoxicity. J Biol Chem 2001; 276(51): 47863-8.
[PMID: 11604391]

[145] Hemming ML, Selkoe DJ. Amyloid beta-protein is degraded by cellular angiotensin-converting enzyme (ACE) and elevated by an ACE inhibitor. J Biol Chem 2005; 280(45): 37644-50.
[http://dx.doi.org/10.1074/jbc.M508460200] [PMID: 16154999]

[146] Papassotiropoulos A, Lambert JC, Wavrant-De Vrièze F, *et al.* Cholesterol 25-hydroxylase on chromosome 10q is a susceptibility gene for sporadic Alzheimers disease. Neurodegener Dis 2005; 2(5): 233-41.
[http://dx.doi.org/10.1159/000090362] [PMID: 16909003]

[147] Puglielli L, Tanzi RE, Kovacs DM. Alzheimers disease: the cholesterol connection. Nat Neurosci 2003; 6(4): 345-51.
[http://dx.doi.org/10.1038/nn0403-345] [PMID: 12658281]

[148] Rademakers R, Dermaut B, Peeters K, *et al.* Tau (MAPT) mutation Arg406Trp presenting clinically with Alzheimer disease does not share a common founder in Western Europe. Hum Mutat 2003; 22(5): 409-11.
[http://dx.doi.org/10.1002/humu.10269] [PMID: 14517953]

[149] Ostojic J, Elfgren C, Passant U, *et al.* The tau R406W mutation causes progressive presenile dementia with bitemporal atrophy. Dement Geriatr Cogn Disord 2004; 17(4): 298-301.
[http://dx.doi.org/10.1159/000077158] [PMID: 15178940]

[150] Lotharius J, Brundin P. Pathogenesis of Parkinsons disease: dopamine, vesicles and alpha-synuclein. Nat Rev Neurosci 2002; 3(12): 932-42.
[http://dx.doi.org/10.1038/nrn983] [PMID: 12461550]

[151] Sidhu A, Wersinger C, Moussa CE, Vernier P. The role of alpha-synuclein in both neuroprotection and neurodegeneration. Ann N Y Acad Sci 2004; 1035: 250-70.
[http://dx.doi.org/10.1196/annals.1332.016] [PMID: 15681812]

[152] Hope AD, Farrer M. Genetics of α-synucleinopathy. In: Kahle PJ, Haass C, Eds. Molecular mechanisms of Parkinson's disease Eureka: Landes Bioscience. 2004; pp. 1-11.

[153] Zimprich A, Biskup S, Leitner P, *et al.* Mutations in LRRK2 cause autosomal-dominant parkinsonism with pleomorphic pathology. Neuron 2004; 44(4): 601-7.
[http://dx.doi.org/10.1016/j.neuron.2004.11.005] [PMID: 15541309]

[154] West AB, Moore DJ, Biskup S, *et al.* Parkinsons disease-associated mutations in leucine-rich repeat kinase 2 augment kinase activity. Proc Natl Acad Sci USA 2005; 102(46): 16842-7.
[http://dx.doi.org/10.1073/pnas.0507360102] [PMID: 16269541]

[155] Shimura H, Hattori N, Kubo Si, *et al.* Familial Parkinson disease gene product, parkin, is a ubiquitin-protein ligase. Nat Genet 2000; 25(3): 302-5.
[http://dx.doi.org/10.1038/77060] [PMID: 10888878]

[156] Valente EM, Abou-Sleiman PM, Caputo V, *et al.* Hereditary early-onset Parkinsons disease caused by mutations in PINK1. Science 2004; 304(5674): 1158-60.
[http://dx.doi.org/10.1126/science.1096284] [PMID: 15087508]

[157] Beilina A, Van Der Brug M, Ahmad R, *et al.* Mutations in PTEN-induced putative kinase 1 associated with recessive parkinsonism have differential effects on protein stability. Proc Natl Acad Sci USA 2005; 102(16): 5703-8.
[http://dx.doi.org/10.1073/pnas.0500617102] [PMID: 15824318]

[158] Bonifati V, Rizzu P, van Baren MJ, *et al.* Mutations in the DJ-1 gene associated with autosomal recessive early-onset parkinsonism. Science 2003; 299(5604): 256-9.
[http://dx.doi.org/10.1126/science.1077209] [PMID: 12446870]

[159] Lockhart PJ, Lincoln S, Hulihan M, *et al.* DJ-1 mutations are a rare cause of recessively inherited early onset parkinsonism mediated by loss of protein function. J Med Genet 2004; 41(3): e22.
[http://dx.doi.org/10.1136/jmg.2003.011106] [PMID: 14985393]

[160] Aran FA. Research on an as yet undescribed disease of the muscular system (progressive muscular atrophy). Arch Gén Méd 1848; 24: 15-35. [French].

[161] Charcot JM. Lectures on the diseases of the nervous system,(New Sydenham Society, London): (ed.Sigerson, G.) 1881; 2: pp. 163-204.2

[162] Andersen PM, Al-Chalabi A. Clinical genetics of amyotrophic lateral sclerosis: what do we really know? Nat Rev Neurol 2011; 7(11): 603-15.
[http://dx.doi.org/10.1038/nrneurol.2011.150] [PMID: 21989245]

[163] Rosen DR, Siddique T, Patterson D, *et al.* Mutations in Cu/Zn superoxide dismutase gene are associated with familial amyotrophic lateral sclerosis. Nature 1993; 362(6415): 59-62.
[http://dx.doi.org/10.1038/362059a0] [PMID: 8446170]

[164] Andersen PM, Sims KB, Xin WW, *et al.* Sixteen novel mutations in the Cu/Zn superoxide dismutase gene in amyotrophic lateral sclerosis: a decade of discoveries, defects and disputes. Amyotroph Lateral Scler Other Motor Neuron Disord 2003; 4(2): 62-73.
[http://dx.doi.org/10.1080/14660820310011700] [PMID: 14506936]

[165] Valdmanis PN, Belzil VV, Lee J, *et al.* A mutation that creates a pseudoexon in SOD1 causes familial ALS. Ann Hum Genet 2009; 73(Pt 6): 652-7.
[http://dx.doi.org/10.1111/j.1469-1809.2009.00546.x] [PMID: 19847927]

[166] Zinman L, Liu HN, Sato C, *et al.* A mechanism for low penetrance in an ALS family with a novel SOD1 deletion. Neurology 2009; 72(13): 1153-9.
[http://dx.doi.org/10.1212/01.wnl.0000345363.65799.35] [PMID: 19332692]

[167] Birve A, Neuwirth C, Weber M, *et al.* A novel SOD1 splice site mutation associated with familial ALS revealed by SOD activity analysis. Hum Mol Genet 2010; 19(21): 4201-6.
[http://dx.doi.org/10.1093/hmg/ddq338] [PMID: 20709807]

[168] Sreedharan J, Blair IP, Tripathi VB, *et al.* TDP-43 mutations in familial and sporadic amyotrophic lateral sclerosis. Science 2008; 319(5870): 1668-72.
[http://dx.doi.org/10.1126/science.1154584] [PMID: 18309045]

[169] Kabashi E, Valdmanis PN, Dion P, *et al.* TARDBP mutations in individuals with sporadic and familial amyotrophic lateral sclerosis. Nat Genet 2008; 40(5): 572-4.
[http://dx.doi.org/10.1038/ng.132] [PMID: 18372902]

[170] Daoud H, Valdmanis PN, Kabashi E, *et al.* Contribution of TARDBP mutations to sporadic amyotrophic lateral sclerosis. J Med Genet 2009; 46(2): 112-4.
[http://dx.doi.org/10.1136/jmg.2008.062463] [PMID: 18931000]

[171] Kirby J, Goodall EF, Smith W, *et al.* Broad clinical phenotypes associated with TAR-DNA binding protein (TARDBP) mutations in amyotrophic lateral sclerosis. Neurogenetics 2010; 11(2): 217-25.
[http://dx.doi.org/10.1007/s10048-009-0218-9] [PMID: 19760257]

[172] Origone P, Caponnetto C, Bandettini Di Poggio M, *et al.* Enlarging clinical spectrum of FALS with TARDBP gene mutations: S393L variant in an Italian family showing phenotypic variability and relevance for genetic counselling. Amyotroph Lateral Scler 2010; 11(1-2): 223-7.
[http://dx.doi.org/10.3109/17482960903165039] [PMID: 19714537]

[173] Benajiba L, Le Ber I, Camuzat A, *et al.* TARDBP mutations in motoneuron disease with frontotemporal lobar degeneration. Ann Neurol 2009; 65(4): 470-3.
[http://dx.doi.org/10.1002/ana.21612] [PMID: 19350673]

[174] Vance C, Rogelj B, Hortobágyi T, *et al.* Mutations in FUS, an RNA processing protein, cause familial amyotrophic lateral sclerosis type 6. Science 2009; 323(5918): 1208-11.
[http://dx.doi.org/10.1126/science.1165942] [PMID: 19251628]

[175]   Ticozzi N, Silani V, LeClerc AL, *et al.* Analysis of FUS gene mutation in familial amyotrophic lateral sclerosis within an Italian cohort. Neurology 2009; 73(15): 1180-5.
[http://dx.doi.org/10.1212/WNL.0b013e3181bbff05] [PMID: 19741215]

[176]   Yan J, Deng HX, Siddique N, *et al.* Frameshift and novel mutations in FUS in familial amyotrophic lateral sclerosis and ALS/dementia. Neurology 2010; 75(9): 807-14.
[http://dx.doi.org/10.1212/WNL.0b013e3181f07e0c] [PMID: 20668259]

[177]   Van Langenhove T, van der Zee J, Sleegers K, *et al.* Genetic contribution of FUS to frontotemporal lobar degeneration. Neurology 2010; 74(5): 366-71.
[http://dx.doi.org/10.1212/WNL.0b013e3181ccc732] [PMID: 20124201]

[178]   Blair IP, Williams KL, Warraich ST, *et al.* FUS mutations in amyotrophic lateral sclerosis: clinical, pathological, neurophysiological and genetic analysis. J Neurol Neurosurg Psychiatry 2010; 81(6): 639-45.
[http://dx.doi.org/10.1136/jnnp.2009.194399] [PMID: 19965854]

[179]   Pfaender S, Grabrucker AM. Characterization of biometal profiles in neurological disorders. Metallomics 2014; 6(5): 960-77.
[http://dx.doi.org/10.1039/c4mt00008k] [PMID: 24643462]

[180]   Guo Z, Cupples LA, Kurz A, *et al.* Head injury and the risk of AD in the MIRAGE study. Neurology 2000; 54(6): 1316-23.
[http://dx.doi.org/10.1212/WNL.54.6.1316] [PMID: 10746604]

[181]   Calderón-Garcidueñas L, Reed W, Maronpot RR, *et al.* Brain inflammation and Alzheimers-like pathology in individuals exposed to severe air pollution. Toxicol Pathol 2004; 32(6): 650-8.
[http://dx.doi.org/10.1080/01926230490520232] [PMID: 15513908]

[182]   Greenough MA, Volitakis I, Li QX, *et al.* Presenilins promote the cellular uptake of copper and zinc and maintain copper chaperone of SOD1-dependent copper/zinc superoxide dismutase activity. J Biol Chem 2011; 286(11): 9776-86.
[http://dx.doi.org/10.1074/jbc.M110.163964] [PMID: 21239495]

[183]   Miller LM, Wang Q, Telivala TP, Smith RJ, Lanzirotti A, Miklossy J. Synchrotron-based infrared and X-ray imaging shows focalized accumulation of Cu and Zn co-localized with beta-amyloid deposits in Alzheimers disease. J Struct Biol 2006; 155(1): 30-7.
[http://dx.doi.org/10.1016/j.jsb.2005.09.004] [PMID: 16325427]

[184]   Camponeschi F, Valensin D, Tessari I, *et al.* Copper(I)-α-synuclein interaction: structural description of two independent and competing metal binding sites. Inorg Chem 2013; 52(3): 1358-67.
[http://dx.doi.org/10.1021/ic302050m] [PMID: 23343468]

[185]   Ermilova IP, Ermilov VB, Levy M, Ho E, Pereira C, Beckman JS. Protection by dietary zinc in ALS mutant G93A SOD transgenic mice. Neurosci Lett 2005; 379(1): 42-6.
[http://dx.doi.org/10.1016/j.neulet.2004.12.045] [PMID: 15814196]

[186]   Trumbull KA, Beckman JS. A role for copper in the toxicity of zinc-deficient superoxide dismutase to motor neurons in amyotrophic lateral sclerosis. Antioxid Redox Signal 2009; 11(7): 1627-39.
[http://dx.doi.org/10.1089/ars.2009.2574] [PMID: 19309264]

[187]   Fox JH, Kama JA, Lieberman G, *et al.* Mechanisms of copper ion mediated Huntingtons disease progression. PLoS One 2007; 2(3): e334.
[http://dx.doi.org/10.1371/journal.pone.0000334] [PMID: 17396163]

[188]   Thakur AK, Srivastava AK, Srinivas V, Chary KV, Rao CM. Copper alters aggregation behavior of prion protein and induces novel interactions between its N- and C-terminal regions. J Biol Chem 2011; 286(44): 38533-45.
[http://dx.doi.org/10.1074/jbc.M111.265645] [PMID: 21900252]

[189]   Brown DR. Prions and manganese: A maddening beast. Metallomics 2011; 3(3): 229-38.
[http://dx.doi.org/10.1039/C0MT00047G] [PMID: 21390367]

[190] Watt NT, Hooper NM. The prion protein and neuronal zinc homeostasis. Trends Biochem Sci 2003; 28(8): 406-10.
[http://dx.doi.org/10.1016/S0968-0004(03)00166-X] [PMID: 12932728]

[191] Watt NT, Taylor DR, Kerrigan TL, *et al.* Prion protein facilitates uptake of zinc into neuronal cells. Nat Commun 2012; 3: 1134.
[http://dx.doi.org/10.1038/ncomms2135] [PMID: 23072804]

[192] Ahmed SS, Santosh W. Metallomic profiling and linkage map analysis of early Parkinsons disease: a new insight to aluminum marker for the possible diagnosis. PLoS One 2010; 5(6): e11252.
[http://dx.doi.org/10.1371/journal.pone.0011252] [PMID: 20582167]

[193] Kanias GD, Kapaki E. Trace elements, age, and sex in amyotrophic lateral sclerosis disease. Biol Trace Elem Res 1997; 56(2): 187-201.
[http://dx.doi.org/10.1007/BF02785392] [PMID: 9164664]

[194] Nadjar Y, Gordon P, Corcia P, *et al.* Elevated serum ferritin is associated with reduced survival in amyotrophic lateral sclerosis. PLoS One 2012; 7(9): e45034.
[http://dx.doi.org/10.1371/journal.pone.0045034] [PMID: 23024788]

[195] Gong G, O'Bryant SE. The arsenic exposure hypothesis for Alzheimer disease. Alzheimer Dis Assoc Disord 2010; 24(4): 311-6.
[http://dx.doi.org/10.1097/WAD.0b013e3181d71bc7] [PMID: 20473132]

[196] Dewji NN, Do C, Bayney RM. Transcriptional activation of Alzheimers beta-amyloid precursor protein gene by stress. Brain Res Mol Brain Res 1995; 33(2): 245-53.
[http://dx.doi.org/10.1016/0169-328X(95)00131-B] [PMID: 8750883]

[197] Yegambaram M, Manivannan B, Beach TG, Halden RU. Role of environmental contaminants in the etiology of Alzheimers disease: a review. Curr Alzheimer Res 2015; 12(2): 116-46.
[http://dx.doi.org/10.2174/1567205012666150204121719] [PMID: 25654508]

[198] Graves AB, White E, Koepsell TD, Reifler BV, van Belle G, Larson EB. The association between aluminum-containing products and Alzheimers disease. J Clin Epidemiol 1990; 43(1): 35-44.
[http://dx.doi.org/10.1016/0895-4356(90)90053-R] [PMID: 2319278]

[199] Crapper DR, Krishnan SS, Dalton AJ. Brain aluminum distribution in Alzheimers disease and experimental neurofibrillary degeneration. Science 1973; 180(4085): 511-3.
[http://dx.doi.org/10.1126/science.180.4085.511] [PMID: 4735595]

[200] Kawahara M, Kato-Negishi M. M. Link between Aluminum and the pathogenesis of Alzheimer's disease: The integration of the aluminum and amyloid cascade hypotheses. Int J Alzheimers Dis 2011; 2011: 276393.

[201] Tolleson CM, Fang JY. Advances in the mechanisms of Parkinsons disease. Discov Med 2013; 15(80): 61-6.
[PMID: 23375015]

[202] Ballard C, Gauthier S, Corbett A, Brayne C, Aarsland D, Jones E. Alzheimers disease. Lancet 2011; 377(9770): 1019-31.
[http://dx.doi.org/10.1016/S0140-6736(10)61349-9] [PMID: 21371747]

[203] Beitz JM. Parkinsons disease: a review. Front Biosci (Schol Ed) 2014; 6: 65-74. [School Ed].
[http://dx.doi.org/10.2741/S415] [PMID: 24389262]

[204] Brown RC, Lockwood AH, Sonawane BR. Neurodegenerative diseases: an overview of environmental risk factors. Environ Health Perspect 2005; 113(9): 1250-6.
[http://dx.doi.org/10.1289/ehp.7567] [PMID: 16140637]

[205] Wanamaker BL, Swiger KJ, Blumenthal RS, Martin SS. Cholesterol, statins, and dementia: what the cardiologist should know. Clin Cardiol 2015; 38(4): 243-50.
[http://dx.doi.org/10.1002/clc.22361] [PMID: 25869997]

[206] Lim WL, Martins IJ, Martins RN. The involvement of lipids in Alzheimers disease. J Genet Genomics 2014; 41(5): 261-74.
[http://dx.doi.org/10.1016/j.jgg.2014.04.003] [PMID: 24894353]

[207] Miklossy J. Emerging roles of pathogens in Alzheimer disease. Expert Rev Mol Med 2011; 13: e30.
[http://dx.doi.org/10.1017/S1462399411002006] [PMID: 21933454]

[208] Wong JC, Hazrati LN. Parkinsons disease, parkinsonism, and traumatic brain injury. Crit Rev Clin Lab Sci 2013; 50(4-5): 103-6.
[http://dx.doi.org/10.3109/10408363.2013.844678] [PMID: 24156652]

[209] Jafari S, Etminan M, Aminzadeh F, Samii A. Head injury and risk of Parkinson disease: a systematic review and meta-analysis. Mov Disord 2013; 28(9): 1222-9.
[http://dx.doi.org/10.1002/mds.25458] [PMID: 23609436]

[210] Ingre C, Roos PM, Piehl F, Kamel F, Fang F. Risk factors for amyotrophic lateral sclerosis. Clin Epidemiol 2015; 7: 181-93.
[PMID: 25709501]

[211] Auld DS, Kornecook TJ, Bastianetto S, Quirion R. Alzheimers disease and the basal forebrain cholinergic system: relations to beta-amyloid peptides, cognition, and treatment strategies. Prog Neurobiol 2002; 68(3): 209-45.
[http://dx.doi.org/10.1016/S0301-0082(02)00079-5] [PMID: 12450488]

[212] Golde TE. Alzheimer disease therapy: can the amyloid cascade be halted? J Clin Invest 2003; 111(1): 11-8.
[http://dx.doi.org/10.1172/JCI200317527] [PMID: 12511580]

[213] Roberds SL, Anderson J, Basi G, *et al.* BACE knockout mice are healthy despite lacking the primary beta-secretase activity in brain: implications for Alzheimers disease therapeutics. Hum Mol Genet 2001; 10(12): 1317-24.
[http://dx.doi.org/10.1093/hmg/10.12.1317] [PMID: 11406613]

[214] Grüninger-Leitch F, Schlatter D, Küng E, Nelböck P, Döbeli H. Substrate and inhibitor profile of BACE (beta-secretase) and comparison with other mammalian aspartic proteases. J Biol Chem 2002; 277(7): 4687-93.
[http://dx.doi.org/10.1074/jbc.M109266200] [PMID: 11741910]

[215] Hong L, Koelsch G, Lin X, *et al.* Structure of the protease domain of memapsin 2 (beta-secretase) complexed with inhibitor. Science 2000; 290(5489): 150-3.
[http://dx.doi.org/10.1126/science.290.5489.150] [PMID: 11021803]

[216] Chang WP, Huang X, Downs D, *et al.* Beta-secretase inhibitor GRL-8234 rescues age-related cognitive decline in APP transgenic mice. FASEB J 2011; 25(2): 775-84.
[http://dx.doi.org/10.1096/fj.10-167213] [PMID: 21059748]

[217] Li Y-M, Xu M, Lai M-T, *et al.* Photoactivated gamma-secretase inhibitors directed to the active site covalently label presenilin 1. Nature 2000; 405(6787): 689-94.
[http://dx.doi.org/10.1038/35015085] [PMID: 10864326]

[218] Wolfe MS, Esler WP, Das C. Continuing strategies for inhibiting Alzheimers gamma-secretase. J Mol Neurosci 2002; 19(1-2): 83-7.
[http://dx.doi.org/10.1007/s12031-002-0015-5] [PMID: 12212799]

[219] Esler WP, Kimberly WT, Ostaszewski BL, *et al.* Transition-state analogue inhibitors of gamma-secretase bind directly to presenilin-1. Nat Cell Biol 2000; 2(7): 428-34.
[http://dx.doi.org/10.1038/35017062] [PMID: 10878808]

[220] De Strooper B, Annaert W, Cupers P, *et al.* A presenilin-1-dependent gamma-secretase-like protease mediates release of Notch intracellular domain. Nature 1999; 398(6727): 518-22.
[http://dx.doi.org/10.1038/19083] [PMID: 10206645]

[221] Weggen S, Eriksen JL, Das P, *et al.* A subset of NSAIDs lower amyloidogenic Abeta42 independently of cyclooxygenase activity. Nature 2001; 414(6860): 212-6.
[http://dx.doi.org/10.1038/35102591] [PMID: 11700559]

[222] Eriksen JL, Sagi SA, Smith TE, *et al.* NSAIDs and enantiomers of flurbiprofen target gamma-secretase and lower Abeta 42 in vivo. J Clin Invest 2003; 112(3): 440-9.
[http://dx.doi.org/10.1172/JCI18162] [PMID: 12897211]

[223] Schenk D, Barbour R, Dunn W, *et al.* Immunization with amyloid-beta attenuates Alzheimer-diseas--like pathology in the PDAPP mouse. Nature 1999; 400(6740): 173-7.
[http://dx.doi.org/10.1038/22124] [PMID: 10408445]

[224] Bard F, Cannon C, Barbour R, *et al.* Peripherally administered antibodies against amyloid beta-peptide enter the central nervous system and reduce pathology in a mouse model of Alzheimer disease. Nat Med 2000; 6(8): 916-9.
[http://dx.doi.org/10.1038/78682] [PMID: 10932230]

[225] Janus C, Pearson J, McLaurin J, *et al.* A beta peptide immunization reduces behavioural impairment and plaques in a model of Alzheimers disease. Nature 2000; 408(6815): 979-82.
[http://dx.doi.org/10.1038/35050110] [PMID: 11140685]

[226] Morgan D, Diamond DM, Gottschall PE, *et al.* A beta peptide vaccination prevents memory loss in an animal model of Alzheimers disease. Nature 2000; 408(6815): 982-5.
[http://dx.doi.org/10.1038/35050116] [PMID: 11140686]

[227] Nicoll JA, Wilkinson D, Holmes C, Steart P, Markham H, Weller RO. Neuropathology of human Alzheimer disease after immunization with amyloid-beta peptide: a case report. Nat Med 2003; 9(4): 448-52.
[http://dx.doi.org/10.1038/nm840] [PMID: 12640446]

[228] Orgogozo JM, Gilman S, Dartigues JF, *et al.* Subacute meningoencephalitis in a subset of patients with AD after Abeta42 immunization. Neurology 2003; 61(1): 46-54.
[http://dx.doi.org/10.1212/01.WNL.0000073623.84147.A8] [PMID: 12847155]

[229] Hock C, Konietzko U, Streffer JR, *et al.* Antibodies against beta-amyloid slow cognitive decline in Alzheimers disease. Neuron 2003; 38(4): 547-54.
[http://dx.doi.org/10.1016/S0896-6273(03)00294-0] [PMID: 12765607]

[230] Gervais F, Chalifour R, Garceau D, *et al.* Glycosaminoglycan mimetics: a therapeutic approach to cerebral amyloid angiopathy. Amyloid 2001; 8 (Suppl. 1): 28-35.
[PMID: 11676287]

[231] Geerts H. NC-531 (Neurochem). Curr Opin Investig Drugs 2004; 5(1): 95-100.
[PMID: 14983981]

[232] Grabrucker AM, Rowan M, Garner CC. Brain-Delivery of Zinc-Ions as Potential Treatment for Neurological Diseases: Mini Review. Drug Deliv Lett 2011; 1(1): 13-23.
[http://dx.doi.org/10.2174/2210303111101010013] [PMID: 22102982]

[233] Duce JA, Bush AI. Biological metals and Alzheimers disease: implications for therapeutics and diagnostics. Prog Neurobiol 2010; 92(1): 1-18.
[http://dx.doi.org/10.1016/j.pneurobio.2010.04.003] [PMID: 20444428]

[234] Bush AI. The metallobiology of Alzheimers disease. Trends Neurosci 2003; 26(4): 207-14.
[http://dx.doi.org/10.1016/S0166-2236(03)00067-5] [PMID: 12689772]

[235] Adlard PA, Parncutt J, Lal V, *et al.* Metal chaperones prevent zinc-mediated cognitive decline. Neurobiol Dis 2014; pii: S0969-9961(14)00383-0

[236] Crouch PJ, Barnham KJ. Therapeutic redistribution of metal ions to treat Alzheimers disease. Acc Chem Res 2012; 45(9): 1604-11.
[http://dx.doi.org/10.1021/ar300074t] [PMID: 22747493]

[237] Andersen OM, Schmidt V, Spoelgen R, *et al.* Molecular dissection of the interaction between amyloid precursor protein and its neuronal trafficking receptor SorLA/LR11. Biochemistry 2006; 45(8): 2618-28.
[http://dx.doi.org/10.1021/bi052120v] [PMID: 16489755]

[238] Anderson JJ, Holtz G, Baskin PP, *et al.* Reductions in beta-amyloid concentrations in vivo by the gamma-secretase inhibitors BMS-289948 and BMS-299897. Biochem Pharmacol 2005; 69(4): 689-98.
[http://dx.doi.org/10.1016/j.bcp.2004.11.015] [PMID: 15670587]

[239] Wischik CM, Edwards PC, Lai RY, Roth M, Harrington CR. Selective inhibition of Alzheimer disease-like tau aggregation by phenothiazines. Proc Natl Acad Sci USA 1996; 93(20): 11213-8.
[http://dx.doi.org/10.1073/pnas.93.20.11213] [PMID: 8855335]

[240] Taniguchi S, Suzuki N, Masuda M, *et al.* Inhibition of heparin-induced tau filament formation by phenothiazines, polyphenols, and porphyrins. J Biol Chem 2005; 280(9): 7614-23.
[http://dx.doi.org/10.1074/jbc.M408714200] [PMID: 15611092]

[241] Chirita C, Necula M, Kuret J. Ligand-dependent inhibition and reversal of tau filament formation. Biochemistry 2004; 43(10): 2879-87.
[http://dx.doi.org/10.1021/bi036094h] [PMID: 15005623]

[242] Bhat RV, Budd Haeberlein SL, Avila J. Glycogen synthase kinase 3: a drug target for CNS therapies. J Neurochem 2004; 89(6): 1313-7.
[http://dx.doi.org/10.1111/j.1471-4159.2004.02422.x] [PMID: 15189333]

[243] Lovestone S, Boada M, Dubois B, *et al.* A phase II trial of tideglusib in Alzheimers disease. J Alzheimers Dis 2015; 45(1): 75-88.
[PMID: 25537011]

[244] Lee VM, Daughenbaugh R, Trojanowski JQ. Microtubule stabilizing drugs for the treatment of Alzheimers disease. Neurobiol Aging 1994; 15(2) (Suppl. 2): S87-9.
[http://dx.doi.org/10.1016/0197-4580(94)90179-1] [PMID: 7700471]

[245] Zhang B, Maiti A, Shively S, *et al.* Microtubule-binding drugs offset tau sequestration by stabilizing microtubules and reversing fast axonal transport deficits in a tauopathy model. Proc Natl Acad Sci USA 2005; 102(1): 227-31.
[http://dx.doi.org/10.1073/pnas.0406361102] [PMID: 15615853]

[246] Schapira AH, Bezard E, Brotchie J, *et al.* Novel pharmacological targets for the treatment of Parkinsons disease. Nat Rev Drug Discov 2006; 5(10): 845-54.
[http://dx.doi.org/10.1038/nrd2087] [PMID: 17016425]

[247] Liu Z, Hamamichi S, Lee BD, *et al.* Inhibitors of LRRK2 kinase attenuate neurodegeneration and Parkinson-like phenotypes in Caenorhabditis elegans and Drosophila Parkinsons disease models. Hum Mol Genet 2011; 20(20): 3933-42.
[http://dx.doi.org/10.1093/hmg/ddr312] [PMID: 21768216]

[248] Chhabra R, Tosi G, Grabrucker AM. Emerging Use of Nanotechnology in the Treatment of Neurological Disorders. Curr Pharm Des 2015; 21(22): 3111-30.
[http://dx.doi.org/10.2174/1381612821666150531164124] [PMID: 26027574]

[249] Grabrucker AM, Ruozi B, Belletti D, *et al.* Nanoparticle transport across the blood brain barrier. Tissue Barriers 2016; 4(1): e1153568.
[http://dx.doi.org/10.1080/21688370.2016.1153568] [PMID: 27141426]

CHAPTER 2

Nanoparticles Targeting Mitochondria in Neurodegenerative Diseases: Toxicity and Challenge for Nanotherapeutics

Michal Cagalinec*

Department of Pharmacology, Faculty of Medicine, University of Tartu, Tartu, Estonia
Centre of Biosciences (former Institute of Molecular Physiology and Genetics), Slovak Academy of Sciences, Bratislava, Slovakia

Abstract: In the past decades, the prevalence of neurodegenerative diseases (NDDs) has risen dramatically with the increasing age of human population. Neurodegeneration is a long-term and complex process resulting in the degeneration of neurons. So far, no causative therapy exists, urging the development of methods for the early diagnostics and efficient therapy. In this respect, nanoparticles (NPs) are considered a promising tool due to their efficient blood-brain barrier penetrance and specific interactions with the cellular components. They can localize to mitochondria, nucleus, and autophagosomes and also interact with the cytoskeletal structures as tubulin and Tau protein. Therefore, as mitochondria represent important target for NPs, the therapeutic potential of NPs together with their toxicity to mitochondria has become an emerging topic. In this review, we describe the current knowledge in targeting NPs into mitochondria in relation to Alzheimer's and Parkinson's disease. Furthermore, we propose a novel idea how to compensate the compromised mitochondrial functioning without the delivery of NPs into the mitochondrial matrix, specifically by the development of NPs targeting either cytoskeleton or the proteins of mitochondrial motility and fusion-fission machinery. As the latter face cytoplasm, this approach does not require targeting NPs into the mitochondrial matrix. At the same time, it could be a significant step to improve the therapy of NDDs, since the movement, fusion, and fission are necessary for mitochondria to exchange their membrane material, mitochondrial DNA, and to remove the damaged mitochondria.

Keywords: Alzheimer's disease, Cytoskeleton, Mitochondria, Mitochondrial dynamics, Mitochondrial fusion and fission, Nanoparticles, Neurodegeneration, Parkinson's disease, Tau protein, β-amyloid.

* **Corresponding author Michal Cagalinec:** Department of Pharmacology, Faculty of Medicine, University of Tartu, Tartu, Estonia; Tel: +372 7374359; E-mail: michal.cagalinec@ut.ee

Giovanni Tosi (Ed)

INTRODUCTION AND OVERVIEW

The development and progression of neurodegenerative diseases (NDDs) is a long-term and complex process. Several factors, such as physiology, environment, and genetics are involved in the manifestation of the NDDs. So far, no causative and thus effective therapy exists, emerging the progress and research in understanding the pathomechanisms of the NDDs. From the NDDs, this review is focused on Alzheimer's (AD) and Parkinson's disease (PD), the two most prevalent NDDs due to their rapidly increasing incidence with the ageing of the world human population [1]. For AD as well as PD, there is the experimental and clinical evidence pointing out the significant contribution of mitochondria to the pathophysiology of both disorders. The importance of mitochondria in the context of NDDs is supported by the fact, that antioxidant therapy, although symptomatic, represents the most effective therapy for AD and PD treatment so far. Therefore, the involvement of mitochondria as the therapeutic target may be a great promise for the successful treatment of the NDDs in general. Herein, the use of nanoparticles (NPs) with their unique physico-chemical properties is a challenging issue. Their highly specific penetrance through biological membranes, focused delivery, and versatility to attach or encapsulate the active substances highlight NPs as a promising tool to target mitochondria and to use them, beyond the early diagnostics, especially for the therapy of the NDDs. In contrary, as any new invention, the toxicity of NPs is of critical note. Since the progress of NDDs is very slow, usually several decades, the long-term treatment is required. Thus, the low-dose long-term toxicity of NPs compared to their acute adverse effects has to be evaluated when using NPs for the treatment of NDDs.

Therefore, this review interconnects three topics: 1. the involvement of mitochondria in AD/PD, with a special focus on the mitochondrial dynamics, 2. the use of NPs as a tool in the treatment of AD/PD, and 3. the direct and indirect relation between NPs and mitochondria (the latter mainly through cytoskeleton and reactive oxygen species). Moreover, the therapeutic potential *vs.* toxicity of NPs is summarized (Fig. **1**) mainly focused on data obtained in experimental condition. At last, the aim of this chapter is to open the readers' mind, to give them inspiration for the future experiments, and to stimulate the paradigm shift in the area of the therapeutic use of nanoparticles in mitochondrial medicine in relation to the neurodegenerative diseases.

ALZHEIMER'S AND PARKINSON'S DISEASE

Progressive increase in the mean age of the human population is a positive end-point showing the progress in the healthcare system, on one side due to success in decreasing the birth deaths and improved care of elderly patients. However, this progress is associated with the increased incidence of diseases associated with the old age – especially the number of patients with neurodegenerative disorders increases non-linearly, sometimes even exponentially [2]. Alzheimer's disease (AD), Parkinson's disease (PD), and related dementias have appeared to be one of the most critical public health problems in the aging population because they are major sources of disabilities, poor quality of life for the patients themselves and their families [3], and of caregiver strain as well [4, 5]. As published in the Delphi consensus study, there is an estimate of 24 million people with AD or dementia worldwide and the prognosis for the year 2040 is 81 million [6]. Regarding the PD, the World Health Organization's has reported estimation of 5 million patients in 2006. Patients suffering from these dementias have increased mortality [7, 8], where in 2006, around half a million of deaths in the world were directly related to AD and dementias, and roughly 100,000 were associated with PD (WHO, 2006). These numbers are proposed to increase significantly in the next years due to increase in the mean age of the world population [9].

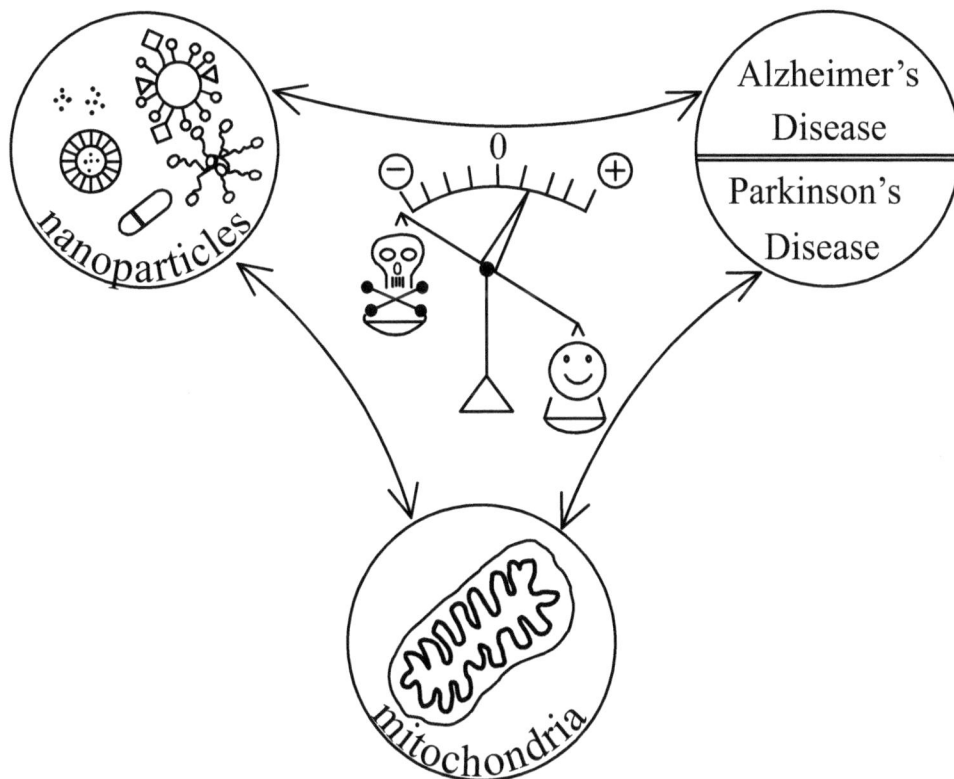

Fig. (1). The aims of the review – to present overlap in the fields of neurodegeneration, mitochondria and nanoparticles focusing on their potential therapeutic use.

Alzheimer's disease. In the United States, it is estimated that one out of every eight persons over the age of 65 suffer from AD, and almost half of those over the age of 85 are affected [10]. AD is mainly recognized by extracellular deposits of polymers of Aβ peptide, forming the neuritic plaques in the later stage of the disease. These plaques are the main hallmark of AD. The next hallmark is presence of Tau neurofibrillary tangles composed of hyperphosphorylated Tau and its truncated forms. Other extensively studied pathologies involve selective neuronal loss, marked synaptic alterations, functional and morphological mitochondrial abnormalities, inflammatory responses and disturbances in calcium homeostasis. These are present in the neocortex, the hippocampus, and other subcortical brain structures involved in cognition [11].

Parkinson's disease (PD) is a neurodegenerative movement disorder [12]. PD is a common age-related disorder that displays infrequent conditions and affects 1%–3% of the population 65 years and older. PD affects the dopaminergic neurons in substantia nigra of the human brain resulting in loss of striatal dopamine release [13]. Therefore the symptoms typically involve bradykinesia (difficulties with slowness of movements), rigidity (stiffness of the muscles), tremor, and ataxia (problems with balance and coordination). In addition, PD is well known to affect cognition and mood. Furthermore, other complaints as disturbances of sleep, bowel dysfunction and pain have negative impact to the patient's quality of life. Although symptomatic treatment has become quite efficient, especially when applied in the early stages, in some cases the subjects do not fulfil expectances. Here, improvements are often selective for some of the symptoms within an individual, such as the relative alleviation of bradykinesia and rigidity but not tremor [14]. Importantly, these therapies are not causal, *i.e.* do not provide neuroprotection. This has prompted the search for molecular therapies that cure or even prevent this disease by targeting PD upstream of the pathophysiology [15].

Although the exact causal factors of Parkinson's disease remain unknown, several research studies point to specific genetic mutations and environmental factors. It has been estimated that around 5–10 in every 100 people suffering from Parkinson's disease are associated with gene mutations. Scientifics have identified at least 13 gene mutations, among which one could highlight those in the genes PINK1 (PTEN-induced putative kinase 1), LRRK2 (leucine-rich repeat kinase 2), SNCA (synuclein, alpha non-A4 component of amyloid precursor), PARK2 (Parkinson's disease autosomal recessive, juvenile 2) and PARK7 (Parkinson's disease autosomal recessive, early-onset 7) [16]. The SNCA gene encodes for the protein alpha-synuclein (α-synuclein), which is a key component of Lewy bodies; the PARK2 gene encodes for the E3 ubiquitin ligase parkin, which is implied in mitochondrial maintenance; the PARK7 gene encodes for the antioxidant protein

DJ-1; PINK 1 gene encodes for a serine/threonine-protein kinase with a protective mitochondrial role. Alterations of SNCA, PARK2, PARK7, and PINK1 genes are involved in the early-onset Parkinson's disease (this is diagnosed before being 50 years old) [17 - 19] and demonstrate strong involvement of mitochondria in pathogenesis of PD. The rest, around 95%, of diagnosed Parkinson's disease cases are sporadic, in which environmental factors such as pesticides and dietary factors, among others, seem to play a crucial role [20].

Reaching the causative therapy, recent strategy is focused on the small protein α-synuclein present in rafts in Lewy bodies and so far seems to be a crucial factor in PD etiology [21]. The use of raft-targeting lipid-based NPs, which could inhibit or even reverse α-synuclein fibrillation *via* endocytosis and the cytosolic release of an α-synuclein-targeting molecule, is therefore one of the possible therapeutic goals and will be discussed in more detail below.

MITOCHONDRIA

Mitochondria are tubular cellular organelles enclosed by two biomembranes, both of them with specific lipid and protein composition. Mitochondria are very versatile organelles, they vary considerably between different species in terms of their metabolism, morphology, ultrastructure, distribution, protein and lipid composition, mitochondrial ion conductance etc. Mitochondria, supply most of energy of the cell through the electron transport chain (ETC) and oxidative phosphorylation localized in the matrix and inner mitochondrial membrane. Next, mitochondria are engaged in calcium homeostasis, apoptosis, ROS balance and thermal regulation. Moreover, mitochondria are the exclusive organelles containing their own DNA and they replicate independently from the cell division cycle. Mitochondrial DNA (mtDNA) is circular and consists of 16,500 base pairs and 37 genes involving the key enzymes of the respiratory cycle. Although some eukaryotic cells are free of mitochondria, vast majority of eukaryotic cells contain usually hundreds of mitochondria, detected either as separated organelles or interconnected to a complex tubular network [22]. The maternal inheritance of mitochondria and close proximity of histone unprotected mtDNA to sources of ROS represents an increased risk for genetic and biological vulnerability. Therefore mitochondrial dysfunction is a hallmark for a spectrum of diseases including NDDs. In neurons, the cells progressively degenerating in NDDs, the number of mitochondria is very high, since they are the major energy generators for the neuronal biological activity. As has been revealed and studied more deeply in the last decade, mitochondria in neurons do not retain their stable position, but are actively transported and distributed around the cell, namely the dendrites, axons and synaptic terminals. Mitochondrial dynamics and distribution involves the movement and continual fusion and fission. This is especially important for highly polarized cells, as neurons where proper distribution of mitochondria guarantees the specific cellular function. Mitochondria use for their movement the cytoskeletal structures, namely the microtubular network. Moreover, as recently reported, mitochondria interact not only with their counterparts, but also with other cellular organelles, especially endoplasmic reticulum, lysosomes, Golgi system and autofagosomes [11, 23].

MITOCHONDRIAL FUSION AND FISSION

In the cells generally, and of special importance for neurons, mitochondria are highly dynamic organelles. Mitochondrial dynamics involves their movement, and in comparison to other organelles, they can fuse together to form one compact organelle and *vice versa*, they can split apart, a process known as mitochondrial fission, out of phase with cell division. Mitochondria can move anterograde, *i.e.* direction from the neuron soma to the periphery and retrograde along the microtubular network. Anterograde movement is kinesin dependent, whereas retrograde movement is carried on by the dynein proteins. To attach to the microtubules, mitochondria use the protein complex composed from Miro1/2 and Milton, where Miro proteins are the key proteins anchoring mitochondria to the cytoskeleton, because Miro possess transmembrane and GTPase domains. Beyond these two, Miro contains also calcium regulating domains, demonstrating the importance and complexity of the process of mitochondrial movement to keep mitochondria in proper shape and distribution. Indeed, mitochondrial movement in axons is precisely controlled, the mitochondria do not move constantly with settled direction, but making a lot of quick runs, stops and backward-forward shifts [24].

Mitochondrial dynamics, involving mitochondrial movement, fusion and fission represents the key mechanism of mitochondria quality control. Mitochondrial dynamics is crucial to delivery mitochondria to specific subcellular regions such as the synapse of a neuron. It is also used to remove damaged mitochondria, through mitophagy mainly occurred in perinuclear area [25]. By fusion, two mitochondria form one composed organelle and mix the mitochondrial matrix and membrane material. Proteins facilitating mitochondrial fusion are mitofusin 1 (Mfn1), mitofusin 2 (Mfn2) able to tether outer mitochondrial membranes by a conformational change after GTP hydrolysis and thus promote membrane fusion. Tethering regions of Mfn1 and Mfn2 facing cytoplasm are compatible for interaction, thus fusion promoted by homo- or heterodimers can be observed. Mitochondrial fission, *i.e.* splitting one mitochondria into two daughter ones is the next step of the mitochondrial fusion-fission cycle. Fission controls optimal mitochondrial length and is involved in separation and elimination of irreversibly damaged mitochondria and mitochondrial content. The main protein facilitating mitochondrial fission is again a GTPase named dynamin-related protein 1 (Drp1, also termed dynamin-like protein 1, Dlp1) and in the cell has been found both in the cytoplasm and on the surface of the outer mitochondrial membrane. Fission occurs when Drp1 oligomerizes into large ring-like complexes and by GTP hydrolysis the diameter of these ring complexes gets smaller and smaller finally followed by separation of the mitochondrial membranes. Drp1 activity and mitochondrial recruitment are regulated by posttranslational modifications such as phosphorylation, S-nitrosylation, sumoylation, and ubiquitination. Fis1, an integral protein of the mitochondrial outer membrane (MOM), was initially suggested to recruit Drp1 to the MOM. However, most recent studies suggest that Fis1 rather plays a regulatory role. Instead, mitochondrial fission factor (Mff), also an MOM protein, directly binds to Drp1 and is necessary and sufficient for mitochondrial recruitment of Drp1 during fission [26]. As we have quantified for the first time [24], the fusion as well as fission events are completed extremely quickly and normally take less than 1 second.

When the mitochondrial dynamics is perturbed, as an outcome fragmented or extremely elongated mitochondria appear. Elongated mitochondria fail to move whereas fragmented mitochondria are associated with reduced respiratory activity and can undergo autophagy [27]. In addition, fusion of mitochondria helps to exchange the membrane material and rescues mitochondria lacking mitochondrial DNA [28]. Thus, perturbations in mitochondrial dynamics, distribution and morphology are proposed to be early indicators of neurodegenerative and neurological disorders happening long time before energetic deficit became prominent [29].

NANOPARTICLES AND THE BRAIN

Nanoparticles (NPs) represent a diverse group objects with size range defined up to 100 nm and they strikingly differ from the particles of bigger size mainly due to their high surface to volume ratio. The NPs can occur naturally or are produced in a synthetic way. In the natural environment, NPs have been detected in terrestrial dust storm, volcanic eruptions, erosion and forest fire. These NPs are heterogeneous in composition, size and properties. In addition, human activities generate NPs as by-products of simple combustion, generated by combustion engines, power plants and other thermodegradation systems. Moreover, the development and application of the synthetically produced NPs is now exposing human to a new category of NPs with controlled chemical composition, shape and size compared to the NPs occurring naturally. Engineered NPs are largely used in the industry and cosmetics, but also application in healthcare and life sciences is increasing rapidly. This includes magnetic resonance imaging, drug delivery, hyperthermia treatment, basic research technologies, biomedical engineering, synthetic NPs are the core components of newly developed biosensors [30]. Thus human exposure to NPs, either by professional exposition or upon release from NP-containing products, is progressively increasing. Therefore, to define the environmental and health impact of engineered NPs and their end-products has become a priority for ensuring health protection. The possible side effects of NPs on human health are still largely unclear, although thousands of different engineered NPs are being developed and have been included in commercial products [31]. In addition, the Food and Drug Administration has approved several nanopharmaceuticals, supposing their low toxicity for the humans [32], but their long-term effects for human health needs to be deeply evaluated. From pivotal studies it is evident that NPs are highly reactive entities, probably due to their large surface area compared to their overall mass. Furthermore, because of their small size, NPs are very good candidates for interaction with cells and subcellular structures in highly efficient but so far poorly characterized ways.

Focusing on the NP and the brain tissue, it has been shown that NPs are capable to enter the central nervous system (CNS) [33]. This capability has a high potential for diagnosis and treatment of CNS disorders, including NDDs, stroke, brain tumours and others. Indeed, NPs-based delivery systems for the diagnosis/ treatment of CNS diseases are under intense investigation. The main obstacle to deliver active substances to neurons is the low permeability of the blood-brain barrier (BBB), a protective system assuring high protection of the neurons against exogenous molecules and particles [34]. The BBB is a protective network of vessels and cells, that strictly regulates the transport from blood to the brain and *vice versa*. Endothelial cells, clamped together by tight junctions, form the main physical barrier of the BBB. The BBB also include astrocyte end-feet and pericytes. Astrocytes can modulate the BBB functionality in several ways: (i) physical barrier, by tightening junctions; (ii) transport barrier, by regulating the expression and polarization of transporters; (iii) metabolic barrier, by modulating specialized enzyme systems [35]. Pericytes seem to regulate the BBB-specific gene expression patterns in endothelial cells and to induce the polarization of astrocyte end-feet [30, 36].

NPs have been shown to penetrate and deliver the active substances through the BBB much more efficiently compared to the classical pharmaceuticals [37]. These NPs have been specifically designed regarding their biocompatibility and biodegradation. Their fate and distribution depends on their composition and structure, thus defining their therapeutic potential in neurology, where both the functionality and toxicity of NPs is on the main interest. Functionality of NPs can be helpful in treatment of NDDs, stroke and brain damage, whereas the targeted toxicity of NPs could be beneficial for the treatment of the brain tumors. However, so far majority of the studies regarding the interaction between CNS and NPs have used metal or metal oxides with selected neuronal cell lines, most probably due to their common use and easy detection within the electron microscopy (EM) samples or with magnetic field. Therefore in the following paragraphs the first is to describe the neurotoxicity of the NPs starting shortly with the metal based NPs and then focus on the toxicity, if known, and the therapeutic potential of the biocompatible and biodegradable NPs follows.

NANOPARTICLES' TOXICITY, FOCUS ON NEUROTOXICITY

Toxicity of the NPs to the brain tissue is still limited and in-depth studies are necessary, especially when considering NPs as drug delivery systems [38]. The penetration of NPs represents a significant risk factor in the case of chronic and accidental exposure as well. For example, nanosized magnetite (Fe_3O_4) or maghemite (Fe_2O_3) associated with senile plaques and Tau filaments, have been detected in human brain implicating their potential risk for development and/or progression of neurodegenerative diseases [39, 40]. Other studies suggest the involvement of NPs in inflammatory processes in the CNS [41]. From the broad spectrum of negative effects of NPs to the brain functioning, here focus is taken on the NPs toxicity to the neurons, mitochondria, cytoskeleton and lysosomes.

In the study of Yang *et al.*, 2014 [42] the authors investigated the effects of silicone NPs in human SK-N-SH and mouse neuro2a (N2a). The diameter of the particles was 15 nm, concentration 10.0 µg/ml and the cells treated for 24 h. Here, the main cellular destination for the Si NPs was cytoplasm. The effect was complex with negative impact on cell survival (decreased viability, decreased cell density, higher cellular apoptosis), cell morphology (loss of dendrite-like processes) and cell metabolism (increased levels of intracellular reactive oxygen species). Moreover, increased deposit of intracellular β-amyloid 1-42 ($Aβ_{1-42}$) and enhanced phosphorylation of Tau at Ser262 and Ser396, two specific pathological hallmarks of Alzheimer's disease were observed in both cell lines. In line with this, the expression of amyloid precursor protein (APP) was up-regulated, while β-amyloid- degrading enzyme neprilysin was down-regulated in Si NP-treated cells. This study shows the penetrance of the 15 nm silicone NPs into the cells and links the exposure of the cells to the pathological signs of AD.

Liu *et al.* 2011 [43] performed the *in vivo* toxicity study in rats of the poly(ethylene glycol)-poly(lactic acid) (PEG-PLA) NPs conjugated to wheat germ agglutinin after repeated intranasal instillation for seven continuous days. NPs showed slight toxicity to brain tissue, as evidenced by increased glutamate level in rat brain and enhanced lactate dehydrogenase activity in rat olfactory bulb. However, the other measured parameters as acetylcholinesterase activity and inflammatory markers were not negatively affected.

In contrary with low toxicity of the polymeric NPs presented above, the exposure to metal NPs seems to have more prominent effects: In mice, after TiO_2 NPs exposure *via* nasal instillation (diameter 80 and 155 nm; 500 µg/mouse; every day up to 30 days), 25–30% of cellular loss has been observed in the cornu ammonis (CA)-1 region and dentate gyrus. Changes in morphology of pyramidal cells were also observed in the CA-4 region after NPs instillation [44]. The Au nanoparticles can also penetrate to the neuronal cells, although here the size of the particles represents the limiting factor: 17 nm Au NPs have been found in cytoplasm of pyramidal cells in the CA region of the hippocampus while 37 nm Au NPs were excluded. This was associated impairment of cognition observed in the mice treated with 17 nm Au NPs [45]. Application of Ag NPs (20 nm, 1, 5, 10 and 50 µg/ml) to primary cultured rat cortical neurons was associated with loss in cell viability, reduced neurite outgrowth of premature neurons, and in mature neurons, degeneration of neuronal processes was demonstrated [46]. Neurotoxicity of Ag NPs may originate from generating free radical-induced oxidative stress and by altering gene expression thereby inducing apoptosis [47]. Both processes mentioned, the oxidative stress and apoptosis are strongly connected with mitochondria, thus associating the Ag NPs with mitochondria.

Once penetrating the plasma membrane the NPs can either be dispersed in the cytoplasm or interact with cellular components [30]. The de Lorenzo study (reviewed in [48]) has shown that intranasally instilled silver-coated colloidal gold particles (50 nm) translocated anterogradely in the axons of the olfactory nerves to the olfactory bulbs in squirrel monkeys. These particles were even able to pass synapses in the olfactory glomerulus to reach mitral cell dendrites. This process occurs rapidly, the particles were transported to the mitral cell dendrites within one hour after beginning of the instillation. Importantly, these silver-coated colloidal gold particles were preferentially located in mitochondria. Direct application of Ag NPs to homogenates from rat brain (diameter from 5 to 45 nm; 10, 25, and 50 mg/l; 1 h) inhibits the activity of mitochondrial respiratory chain complexes I and III tissue [49]. Moreover, after application of 40 and 80 nm Ag NPs (2 and 5 µg/mg protein) the oxidative phosphorylation capacity was impaired and permeability of the inner membrane to protons was increased in isolated rat liver mitochondria [50]. Ag NPs (6–20 nm; 25–100 µg/ml; 48 h) in a dose dependent manner reduced cell ATP content, caused damage to mitochondria followed by increased ROS production and interrupted ATP synthesis in human lung fibroblasts (IMR-90) and human glioblastoma cells (U251), and presence of Ag NPs in mitochondria was shown by electron microscopy [51]. Relation between Fe NPs and mitochondria was shown by application of iron oxide superparamagnetic NPs in concentration of 10 µg/ml for 6 h in astrocyte cell cultures. Treatment resulted in increase of mitochondrial activity, most probably because of mitochondrial uncoupling [52]. Aluminium NPs were also tested for their effect on mitochondria in a microvascular endothelial cell line from human brain. Treatment with Al NPs (8–12 nm; 10 μM; up to 24 h) has decreased mitochondrial membrane potential [53]. Finally, ultrafine particles (diameter <150 nm) collected from ambient particle concentrators in the US and mostly composed from organic carbon localize in mitochondria, where they induce major structural damage and increase the ROS production in the cultured RAW 264.7 cells, a murine macrophage cell line [54]. However, as presented by evidence above, the NPs' effects on mitochondria have been widely studied in *in vitro* models with transformed cell lines [55], but *in vitro* neuronal models and *in vivo* studies analysing the impact of the NPs on respiratory chain and mitochondrial condition are still largely absent.

Beyond mitochondria, NPs target the autophagic-lysosomal system as well. Why the lysosomal system has to be mentioned at this point is, that mitochondria and lysosomes share the same cargo system of microtubules. In mice treated with micro Ag particles (<20 μm; 1.18 mg/mouse; observation after 7, 14 days and 9 months), heavy accumulations of Ag-sulphur nanocrystals was observed in endothelial cells, neurons and glial cells, in particular astroglia, within lysosome-like organelles [56]. Beyond this study, the information available for neurons is mostly absent here, therefore studies performed on non-neuronal cells are present below. In summary, they report accumulation of NPs in lysosomes, impairment of the lysosome-degradation capacity and induction of autophagy after treatment with different NPs [57]. NPs are present in endosomes and lysosomes in the model of human BBB composed of endothelial cells exposed to 50 nm SiO_2 NPs (100 μg/ml; 1 h) [58]. In line with this, the carboxylated polystyrene NPs (100 nm, 100–300 μg/ml, up to 48 h exposure) accumulated within lysosomes without degradation in the same BBB model [59]. Treatment with iron oxide NPs Resovist® (58.7 nm, 1–50 μg Fe/ml, 30 min) caused accumulation of NPs in endosomes and lysosomes in murine primary microglial cells [60]. Uncoated and oleic acid-coated iron oxide (core 8 nm; 15 μg/cm^2), TiO_2 (21 nm; 15 μg/cm^2) and SiO_2 (25–50 nm; 75 μg/cm^2) NPs were transported in endosomes and lysosomes in human brain-derived endothelial cells treated for 24 h with mentioned NPs [61]. LC3-II induction, the double membrane vacuoles and autolysosomes containing NPs were detected in these cells, in particular when exposed to 50 nm SiO_2.

The other issue concerning the autophagic-lysosomal system is that the lysosomal system is not completely efficient, a fact of special importance for long living cells as neurons. Here, in the lysosomal compartment an accumulation of non-degradable material has been observed within increased age. This material ends up concentrated in neurons in conjugates of neuromelanin, dark pigment droplets well defined from histological brain slice preparations. Although observed already long time ago, its chemical composition is still not completely clarified. What is known shown so far, neuromelanin consists of 30 nm nanoparticulate subunits together with lipid droplets localized in cell organelles [62]and accumulates during ageing in autophagic vacuoles [63]. Importantly, it has to be noticed that material from damaged and non-functional mitochondria represents a significant content of these autophagic vacuoles and mitochondrial membrane material thus could interfere with the degradation products of the NPs. Moreover, as excellently reviewed in Wischik *et al.*, 2014 [64], toxic forms of Tau protein able to start the nucleation process of the Tau monomers resulting in paired helical filaments are also located in autophagy-lysosomal vacuoles. Thus it is logical to deduce, that NPs or their metabolites could be, at least in part, responsible for Tau aggregation, one of the main hallmarks of the AD. Considering the presence of NPs in autophagosome-lysosomal system, on the other side one can hypothesize, that NPs with long-span lifetime specifically targeted to lysosomes and blocking the Tau nucleation and/or oligomerization could be of great interest in the efficient treatment of the AD in the future [64].

NPs can influence mitochondria also indirectly – to interfere with the cytoskeletal structures which in turn impacts their movement, distribution, fusion-fission and later their respiratory function. Indeed, interaction of NPs, mainly the solid ones with cytoskeleton is documented by following evidence: assembly/disassembly of cytoskeletal components, namely beta-tubulin and actin was disturbed in primary cultured rat cortical neurons after treatment of Ag NPs (20 nm, 1, 5, 10 and 50 µg/ml) in a dose dependent manner. This was followed by disturbances of synaptic structure and function: number of synaptic clusters of the presynaptic vesicle protein synaptophysin, and the postsynaptic receptor density protein PSD-95 were reduced in presence of Ag NPs [46]. In human brain endothelial cell lines the exposure to Al NPs with lower concentration (10 µM; up to 24 h) induced rearrangement of F-actin towards its deposition at the cell–cell borders. The higher concentration of 1 mM together with cytoskeleton rearrangements induced alterations in protein expression of tight junctions, namely loss of F-actin, zonula occludens proteins 1 and 2 and the junctional adhesion molecules-A [53]. In microtubules isolated from sheep brain TiO_2 NPs (20 nm; 0–50 µg/ml) negatively affect tubulin polymerization [65, 66] what can have in the cell negative effect on organelle trafficking.

Finally, NPs were shown to interact with the Tau protein: silica NPs (Si NPs) used in medicine are also able to increase P-Tau phosphorylation at Ser262 and Ser396, two phosphorylation sites characteristic of AD [42]. Summary of the toxicity of the NPs described above is given in Table **1**.

Table 1. Toxicity of the nanoparticles in experimental models: summary.

Nanoparticle	Cell type/ Exp. model	Organelle	Effect	Ref.
SiO_2, 15 nm	Human SK-N-SH, mouse neuro2a	Cytoplasm	Decreased viability, decreased cell density, higher cellular apoptosis, loss of dendrite-like processes, increased deposit of intracellular β-amyloid 1-42, enhanced phosphorylation of Tau at Ser262 and Ser396	[42]
SiO_2, 50 nm	Endothelial cells	Endosomes and lysosomes	Localization in endosomes and lysosomes	[58]
SiO_2, 25-50 nm	Human brain-derived endothelial cells	Endosomes and lysosomes	LC3-II induction, the double membrane vacuoles and autolysosomes containing NPs	[61]
TiO_2, 80 and 155 nm	Mouse, instillation	CA-1, dentate gyrus, CA-4	Changes in morphology of pyramidal cells	[44]
TiO_2, 21 nm	Human brain-derived endothelial cells	Endosomes and lysosomes	Transported in endosomes and lysosomes	[61]
TiO_2, 20 nm	Isolated microtubules	-	Disturbance in tubulin polymerization	[65]
Au, 17 nm	Mouse, injection	CA region, cytoplasm	Cognition impairment	[45]
Ag, 20 nm	Primary cultured rat cortical neurons	Cytoplasm, cytoskeleton	Loss in cell viability, reduced neurite outgrowth of premature neurons, and in mature neurons, degeneration of neuronal processes	[46]
Ag, 25 nm	Mouse, i.p. injection	-	Generating free radical-induced oxidative stress and by altering gene expression	[47]
Ag-coated Au, 50 nm	Squirrel monkeys, intranasal instillation	Axons, cell dendrites, mitochondria	Synapse penetration, localization in mitochondria	[48]
Ag, 5-45 nm	Homogenates from rat brain	-	Activity of mitochondrial respiratory chain complexes I and III inhibited	[49]
Ag, 40 and 80 nm	Isolated rat liver mitochondria	Mitochondria	Loss in oxidative phosphorylation capacity, increase in permeability of the inner membrane to protons	[50]
Ag, 6-20 nm	Human lung fibroblasts (IMR-90), human glioblastoma cells (U251)	Mitochondria	Reduced cell ATP content, damage to mitochondria, increased ROS production, interrupted ATP synthesis, presence in mitochondria	[51]

(Table 1) contd.....

Nanoparticle	Cell type/ Exp. model	Organelle	Effect	Ref.
Ag, <20 nm	Mouse, endothelial cells, neurons and glial cells	Lysosome-like organelles	accumulations of Ag-sulphur nanocrystals	[56]
Ag, 20 nm	Primary cultured rat cortical neurons	Cytoskeleton	Assembly/disassembly of beta-tubulin and actin, disturbances of synaptic structure and function	[46]
Fe, SPION, Fe₃O₄ or γ-Fe₂O₃	Astrocyte cell cultures	-	Increase of mitochondrial activity	[52]
Fe, Resovist®,58.7 nm	Murine primary microglial cells	Endosomes and lysosomes	Localization in endosomes and lysosomes	[60]
Fe, core 8 nm	Human brain-derived endothelial cells	Endosomes and lysosomes	Transported in endosomes and lysosomes	[61]
Al, 8-12 nm	Human endothelial cell line	Mitochondria Cytoskeleton	Loss in mitochondrial membrane potential, rearrangement of F-actin, alterations in protein expression of tight junctions	[53]
Organic carbon, <150 nm	Murine macrophage cell line (RAW 264.7)	Mitochondria	Structural damage and increase the ROS production	[54]
Carboxylated polystyrene, 100 nm	Endothelial cells	Lysosomes	Localization in lysosomes	[59]
PEG-PLA conjugated to wheat germ agglutinin	Rat, instillation	Brain tissue, olfactory bulb	Increased glutamate level, enhanced lactate dehydrogenase activity, no change in acetylcholinesterase activity and inflammatory markers	[43]

NANOPARTICLES AS DRUG CARRIERS

NPs, due to their unique physico-chemical properties and advantageous penetration through biological tissue are challenging tools for clinical medicine, starting with screening and ending with treatment of various diseases. NPs can be engineered with controlled properties of pharmacokinetics, encapsulation and cargo release when delivered to the target environment all of these with low level of systemic toxicity [67]. Progress in NP engineering allows to carry insoluble or poorly soluble drugs with higher efficiency and can have capacity to stabilize labile molecules. Moreover, NPs can have potential to reduce the immune response in case where proteins, peptides or DNA need to be delivered to desired destination. NPs may improve drug concentration to a therapeutically desirable range by increasing half-life, solubility, stability and permeability of drugs.

The use of NPs for medical purposes has started with organometallic compounds as Fe_3O_4 NPs, quantum dots, in some cases coated with inorganic or organic shells. Nowadays, the theranostic nanoparticular systems are coming closer to compatibility with the biological environment: significant approach has been reached in lipidic NPs as solid lipid NPs, liposomes, nanoemulsions. The other group with low adverse effects are the polymeric NPs including also dendrimers, polymeric micelles and nanogels. Here, the active drug may be entrapped into the NPs (encapsulated or dissolved), adsorbed on the surface or covalently bound to the matrix or [30].

Well known polymer for the preparation of NPs for CNS drug delivery is polyalkylcyanoacrylate [68], currently in phase II of clinical trial. Polymeric micelles are amphiphilic polymers with a hydrophilic core and hydrophobic shell able to self-assemble in an aqueous solution [69]. This arrangement avoids rapid uptake by the reticuloendothelial system and thus improving circulation time in the body [70]. The hydrophobic core serves allows the encapsulation of compounds poorly soluble in water, protecting these molecules from degradation, thus prolonging therapeutic activity. Poly(ethylene glycol) (PEG), due to its minimal adverse effects represents the most commonly used hydrophilic polymer. PEG is biocompatible, water soluble and highly hydrated. The next example applicable for delivery to the CNS are relatively thin rods (diameter 12 nm). These have a more favourable morphology for the internalization in neurons, while urchin NPs (size roughly 70 nm) are easily phagocytosed by microglia [71].

The drugs desired to be transported by NPs in case of NDDs are so far mostly neuroprotective compounds of a natural origin and synthetic derivatives present encouraging results with minimal adverse effects, and some of these molecules are currently in different phases of clinical trials [72]. Alkaloids are effective in alleviating the symptoms of neurodegenerative diseases, such as AD. Large numbers of natural alkaloids and synthetic derivatives have exhibited neuroprotective effects. Polyphenols are one of the most significant secondary metabolites, exhibiting natural antioxidant properties. Polyphenols, both flavnoids and nonflavnoids, are effective in alleviating and protecting against the neurodegenerative diseases in various cell culture and animal models as well as multiple epidemiological studies. Moreover, various polyphenols and nutrients with respect to application of nanotechnology have recently reported as cancer chemotherapeutic agents in the literature [73]. Epidemiological evidence has shown that the Mediterranean diet, which is rich in antioxidants, is effective in the prevention of age-related diseases, such as AD [74].

Another approach in the treatment of AD is to remove the transitional metal ions due to their catalyzing properties to form the ROS. A number of metal chelators were delivered in the form of nanoparticles to facilitate BBB penetration of the metal ions [75, 76]. These studies showed promising results: low toxicity, sufficient level of stability and significant metal chelating properties of the nanodelivery systems used.

In case of PD, current PD nanomedicine approach has been primarily focused on the gene therapy. As an example, nanoparticles containing condensed DNA were tested for the purpose to increase production of dopamine. Significant improvement from baseline was observed in both clinical symptoms and abnormal brain metabolism as measured by tomography [77]. The use of NPs as drug delivery systems, specifically for the treatment of AD and PD, will be discussed in more detail in the sections Nanoparticles and Alzheimer's disease treatment and Nanoparticles and Parkinson's disease treatment.

MITOCHONDRIA AND ALZHEIMER'S DISEASE

Mitochondrial dysfunction is a prominent feature of various NDDs. Increased oxidative damage is often associated with AD, PD, HD, ALS and several other neurodegenerative disorders suggesting that oxidative stress may play an important role in the pathophysiology of these diseases [78]. Other evidence linking mitochondria with NDDs includes defects and/or reduced amount of mtDNA, apoptosis, altered mitochondrial calcium handling and morphological ultrastructural changes. Finally, widely increasing research reveals influence of the mitochondrial dynamics in NDDs [79].

In case of AD, the prevalent data in the literature are linking mitochondria with increased ROS production and Aβ. Indeed, mitochondrial dysfunction is one of the symptoms of AD, where accumulation of Aβ within mitochondria has been shown. The Aβ is transported into mitochondria *via* the translocase of the outer membrane and accumulates at the mitochondrial cristae. Aβ inhibits enzymes of respiratory chain such as cytochrome oxidase (complex IV) and the key Krebs-cycle enzymes (α-ketoglutarate and pyruvate dehydrogenase) impairing mitochondrial membrane potential, ETC, ATP production and oxygen consumption. These data were demonstrated both in the brain and isolated mitochondria. As a consequence, level of mitochondrial ROS increase significantly inducing oxidative damage of mtDNA [80]. In line with this, current evidence demonstrates that mitochondria-derived ROS are sufficient to trigger Aβ production both *in vitro* and *in vivo*, and thereby initiate a vicious cycle [81]. In the mitochondria, Aβ also interacts with the 17-β-hydroxysteroid dehydrogenase X (HSD17B10) also known as Aβ peptide-binding alcohol dehydrogenase, an enzyme that catalyzes the oxidation of a wide variety of fatty acids, alcohols, and steroids, exacerbating neuronal cell death by mitochondrial dysfunction and oxidative stress [82]. Next, impaired Ca^{2+} homeostasis induced by Aβ can trigger ROS formation and mitochondrial dysfunction. Mitochondrial Aβ is also reported to interact with cyclophilin D, an integral part of mitochondrial permeability transition pore (mPTP). Potentiation of free radical production, as well as synaptic failure promotes the cyclophilin D mediated opening of the mPTP resulting in apoptotic response.

Although a lot of effort has been taken to the combat the Aβ plaques, in the clinical trials this approach resulted to be not efficient (see [64] for review). Therefore for effective therapy also other known mechanisms need to be considered, including the Tau hypothesis in the AD. Indeed, hyperphosphorylated Tau negatively impacts axonal organellar transport including mitochondria resulting in energy deficit, oxidative stress at the synapse and finally to neurodegeneration [22]. Under yet unresolved triggering conditions, Tau is hyperphosphorylated at 'pathological' sites leading to microtubules depolymerization, axonal transport disruption and aggregation [83].

Next evidence connecting mitochondria and AD includes decreased glucose and oxygen use by the AD brain, although there is no evidence in humans or animal models for a massive disruption of the BBB in AD [84]. Next, Pioglitazone attenuates mitochondrial dysfunction by improving mitochondrial DNA (mtDNA) content, nuclear-encoded electron transport chain subunit proteins, increased oxygen consumption, and elevated complex I and complex IV V_{max} activities. The other example is resveratrol - with antioxidant, anti-inflammatory, and metal-chelating effects prevents mitochondrial dysfunction as well [79]. In addition, in AD, accumulation of autophagosomes within large swellings along degenerating neurones has been described. This can be explained as deficit in the maturation of autophagosomes and their retrograde transport toward the cell body implicating the misleading of the mitochondrial transport as well. Finally, perturbations in iron homeostasis have also been reported in AD: levels of iron, ferritin, and transferrin receptor in the hippocampus and cerebral cortex are changed. Iron ions are associated with the deposits of Aβ and the promotion of oxidative stress. Indeed, Aβ deposits are enriched with zinc, iron, and copper linking the mitochondria and AD both directly (ROS) and indirectly (through the autophagy and mitophagy clearance system) [15].

A clever alternative how to bypass the BBB is retrograde axonal transport through the nasal instillation. Curcumin-incorporated nanoparticles coupled with Tet-1 peptide penetrate into neurons and are able to travel towards cell soma by axonal retrograde transport. These NPs were capable to destroy Aβ aggregates and exhibited antioxidative activity. However, efficiency of this system needs to be tested *in vivo* [37].

With the presented success of NPs to target axonal transport, mitochondrial dynamics in AD needs to be reviewed at this place. Axonal mitochondria are actively transported retrogradely and anterogradely, frequently fusing and splitting apart, ensuring proper mtDNA/membrane material exchange and controlling mitochondrial distribution. In mouse model of AD disbalance toward mitochondrial retrograde movement, promoted mitochondrial fission, loss of fusion and defective mitochondrial metabolism were detected [29, 85]. This is consistent with the finding that mitochondria accumulate in the soma and are reduced in neuronal processes *in vivo* in AD pyramidal neurons [86, 87]. Exposure of primary neurons to preaggregated $A\beta_{25-35}$ found in AD induces acute impairment in mitochondrial axonal transport [88] and mitochondrial fragmentation prior to neuronal death [89]. This is presumably through enhanced nitric oxide (NO) production and increased S-nitrosylated Drp1 formation which activates GTPase activity and mitochondrial fission [90]. In line with this, increased S-nitrosylation of Drp1 is also found in AD brain tissues [87, 90]. Enhanced nitrosylation of Drp1 is caused also by phosphorylated Tau and $A\beta$, which leads to increased mitochondrial fission and neurodegeneration [85]. To link mitochondrial fragmentation and fusion-fission in AD, authors Wang *et al.* have immunoblotted key fusion-fission proteins in hippocampal tissues from AD patients compared with age-matched controls. Analysis has revealed that expression of Drp1, OPA1, Mfn1, and Mfn2 was reduced, whereas expression of Fis1 was significantly increased [91]. Calkins *et al.* [29] also found altered expression of Mfn1 and Fis1 (but not Mfn2, OPA1 and Drp1) in primary cultured neurons isolated from Tg2576 transgenic mice expressing the human APP Swedish mutation and has shown increased interaction between Drp1 and $A\beta$ in the brain tissues. Altogether, presented facts evidence perturbations in mitochondrial dynamics in AD and potentiate the use of NPs in this recently developing field of research.

MITOCHONDRIA AND PARKINSON'S DISEASE

Regarding current knowledge, aging represents the most important risk factor of PD together with a few genetic mutations involving genes encoding α-synuclein and LRRK2 (leucine-rich repeat kinase 2) causing autosomal dominant forms of PD. Next dysfunctional proteins in PD are originating from loss-of-function mutations in the genes encoding Parkin and PINK1, both of them related with mitochondrial quality control and mitochondrial fission and fusion. DJ-1 loss of function mutants result in abnormal mitochondria and early onset PD and mediate the autosomal recessive form of PD. All of proteins coded by these genes are associated or directly localized to mitochondria. Mutations in the gene encoding LRRK2 are related to autosomal dominant Parkinson's disease form. The most common mutation is G2019S that accounts for 5-6% of familial cases of

Parkinson's disease. Experimental studies have identified different pathogenic mechanisms for altered LRRK2 that involve inflammation processes, oxidative stress and mitochondrial dysfunction, among others. Focusing on the last mentioned, mutations in LRRK2 cause mitochondrial fragmentation and a downregulation in mitochondrial homeostasis (reduction in mitochondrial membrane potential and ATP production) [20]. Mutations in the protein α-synuclein, particularly two mutations in the α-synuclein gene (A30P and A53T) have been identified which lead to the formation of pathogenic pore-like annular and tubular protofibrils. These mutations inhibit the activity of complex I and induce mitochondrial fragmentation [22]. Mutations in the Parkin gene product, which is an ubiquitin ligase, lead to an early-onset familial Parkinson's disease. Experimental studies have determined that the pathology of Parkin is associated with alterations in the mitochondrial recognition, transportation, and ubiquitination and with mitophagy impairment. Mutations in the mitochondrial serine/threonine-protein kinase PINK1 result in alterations in the mitochondrial morphology and function (defects in complex I activity) and they are strongly associated with a form of autosomal recessive early-onset Parkinson's disease. DJ-1 mutations on chromosome 1p36 cause autosomal recessive early-onset PD and its pathological mechanism seems to be linked with mitochondrial fragmentation and mitochondrial structural damage and consequently defects in the mitochondrial function of dopaminergic cells.

Compared to well defined gene mutation forms of PD, the sporadic case of PD represents around 95% of all cases. In focus to mitochondria, here the association is not so clearly linked: one of the proposed mechanisms for the dopaminergic neurons degeneration in sporadic Parkinson's disease cases is related to an excessive production of ROS that leads to oxidative stress. An excess of ROS causes the oxidative modification of macromolecules (lipids, proteins, and DNA) leading to cell damage and finally to cell death. The pathological effect of ROS is also involved in a reduction of ATP production, in an increase of iron levels, and in an increase of intracellular calcium levels and alterations in mitochondrial respiratory chain complexes function. Moreover, evidence implicating the role of mitochondria in PD prognosis are validated with the discovery that rotenone and 1-methyl-4-phenyl-1,2,3,6-tetrahydropyridine (MPTP), both specifically affecting mitochondrial metabolism can induce PD in animal models and humans. MPTP is highly lipophilic and easily crosses the BBB and is further oxidized to the toxic molecule MPP$^+$, which can be taken up by dopaminergic neurons *via* dopamine transporters. MPP$^+$ is highly concentrated in neuronal mitochondria and can inhibit complex I of the ETC, which results in impaired electron flux *via* complex I, subsequently resulting in decreased in mitochondrial ATP production and elevated ROS levels [22]. In addition, a dysfunction in the ubiquitin-proteasom--system (UPS) and the autophagy-lysosomal pathway (ALP) as evidenced in a reduction of proteasome and autophagy activities and in postmortem brains of PD patients has been demonstrated [20].

Focusing more specifically to mitochondrial dynamics in PD, it has been demonstrated recently that Drp1-dependent mitochondrial fragmentation occurs after MPP$^+$ treatment. This was connected with ATP depletion in both primary rat dopaminergic midbrain neurons and SH-SY5Y neuroblastoma cells. Evidence, that this pathway is mediated by the Drp1 ensures the fact, that Drp1 knockdown reduces mitochondrial fragmentation induced by MPP$^+$ [92]. Increasing evidence is directly linking the α-synuclein and mitochondrial dynamics: overexpression of α-synuclein fragments mitochondria independently from fission/fusion system. In addition, mitochondrial fragmentation induced by α-synuclein could be partially rescued by PINK1, Parkin, or DJ-1 overexpression [93]. Moreover, a A53T mutation of α-synuclein, when inserted into the genome of transgenic mice, induced mitochondrial fragmentation [94]. The authors of this study describe also significant changes in the expression of Drp1, Mfn1 and Mfn2 in this mouse model, indicating involvement of the mitochondrial fusion-fission machinery. Next, it seems PINK1/parkin can directly influence mitochondrial dynamics: when PINK1 was knocked down in COS-7 cells, elongation of mitochondria was observed. This mitochondrial elongation can be inhibited by Drp1 or Fis1 overexpression [95]. Moreover, expression of PINK1 and Parkin in primary cultured neurons resulted in mitochondrial fragmentation and vice versa, increased mitochondrial length was observed after their silencing [96]. Degradation of Parkin ubiquitinated Mfn2 may explain diminished mitochondrial fusion, as Mfn2 is well known substrate for PINK1-Parkin pathway [97]. In contrast, several groups demonstrated that PINK1 or Parkin silencing by siRNA leads to mitochondrial fragmentation, which could be exacerbated by Drp1 overexpression and rescued by expression of a dominant negative Drp1 mutant in HeLa cells or neuroblastoma cells [98 - 100]. Finally, overexpression of wild-type LRRK2 causes mitochondrial fragmentation; the mechanism behind could be the interaction between LRRK2 and Drp1, where LRRK2 facilitates Drp1 translocation to mitochondria triggering mitochondrial fission [101]. Therefore pharmaceutical intervention or nanotherapeutics targeting Mfn1, Mfn2, Mff1, Fis1, Drp1, PINK1 and Parkin could provide novel treatment strategies for neurodegenerative diseases.

NANOPARTICLES AND ALZHEIMER'S DISEASE TREATMENT

Up to date, the therapy of AD by pharmacological treatment can reduce the symptoms of the disease, unfortunately it cannot reverse disease progression. Therefore, researchers are trying to treat AD where implementing the NP delivery systems is of great interest for effective drug delivery. For the AD treatment, there are several promising drugs in development, including use of enzymes to decrease Aβ production, substances that increase Aβ clearance, immunotherapy, enzymatic degradation of Aβ, inhibition of Aβ aggregation, antioxidants, anti-inflammatory drugs, statins, etc. From among these approaches, in this review we concentrate on two of them due to their implications for improvement in mitochondrial functioning: the inhibition of Aβ assembly which has been considered the primary therapeutic strategy for AD *via* NPs and the management of oxidative stress in AD using nanoparticle delivery systems, as currently the clinical outcomes of the antioxidant therapy seem to be the most effective ones [102].

Aβ Strategy: To date, the most effort was to combat the Aβ fibrils. Reducing the concentration of Aβ fibrils can be achieved by decreasing monomer nucleation and thus blocking aggregation. This is expected to reduce formation of Aβ derived diffusible ligands (ADDLs), fibrils, and plaques. The alternative strategy is disintegration of already existing amyloid plaques or fibrils. It has to be noticed here, that the results performed to understand the process of Aβ oligomerization inhibition are controversial and extremely dependent on the experimental protocols. As an example, it was demonstrated that cationic surfactants had a dual effect on Aβ fibrillation: depending on the surfactant concentration they could either accelerate or inhibit fibrillation. A similar story has been reported for liposomes and for cationic polystyrene NPs. For example, Mahmoudi *et al.*, 2013 [103] reports the effect of magnetic nanoparticles on Aβ fibrillation. Depending on the surface coating charge, lower concentrations of superparamagnetic iron oxide nanoparticles (SPIONs) have inhibited fibrillation, whereas the rate of Aβ fibrillation was higher when higher concentration of SPIONs was applied. Moreover, it seems the surface charge coating plays important role: authors proclaim that SPIONs designed for *in vivo* medical imaging and therapy should be negatively charged on the surface or the generally electroneutral [102]. Other groups have reported that NPs such as carbon nanotubes, quantum dots, cerium oxide particles and copolymer particles enhance the possibility for fibril nucleation. Indeed, nanoparticles with their high surface-to-volume ratio facilitate protein binding onto the particle surface. This may enhance the probability of partially unfolded proteins coming into close contact, interact together and thus accelerate Aβ formation.

Towards to the complex therapy bridging the Aβ and antioxidants strategy could be the CeO (cerium oxide) NP used to treat or prevent both AD and PD. In particular, CeO NPs with diameter ranging from 2 to 100 nm can be administered in concentrations capable to block lipid peroxidation, block free radical production induced by $A\beta_{1-42}$, production of hydroxyl or superoxide radicals, reduce neuronal death, decrease loss of dopaminergic neurotransmission, balance $[Ca^{2+}]$ dysfunction in neurons, or reduce mitochondrial dysfunction. In general,

CeO NPs can be effective in treatment of toxic compounds inducing mitochondrial dysfunction, such as cyanide, rotenone and other mitochondrial toxins [104].

Antioxidants Strategy: The use of antioxidants is the next strategy in AD therapeutics because oxidative damage is an early symptom of AD. Antioxidants as ferulic acid, glutathione, nanoceria and fullerenes were able to inhibit the Aβ fibrillization and/or neuronal oxidative stress [37]. Cheng *et al.*, 2013 tested the efficacy of curcumin PEG-PLGA-polyvinylpirrolidone NP freeze dried with β-cyclodextrin orally administered in AD Tg2576 mice. Treated animals showed better cue memory in the contextual fear conditioning test after three month treatment [105]. A naturally occuring flavonoid quercetin, possessing antioxidant activity was encapsulated into solid lipid NP and administered in rats with aluminium induced dementia. Animals improved memory retention in the spatial navigation task and in the elevated plus maze paradigm compared to free quercetin administration [106]. Trimethylated chitosan (TMC) poly(d,l-lactide--o-glycolide) (PLGA) nanoparticles (TMC/PLGA-NP) were synthesized as a drug carrier for brain delivery. Then, 6-coumarin loaded PLGA-NP and TMC/PLGA-NP were applied intravenously into the mice. Higher accumulation of TMC/PLGA-NP was detected in the cortex, paracoele, the third ventricle and choroid plexus epithelium, while no brain uptake of PLGA-NP was observed. Coenzyme Q10 loaded TMC/PLGA-NP greatly improved memory impairment in the injected mice, whereas the efficacy was slight for loaded PLGA-NP (no TMC conjugation. Coenzyme Q10 carried to the brain through the TMC/PLGA-NP has also positively influenced senile plaque formations and selected biochemical parameters. Results of this study demonstrate that TMC/PLGA-NPs Thus TMC effectively cross the BBB and appear to be a low toxicity brain drug delivery carrier [107].

Chonpathompikunlert *et al.*, 2015 [108] studied the redox polymer nanotherapeutics in a model of senescence-accelerated prone mice, a model mimicking the AD condition. The NPs prepared of redox polymers possessing antioxidant nitroxide radicals were delivered to the brain after oral administration for one month. The oxidative stress in the brain was remarkably reduced by treatment with redox nanoparticles, compared to low-molecular-weight nitroxide radicals treatment. This was associated with amelioration of cognitive impairment with increased numbers of surviving neurons. Importantly, any signs of toxicity were observed after treatment by these NPs.

Finally, neuroprotective agents as S14G-humanin and NAP (octapeptide NAPVSIPQ) represent a promising tool to combat AD because they block AD-related neuron death and improve neuronal activity. S14G-humanin is a neuroprotective peptide, its digestion by peptidase was minimized by packaging S14G-humanin into polymersomes. As next step, lactoferrin was implemented into these polymersomes to promote BBB transport. Created nanoparticular system transported 3.32-fold more cargo to the brain than did unmodified polymersomes. Moreover, these polymersomes reversed the decrease of choline acetyltransferase (ChAT) activity caused by intracranial injection of $A\beta_{25-35}$ [109]. The NAP derived from the Activity Dependent Neuroprotective Protein has been identified as a next promising neuroprotective agent against AD. Liu *et al.*, proposed nanoencapsulation of NAP in B6 peptide- modified PEG-PLGA NP to enhance its brain delivery and to protect the neuropeptide from degradation [110]. Indeed, B6 peptide was significantly enhancing brain targeting of the NPs resulting in higher brain accumulation. Moreover, B6-NP-NAP significantly ameliorated the spatial learning deficit, the cholinergic dysfunction and the loss of hippocampal neurons in mice stereotactically coinjected with $A\beta_{1-40}$ and ibotenic acid.

NANOPARTICLES AND PARKINSON'S DISEASE TREATMENT

Current approaches for PD treatment are mostly addressing motor symptoms enhancing dopamine (DA) levels and far fewer are focused the the causal therapy or on modifying disease progression. DA agonists as well as Levodopa (L-DOPA) exhibit very low brain uptake and low oral bioavailability, where NPs could be beneficial to increase penetration of these substances. Beyond this, there is urgent need for more efficient therapy combating the cause of the PD. The current research aiming more specific targets in PD therapy is based on: 1) the delivery of specific genes to improve the enzymatic activity *via* the NPs, 2) using the cytoprotectants and 3) blocking α-synuclein aggregation [111].

Gene therapy, by increasing level of dopamine synthetic enzymes or improving function of dopaminergic neurons has potential to reverse the progression of PD. In more detail, one of the goals of PD gene therapy is restoring tyrosine hydroxylase (TH) in the striatum. TH-encoding plasmids were encapsulated in liposomes that were targeted *via* OX26. This treatment normalized the striatal TH activity and also reversed the apomorphine-induced rotational behaviour. Other system to deliver genes into the brain are lactoferrin-modified NPs. With this NP carrier system 4.2-fold higher gene expression in the brain was observed when compared to unmodified nanoparticles [112]. Lactoferrin-modified NPs loaded with the GDNF gene reduced the dopaminergic neuronal loss, improved locomotor activity, and enhanced the levels of monoamine neurotransmitters in the unilaterally 6-hydroxydopamine (6-OHDA)-lesioned rat model [37, 113].

Urocortin, featuring cytoprotective effects, has diminished the PD phenotype in 6-OHDA model after intracerebral administration. Unfortunately, when applied intravenously, the penetration to brain was very low. Lactoferrin-modified nanoparticles deliver 1.98-fold more cargo to brain than do unmodified nanoparticles. Therefore Hu *et al.* [114] encapsulated urocortin in lactoferrin-modified nanoparticles to enhance the penetration of urocortin through the BBB. As indicated by the results of behavioral tests, immunohistochemistry and a striatal transmitter assay, the urocortin-loaded NPs attenuated the striatal lesions in rats treated with 6-OHDA. These authors have successfully used also an alternative way for targeted delivery of urocortin *via* odorranalectin-modified NPs administered intranasally: significant reduction of the rotational behaviour and alleviation of the reduction of neurotransmitters and reduction of the loss of dopaminergic cells caused by 6-OHDA is reported in this study.

Failure of the ubiquitin-proteasome system to degrade abnormal proteins may underlie the accumulation of α-synuclein and dopaminergic neuronal degeneration that occurs in Parkinson's disease. VP025 (Vasogen Inc.) is a preparation of phospholipid nanoparticles incorporating phosphatidylglycerol that has been shown to have neuroprotective effects [37]. Fitzgerald *et al.*, 2008 [13] has shown that VP025 prevents the deficits in motor coordination and dopamine observed in a proteasome inhibitor rat model of PD.

Joshi *et al.*, 2015 [115] have been studying early stage of aggregation of α-synuclein in the presence of Fe_3O_4 NPs using confocal microscopy and fluorescence correlation spectroscopy (FCS). Pure Fe_3O_4 NPs accelerated the rate of early α- synuclein aggregation, without binding the monomeric α- synuclein. In contrast, when Fe_3O_4 NPs were coated with L-lysine (Lys), the authors noticed strong binding of these coated NPs with the monomeric α- synuclein preventing its aggregation. Moreover, Lys-coated Fe_3O_4 NPs toxicity was significantly lower compared to pure Fe_3O_4 NPs. Dendrimers represent other strategy to combat α-synuclein aggregation: authors Milowska *et al.*, 2013 [116] have been studying the impact of viologen-phosphorus dendrimers to the α-synuclein fibrillation. Depending on the dendrimer concentration and the surface modification by defined reactive groups, α-synuclein fibrillation kinetics can be significantly reduced, and even, the fibrillation can be totally inhibited at low concentrations in case of phosphonate dendrimers.

NANOPARTICLES AS DRUG DELIVERY SYSTEMS TO MITO-CHONDRIA

Due to the well documented evidence of mitochondrial dysfunction in AD and PD, mitochondria are promising targets for therapeutic intervention. As mitochondria are the main energy suppliers for the majority of the human cells, design of therapeutics targeted to mitochondria represents a promising therapeutic strategy. In this concept, nanoparticles could be a promising avenue of drug delivery. Their surface modification may help against rapid immune clearance, they can carry pharmaceuticals with low solubility in water and their high surface-volume ratio increases the active surface for chemical reactions and adsorption capacity [22]. Here, the transport of NPs into mitochondria is limited due to several barriers at the organism level, tissue level (crossing the BBB), cellular level (penetration of the plasmatic membrane), and organelle level (delivery of NPs to mitochondrial matrix). Therefore, advanced strategies have to be implemented when mitochondria is the desired destination in the cell. Up to date, the main progress in this field involves use of substrates for mitochondrial receptors, delocalized cations, conjugation of mitochondrial peptides to the drugs or use of specific compounds able to accumulate at the mitochondrial surface [22].

So far, the main attention is concentrated to DQAsomes; self-assembling nanocarriers of the bolaamphiphile dequalinium chloride targeted to mitochondria. Dequalinium is a molecule with two symmetrical cationic charge centres separated by a hydrophobic carbon chain. Dequalinium aggregates in aqueous solution forming cationic colloidal nanoparticles, which can form complexes with DNA. These DQAsome/DNA complexes (DQAplexes) were shown to penetrate the cells *via* endocytosis. DQAplexes were attracted to the depolarized mitochondria by electrostatic interactions and selectively released DNA into mitochondria [117]. In addition, the authors of this study demonstrated that DQAsomes were able to deliver plasmid-sized DNA into mitochondria selectively through a mitochondrial targeting sequence of 20–40 amino acids recognized by receptors on the surface of mitochondria. This success inspired other group to engage the DQAsomes to deliver a non-DNA cargo, in this case paclitaxel. Created nanoparticular complex was shown to partially co-localize with mitochondria and pro-apoptotic effect of paclitaxel was shown to be remarkably increased [118]. Although unexplored yet, this approach shows a perspective how to deliver pharmacologically active molecules which, otherwise, would be unable to enter mitochondria for treatment of neurodegenarative diseases. Unfortunately, most of the liposomes made from dequalinium analogs are cytotoxic and this approach to deliver drugs or DNA into mitochondria is still in the early stage of development [2].

The next investigated carrier system targeting mitochondria is called MITO-Porter, a liposome-based nanoparticular system, able to carry various agents into mitochondria *via* membrane fusion [119]. MITO-Porter is an extension of a concept based on envelope-type viruses, composed from a condensed core, as plasmid DNA or proteins, and a lipid envelope with attached specific functional groups. As example, incorporation of octa-arginine moieties (R8) allows to deliver green fluorescent protein to mitochondria, cytosol and nucleus of HeLa cells [119, 120]. In addition, R8 decorated liposomes were introduced by Furukawa *et al.* to enhance SOD's anti-oxidant effect of superoxide dismutase (SOD), since SOD is unable to cross cellular membranes by itself [2]. The authors observed by increasing cellular uptake and accumulation of SOD [121].

Other approach is based on oxide nanoparticles. For instance, titanium dioxide NPs are able to penetrate mitochondria [122]. Therefore TiO_2 NPs have been used for mitochondrial drug delivery with and without specific mitochondrial targeting surface modifications (mitochondrial targeted oligonucleotides). Metallic nanoparticles represent another candidate for mitochondrial drug delivery: gold nanoparticles can enter mitochondria, although here the size of the gold particle is the limiting factor. 3 nm particles pass mitochondrial membranes, but 6 nm ones are not able to penetrate indicating involvement of a ion channel mediated entry [123]. Carbon nanotubes is other carrier system for mitochondrial drug delivery. Multi-walled carbon nanotubes fused with cationic rhodamine-110 as the mitochondria specific targeting ligand have been developed to deliver the anti-cancer cisplatin prodrug and 3-BP to mitochondria [124]. Finally, polymeric nanoparticles are ideal vectors for mitochondrial drug delivery applications due to their biocompatibility and flexibility. A prime example that demonstrates the flexibility of polymeric platforms is the development of triphenylphosphonium (TPP) conjugated particles [125]. The platform of this system consisted of a poly(D,L-lactic-co-glycolic acid)-block (PLGA-b)-poly(ethylene glycol) (PEG)-TPP polymer (PLGA-b-PEG-TPP) that could be blended with other PLGA-PEG polymers to control the physicochemical properties and optimize mitochondrial delivery [22, 125].

COULD NANOPARTICLES IMPROVE MITOCHONDRIAL FUNCTION-ING WITHOUT ENTERING MITOCHONDRIA?

As mentioned above, the research idea to improve mitochondrial functioning by targeting NPs directly to mitochondria was stated time ago, with special interest to deliver NPs into the matrix of mitochondria. This means, in case of neurons, degenerating in AD and PD, to penetrate three membranes after crossing the BBB – first the cytoplasmic membrane of neurons and subsequently the outer and inner mitochondrial membrane. This classical approach has a logical background, as the main place of ROS production and location of mitochondrial respiratory enzymes are located in the matrix close to MIM, similarly with the mtDNA located in the mitochondrial matrix. By advanced delivery systems, the localization of NPs in mitochondria is already reported, however, especially for therapeutic treatment, to internalize NPs into mitochondrial matrix with the current state of knowledge is limited. Thus, we would like to emphasize an alternative approach, how to potentially improve mitochondrial quality control bypassing NPs internalization into mitochondria: a potential novel approach based on effective controlling of mitochondrial movement and fusion-fission dynamics. As we already have reviewed above, mitochondrial dynamics is critically important for proper mitochondrial distribution in neurons. Moreover, stimulating of mitochondrial movement with Miro1 (small Rho GTPase responsible for attaching mitochondria to cytoskeletal tracks) was able to reverse loss in mitochondrial movement in neuronal model of AD and PD [24]. Could it be possible to design NPs penetrating only through cytoplasmic membrane and specifically interacting with the proteins involved in mitochondrial movement? The first step, the integration of NPs into neuronal cytoplasm is already quite well developed and described in several cellular as well as experimental models. The more challenging part represents the progress regarded to study the interaction of NPs with proteins involved in mitochondrial dynamics and facing the cytosol, namely the Mfn1/2 executing mitochondrial fusion; Drp1, Fis1, Mff1 facilitating mitochondrial fission and the proteins involved in the mitochondrial movement directly: Miro1/2, Tau and microtubules. Recent studies demonstrate, that these proteins, so far mostly the Tau protein can be effectively targeted by NP: cationic phosphorus-containing dendrimers of generation 3 and generation 4 (CPDG3, CPDG4) are able to affect Tau aggregation process in a neuro-2a cell line and have a beneficial effect by reducing cell toxicity [126]. In an *in vitro* study, Dadras *et al.*, 2013 [127] predicted three possible binding sites on Tau protein and one possible binding site on tubulin dimer for magnetite (Fe_3O_4) nanoparticles. Magnetite also caused here abnormal morphology of PC12 cells. Authors have also analysed the PC12 cells viability for several magnetite NPs concentrations showing clear dose-dependent manner. If the low concentrations of NPs were used (up to 10 µg/ml), the toxic effect was minimal, opening a plausible therapeutic availability to positively modulate Tau interaction with microtubules and thus mitochondrial movement. In other recent paper [128], TiO_2 nanoparticle-modified membranes were able to selectively attract the phosphorylated form of Tau, but not the non-phosphorylated one, from the solutions passing through the membrane. The TiO_2 selectivity of these NPs for phosphorylated forms of Tau clearly indicates their potential as a tool for targeting the cytoskeletal structures and thus positively modulating their structure and function. The other approach, we aim to highlight in this review, is to deliver the functional proteins into neurons using NPs. This idea is based on the work of Hasadsri and co-workers in 2009 [129], who used the polybutylcyanoacrylate nanoparticles for delivery of intact, functional proteins into neurons and neuronal cell lines. Here, delivery of the small GTPase RhoG has induced neurite outgrowth and differentiation in

PC12 cells. Regarding to Aspenström *et al.*, 2004 [130], RhoG is located in mitochondria and belongs to the same family of Rho GTPases as Miro1/2, although Miro subfamily are bigger than the classical small GTPases, and they contain additional domain structures including the GTP-binding domains. Thus we suppose, with presented nanoparticle delivery system it would be possible to transport the Miro1/2 protein into neurons with beneficial effects on neurite outgrowth based on improvement in mitochondrial movement in neurons.

FUTURE DIRECTIONS, PERSPECTIVES, LIMITATIONS

Targeted drug delivery by the NPs to effectively treat neurodegenerative diseases is a promising branch of pharmacological research. NPs have been shown to penetrate the BBB much more effectively compared to classical pharmaceuticals, and the specific cellular targeting of some types of NPs has already been demonstrated. However, the majority of research connected with mitochondria and cytoskeleton has been performed on solid state NPs so far, most probably for their wide-spread use in everyday life as well as because they are easy-to-detect in EM biological samples. For example, NPs with Ag ions, generally inducing the oxidative stress and apoptosis, inhibit activity of I and III mitochondrial respiration complexes in rat brain tissue *in vitro*, reduce oxidative phosphorylation capacity in isolated mitochondria, and induce beta-tubulin and actin loss in neurons. Al NPs induce rearrangement of F-actin and general cytoskeleton rearrangement in endothelial cells. TiO_2 NPs have an inhibitory effect on tubulin polymerization in microtubules of the sheep brain. Recently, the interaction between TiO_2 and Tau protein has been observed. The latter two point out the strong impact of NPs on the mitochondrial dynamics, supported by the recent observation of Chin-Chan and colleagues [131], where TiO_2 NPs reduced the expression of Parkin (E3 ubiquitin ligase), a protein involved in mitochondria quality control and associated with PD. However, a similar approach with biocompatible and biodegradable NPs involving the polymeric NPs (PLGA, PEG, polybutylcyanoacrylate), liposomes, and dendrimers is largely absent at the moment. Although these NP delivery systems were already approved by Food and Drug Administration in some limited cases, their potential benefit to risk ratio needs to be deeply evaluated at the level of isolated mitochondria, neurons, as well as experimental models (Fig. **2**). The benefit *vs.* risk appears to be of critical importance especially in case of effective treatment of AD and PD, in which the long-term exposition with low doses of NPs will occur. Unfortunately, the chronic prolonged exposures, as they may occur in humans *in vivo*, are extremely difficult to reproduce in experimental systems. Animal models are required in order to assess the role of NPs in the development and progression of neurodegenerative diseases, and the long-term treatment and/or long-term effects after a single injection, both functional and morphological, have to be analysed at this state.

More specifically, we would like to draw the attention to the design of the future experiments, mainly the beneficial effects of NPs to mitochondria for Alzheimer's and Parkinson's disease treatment, as this area of research is currently not sufficiently evidenced (Fig. **2**). For the scientific community, we also emphasize a novel idea to treat mitochondria without targeting their matrix, thus, to develop NPs specifically interacting with cytoskeletal structures or fusion-fissio--movement proteins to control and improve mitochondrial functioning, motility and distribution. Moreover, targeting mitochondria by NPs from cytoplasm can be tailored as 'multitargeting' in the future, as NPs could target mitochondria surface proteins (Mfn, Miro) and cytoplasmic proteins (Drp, Tau, tubulin) at the same time, when specifically designed.

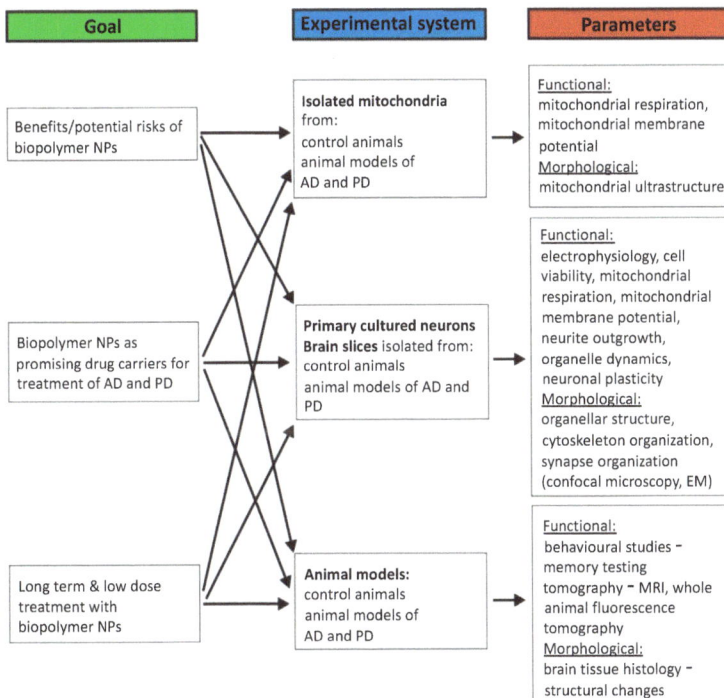

Fig. (2). Summary of perspective experiments regarding the potential use of biopolymer NPs targeting mitochondria for treatment of AD and PD. Three main goals are presented, in all of them the NPs should be tested on the level of isolated mitochondria, cellular and experimental model. More detailed description in the text.

In general, despite the growing interest in the field of nanoscience, caution needs to be employed when interpreting the effects of the nanoparticle delivery systems. An in-depth investigation and clear evidence of the long-term biocompatibility of these nano-carriers are needed, together with fully standardized protocols for the preparation of the NPs, so that the results become more reproducible and comparable among different research groups, and with the extensive *in vitro* and *in vivo* experiments to understand the intracellular trafficking and mechanisms of actions responsible for the therapeutic properties of NPs in AD and PD.

CONCLUSIONS

This chapter aims to bridge the available information for the current and future therapy of Alzheimer's and Parkinson's diseases, the most common neurodegenerative diseases, with nanoparticles targeting mitochondria and mitochondrial dynamics. So far, robust evidence exists regarding both the involvement of mitochondria in AD and PD and the impact of nanoparticles on mitochondria, namely the toxicity of solid nanoparticles. However, less information is available for the treatment of AD and PD with nanoparticles; and especially their inter-connection with mitochondria is still very limited. In this review, we tried to draw reader's attention/ presented the information ranging from the toxicity of nanoparticles to nanoparticles as therapeutic target of AD/PD and (tried) to provide a brainstorming for the future experiments designed to target mitochondria with nanoparticles. We also pointed out the gaps in the current knowledge and summarized the experiments essential in the nearest future: 1. to evaluate the impact of biodegradable nanoparticles on the cells and animals; 2. to (pay attention to) analyze the long-term treatment with low doses of nanoparticles; 3. to test the nanoparticles on primary cultured neurons, brain slices, and the animal models. Finally, a special attention should be taken to mitochondrial dynamics, distribution, morphology, and ultrastructure when considering nanoparticles as a potential therapeutic target in the future. For the first time, we also propose a novel idea to target mitochondria with nanoparticles without their internalization to mitochondrial matrix. By documenting the interaction of nanoparticles with the cytoskeleton, we emphasize to develop nanoparticles positively modulating the cytoplasmic proteins involved in mitochondrial movement and fusion-fission.

CONFLICT OF INTEREST

The authors confirm that they have no conflict of interest to declare for this publication.

ACKNOWLEDGEMENTS

This review has originated with the author's support from the SASPRO grant No. 0063/01/02 financed by the Slovak Academy of Sciences and by a "Co-financing of regional, national and international programs (COFUND)", which is part of the Marie Curie Action of the EU 7th Framework Programme, under Grant Agreement No. 609427; grant 2/0169/16 from the Scientific Grant Agency

VEGA; bi-lateral Czech-Estonian collaborative grant ETA 15-04-4 and Estonian Research Council personal grant PUT771.

ABBREVIATIONS

Aβ	= β-amyloid		**MPTP 1**	= methyl-4-phenyl-1,2,3,6-tetrahydro-pyridine
AD	= Alzheimer's disease		**mtDNA**	= mitochondrial DNA
APP	= amyloid precursor protein		**NAP**	= octapeptide NAPVSIPQ
BBB	= blood-brain barrier		**NDD**	= neurodegenerative disease
CA	= cornu ammonis		**NP**	= nanoparticle
CNS	= central nervous system		**OPA1**	= optic atrophy protein 1
DA	= dopamine		**PD**	= Parkinson's disease
Drp1	= Dynamin-related protein 1		**PEG**	= poly(ethylene glycol)
EM	= electron microscopy		**PINK1**	= PTEN induced putative kinase 1
ETC	= electron transport chain		**PLA**	= poly(lactic acid)
L-DOPA	= levodopa		**PLGA**	= poly(d,l-lactide-co-glycolide)
LRRK2	= leucine-rich repeat kinase 2		**ROS**	= reactive oxygen species
Mff	= mitochondrial fission factor		**SOD**	= superoxide dismutase
Mfn	= Mitofusin		**SPION**	= superparamagnetic iron oxide nanoparticle
MIM	= mitochondrial inner membrane		**TH**	= tyrosine hydroxylase
MOM	= mitochondrial outer membrane		**TMC**	= trimethylated chitosan
mPTP	= mitochondrial permeability transition pore		**6-OHDA**	= 6-hydroxydopamine

REFERENCES

[1]　Breuer ME, Koopman WJ, Koene S, *et al.* The role of mitochondrial OXPHOS dysfunction in the development of neurologic diseases. Neurobiol Dis 2013; 51: 27-34.
[http://dx.doi.org/10.1016/j.nbd.2012.03.007] [PMID: 22426394]

[2]　Gruber J, Fong S, Chen CB, *et al.* Mitochondria-targeted antioxidants and metabolic modulators as pharmacological interventions to slow ageing. Biotechnol Adv 2013; 31(5): 563-92.
[http://dx.doi.org/10.1016/j.biotechadv.2012.09.005] [PMID: 23022622]

[3]　Murray LM, Boyd S. Protecting personhood and achieving quality of life for older adults with dementia in the U.S. health care system. J Aging Health 2009; 21(2): 350-73.
[http://dx.doi.org/10.1177/0898264308329017] [PMID: 19114609]

[4]　Agüero-Torres H, Fratiglioni L, Guo Z, Viitanen M, von Strauss E, Winblad B. Dementia is the major cause of functional dependence in the elderly: 3-year follow-up data from a population-based study. Am J Public Health 1998; 88(10): 1452-6.
[http://dx.doi.org/10.2105/AJPH.88.10.1452] [PMID: 9772843]

[5]　Jalbert JJ, Daiello LA, Lapane KL. Dementia of the Alzheimer type. Epidemiol Rev 2008; 30: 15-34.
[http://dx.doi.org/10.1093/epirev/mxn008] [PMID: 18635578]

[6]   Ferri CP, Prince M, Brayne C, *et al.* Global prevalence of dementia: a Delphi consensus study. Lancet 2005; 366(9503): 2112-7.
[http://dx.doi.org/10.1016/S0140-6736(05)67889-0] [PMID: 16360788]

[7]   Dewey ME, Saz P. Dementia, cognitive impairment and mortality in persons aged 65 and over living in the community: a systematic review of the literature. Int J Geriatr Psychiatry 2001; 16(8): 751-61.
[http://dx.doi.org/10.1002/gps.397] [PMID: 11536341]

[8]   Helmer C, Joly P, Letenneur L, Commenges D, Dartigues JF. Mortality with dementia: results from a French prospective community-based cohort. Am J Epidemiol 2001; 154(7): 642-8.
[http://dx.doi.org/10.1093/aje/154.7.642] [PMID: 11581098]

[9]   Désesquelles A, Demuru E, Salvatore MA, *et al.* Mortality from Alzheimers disease, Parkinsons disease, and dementias in France and Italy: a comparison using the multiple cause-of-death approach. J Aging Health 2014; 26(2): 283-315.
[http://dx.doi.org/10.1177/0898264313514443] [PMID: 24667337]

[10]   2011 Alzheimers disease facts and figures. Alzheimers Dement 2011; 7(2): 208-44.
[http://dx.doi.org/10.1016/j.jalz.2011.02.004] [PMID: 21414557]

[11]   Baloyannis SJ. Mitochondria are related to synaptic pathology in Alzheimer's disease Int J Alzheimers Dis 2011;2011:305395.
[http://dx.doi.org/10.4061/2011/305395]

[12]   Goetz CG. The history of Parkinsons disease: early clinical descriptions and neurological therapies. Cold Spring Harb Perspect Med 2011; 1(1): a008862.
[http://dx.doi.org/10.1101/cshperspect.a008862] [PMID: 22229124]

[13]   Fitzgerald P, Mandel A, Bolton AE, Sullivan AM, Nolan Y. Treatment with phosphotidylglycerol-based nanoparticles prevents motor deficits induced by proteasome inhibition: implications for Parkinsons disease. Behav Brain Res 2008; 195(2): 271-4.
[http://dx.doi.org/10.1016/j.bbr.2008.08.041] [PMID: 18817814]

[14]   Lewis SJ, Barker RA. Understanding the dopaminergic deficits in Parkinsons disease: insights into disease heterogeneity. J Clin Neurosci 2009; 16(5): 620-5.
[http://dx.doi.org/10.1016/j.jocn.2008.08.020] [PMID: 19285870]

[15]   Iqbal A, Ahmad I, Khalid MH, Nawaz MS, Gan SH, Kamal MA. Nanoneurotoxicity to nanoneuroprotection using biological and computational approaches. J Environ Sci Health C Environ Carcinog Ecotoxicol Rev 2013; 31(3): 256-84.
[http://dx.doi.org/10.1080/10590501.2013.829706] [PMID: 24024521]

[16]   Gandhi S, Wood NW. Genome-wide association studies: the key to unlocking neurodegeneration? Nat Neurosci 2010; 13(7): 789-94.
[http://dx.doi.org/10.1038/nn.2584] [PMID: 20581814]

[17]   Björkblom B, Adilbayeva A, Maple-Grødem J, *et al.* Parkinson disease protein DJ-1 binds metals and protects against metal-induced cytotoxicity. J Biol Chem 2013; 288(31): 22809-20.
[http://dx.doi.org/10.1074/jbc.M113.482091] [PMID: 23792957]

[18]   Morais VA, Verstreken P, Roethig A, *et al.* Parkinsons disease mutations in PINK1 result in decreased Complex I activity and deficient synaptic function. EMBO Mol Med 2009; 1(2): 99-111.
[http://dx.doi.org/10.1002/emmm.200900006] [PMID: 20049710]

[19]   Veeriah S, Taylor BS, Meng S, *et al.* Somatic mutations of the Parkinsons disease-associated gene PARK2 in glioblastoma and other human malignancies. Nat Genet 2010; 42(1): 77-82.
[http://dx.doi.org/10.1038/ng.491] [PMID: 19946270]

[20]   Fernandez-Moriano C, Gonzalez-Burgos E, Gomez-Serranillos MP. Mitochondria-Targeted Protective Compounds in Parkinson's and Alzheimer's Diseases Oxid med cell longev 2015;2015:408927.
[http://dx.doi.org/10.1155/2015/408927]

[21]   Scott DA, Tabarean I, Tang Y, Cartier A, Masliah E, Roy S. A pathologic cascade leading to synaptic dysfunction in alpha-synuclein-induced neurodegeneration. J Neurosci 2010; 30(24): 8083-95.
[http://dx.doi.org/10.1523/JNEUROSCI.1091-10.2010] [PMID: 20554859]

[22]   Milane L, Trivedi M, Singh A, Talekar M, Amiji M. Mitochondrial biology, targets, and drug delivery. J Control Release 2015; 207: 40-58.
[http://dx.doi.org/10.1016/j.jconrel.2015.03.036] [PMID: 25841699]

[23]   Apostolova N, Victor VM. Molecular strategies for targeting antioxidants to mitochondria: therapeutic implications. Antioxid Redox Signal 2015; 22(8): 686-729.
[http://dx.doi.org/10.1089/ars.2014.5952] [PMID: 25546574]

[24]   Cagalinec M, Safiulina D, Liiv M, *et al.* Principles of the mitochondrial fusion and fission cycle in neurons. J Cell Sci 2013; 126(Pt 10): 2187-97.
[http://dx.doi.org/10.1242/jcs.118844] [PMID: 23525002]

[25]   Braschi E, McBride HM. Mitochondria and the culture of the Borg: understanding the integration of mitochondrial function within the reticulum, the cell, and the organism. BioEssays 2010; 32(11): 958-66.
[http://dx.doi.org/10.1002/bies.201000073] [PMID: 20824657]

[26]   Yan MH, Wang X, Zhu X. Mitochondrial defects and oxidative stress in Alzheimer disease and Parkinson disease. Free Radic Biol Med 2013; 62: 90-101.
[http://dx.doi.org/10.1016/j.freeradbiomed.2012.11.014] [PMID: 23200807]

[27]   Dagda RK, Gusdon AM, Pien I, *et al.* Mitochondrially localized PKA reverses mitochondrial pathology and dysfunction in a cellular model of Parkinsons disease. Cell Death Differ 2011; 18(12): 1914-23.
[http://dx.doi.org/10.1038/cdd.2011.74] [PMID: 21637291]

[28]   Twig G, Elorza A, Molina AJ, *et al.* Fission and selective fusion govern mitochondrial segregation and elimination by autophagy. EMBO J 2008; 27(2): 433-46.
[http://dx.doi.org/10.1038/sj.emboj.7601963] [PMID: 18200046]

[29]   Calkins MJ, Manczak M, Mao P, Shirendeb U, Reddy PH. Impaired mitochondrial biogenesis, defective axonal transport of mitochondria, abnormal mitochondrial dynamics and synaptic degeneration in a mouse model of Alzheimers disease. Hum Mol Genet 2011; 20(23): 4515-29.
[http://dx.doi.org/10.1093/hmg/ddr381] [PMID: 21873260]

[30]   Cupaioli FA, Zucca FA, Boraschi D, Zecca L. Engineered nanoparticles. How brain friendly is this new guest? Prog Neurobiol 2014; 119-120: 20-38.
[http://dx.doi.org/10.1016/j.pneurobio.2014.05.002] [PMID: 24820405]

[31]   Pautler M, Brenner S. Nanomedicine: promises and challenges for the future of public health. Int J Nanomedicine 2010; 5: 803-9.
[PMID: 21042425]

[32]   Bawarski WE, Chidlowsky E, Bharali DJ, Mousa SA. Emerging nanopharmaceuticals. Nanomedicine (Lond) 2008; 4(4): 273-82.
[PMID: 18640076]

[33]   Kreuter J. Drug targeting with nanoparticles. Eur J Drug Metab Pharmacokinet 1994; 19(3): 253-6.
[http://dx.doi.org/10.1007/BF03188928] [PMID: 7867668]

[34]   Andrieux K, Couvreur P. [Nanoparticles for brain delivery of drugs or contrast agents. Application to Alzheimers disease]. Biol Aujourdhui 2012; 206(3): 185-90. [Nanoparticles for brain delivery of drugs or contrast agents. Application to Alzheimer's disease].
[http://dx.doi.org/10.1051/jbio/2012019] [PMID: 23171841]

[35]   Abbott NJ, Rönnbäck L, Hansson E. Astrocyte-endothelial interactions at the blood-brain barrier. Nat Rev Neurosci 2006; 7(1): 41-53.
[http://dx.doi.org/10.1038/nrn1824] [PMID: 16371949]

[36]   Armulik A, Genové G, Mäe M, *et al.* Pericytes regulate the blood-brain barrier. Nature 2010; 468(7323): 557-61.
        [http://dx.doi.org/10.1038/nature09522] [PMID: 20944627]

[37]   Gao H, Pang Z, Jiang X. Targeted delivery of nano-therapeutics for major disorders of the central nervous system. Pharm Res 2013; 30(10): 2485-98.
        [http://dx.doi.org/10.1007/s11095-013-1122-4] [PMID: 23797465]

[38]   Costantino L, Boraschi D. Is there a clinical future for polymeric nanoparticles as brain-targeting drug delivery agents? Drug Discov Today 2012; 17(7-8): 367-78.
        [http://dx.doi.org/10.1016/j.drudis.2011.10.028] [PMID: 22094246]

[39]   Dobson J. Nanoscale biogenic iron oxides and neurodegenerative disease. FEBS Lett 2001; 496(1): 1-5.
        [http://dx.doi.org/10.1016/S0014-5793(01)02386-9] [PMID: 11343696]

[40]   Hautot D, Pankhurst QA, Khan N, Dobson J. Preliminary evaluation of nanoscale biogenic magnetite in Alzheimers disease brain tissue. Proc Biol Sci 2003; 270 (Suppl. 1): S62-4.
        [http://dx.doi.org/10.1098/rsbl.2003.0012] [PMID: 12952638]

[41]   Sharma HS, Hussain S, Schlager J, Ali SF, Sharma A. Influence of nanoparticles on blood-brain barrier permeability and brain edema formation in rats. Acta Neurochir Suppl (Wien) 2010; 106: 359-64.
        [http://dx.doi.org/10.1007/978-3-211-98811-4_65] [PMID: 19812977]

[42]   Yang X, He C, Li J, *et al.* Uptake of silica nanoparticles: neurotoxicity and Alzheimer-like pathology in human SK-N-SH and mouse neuro2a neuroblastoma cells. Toxicol Lett 2014; 229(1): 240-9.
        [http://dx.doi.org/10.1016/j.toxlet.2014.05.009] [PMID: 24831964]

[43]   Liu Q, Shao X, Chen J, *et al. In vivo* toxicity and immunogenicity of wheat germ agglutinin conjugated poly(ethylene glycol)-poly(lactic acid) nanoparticles for intranasal delivery to the brain. Toxicol Appl Pharmacol 2011; 251(1): 79-84.
        [http://dx.doi.org/10.1016/j.taap.2010.12.003] [PMID: 21163285]

[44]   Wang J, Chen C, Liu Y, *et al.* Potential neurological lesion after nasal instillation of TiO(2) nanoparticles in the anatase and rutile crystal phases. Toxicol Lett 2008; 183(1-3): 72-80.
        [http://dx.doi.org/10.1016/j.toxlet.2008.10.001] [PMID: 18992307]

[45]   Chen YS, Hung YC, Lin LW, Liau I, Hong MY, Huang GS. Size-dependent impairment of cognition in mice caused by the injection of gold nanoparticles. Nanotechnology 2010; 21(48): 485102.
        [http://dx.doi.org/10.1088/0957-4484/21/48/485102] [PMID: 21051801]

[46]   Xu F, Piett C, Farkas S, Qazzaz M, Syed NI. Silver nanoparticles (AgNPs) cause degeneration of cytoskeleton and disrupt synaptic machinery of cultured cortical neurons. Mol Brain 2013; 6: 29.
        [http://dx.doi.org/10.1186/1756-6606-6-29] [PMID: 23782671]

[47]   Rahman MF, Wang J, Patterson TA, *et al.* Expression of genes related to oxidative stress in the mouse brain after exposure to silver-25 nanoparticles. Toxicol Lett 2009; 187(1): 15-21.
        [http://dx.doi.org/10.1016/j.toxlet.2009.01.020] [PMID: 19429238]

[48]   Oberdörster G, Oberdörster E, Oberdörster J. Nanotoxicology: an emerging discipline evolving from studies of ultrafine particles. Environ Health Perspect 2005; 113(7): 823-39.
        [http://dx.doi.org/10.1289/ehp.7339] [PMID: 16002369]

[49]   Costa CS, Ronconi JV, Daufenbach JF, *et al. In vitro* effects of silver nanoparticles on the mitochondrial respiratory chain. Mol Cell Biochem 2010; 342(1-2): 51-6.
        [http://dx.doi.org/10.1007/s11010-010-0467-9] [PMID: 20411305]

[50]   Teodoro JS, Simões AM, Duarte FV, *et al.* Assessment of the toxicity of silver nanoparticles *in vitro*: a mitochondrial perspective. Toxicol In Vitro 2011; 25(3): 664-70.
        [http://dx.doi.org/10.1016/j.tiv.2011.01.004] [PMID: 21232593]

[51] AshaRani PV, Low Kah Mun G, Hande MP, Valiyaveettil S. Cytotoxicity and genotoxicity of silver nanoparticles in human cells. ACS Nano 2009; 3(2): 279-90.
[http://dx.doi.org/10.1021/nn800596w] [PMID: 19236062]

[52] Au C, Mutkus L, Dobson A, Riffle J, Lalli J, Aschner M. Effects of nanoparticles on the adhesion and cell viability on astrocytes. Biol Trace Elem Res 2007; 120(1-3): 248-56.
[http://dx.doi.org/10.1007/s12011-007-0067-z] [PMID: 17916977]

[53] Chen L, Yokel RA, Hennig B, Toborek M. Manufactured aluminum oxide nanoparticles decrease expression of tight junction proteins in brain vasculature. J Neuroimmune Pharmacol 2008; 3(4): 286-95.
[http://dx.doi.org/10.1007/s11481-008-9131-5] [PMID: 18830698]

[54] Li N, Sioutas C, Cho A, *et al.* Ultrafine particulate pollutants induce oxidative stress and mitochondrial damage. Environ Health Perspect 2003; 111(4): 455-60.
[http://dx.doi.org/10.1289/ehp.6000] [PMID: 12676598]

[55] Schrand AM, Rahman MF, Hussain SM, Schlager JJ, Smith DA, Syed AF. Metal-based nanoparticles and their toxicity assessment. Wiley Interdiscip Rev Nanomed Nanobiotechnol 2010; 2(5): 544-68.
[http://dx.doi.org/10.1002/wnan.103] [PMID: 20681021]

[56] Locht LJ, Pedersen MO, Markholt S, *et al.* Metallic silver fragments cause massive tissue loss in the mouse brain. Basic Clin Pharmacol Toxicol 2011; 109(1): 1-10.
[http://dx.doi.org/10.1111/j.1742-7843.2010.00668.x] [PMID: 21205224]

[57] Cengelli F, Voinesco F, Juillerat-Jeanneret L. Interaction of cationic ultrasmall superparamagnetic iron oxide nanoparticles with human melanoma cells. Nanomedicine (Lond) 2010; 5(7): 1075-87.
[http://dx.doi.org/10.2217/nnm.10.79] [PMID: 20874022]

[58] Ragnaill MN, Brown M, Ye D, *et al.* Internal benchmarking of a human blood-brain barrier cell model for screening of nanoparticle uptake and transcytosis. Eur J Pharm Biopharm 2011; 77(3): 360-7.
[http://dx.doi.org/10.1016/j.ejpb.2010.12.024] [PMID: 21236340]

[59] Raghnaill MN, Bramini M, Ye D, *et al.* Paracrine signalling of inflammatory cytokines from an *in vitro* blood brain barrier model upon exposure to polymeric nanoparticles. Analyst (Lond) 2014; 139(5): 923-30.
[http://dx.doi.org/10.1039/C3AN01621H] [PMID: 24195103]

[60] Wu HY, Chung MC, Wang CC, Huang CH, Liang HJ, Jan TR. Iron oxide nanoparticles suppress the production of IL-1beta *via* the secretory lysosomal pathway in murine microglial cells. Part Fibre Toxicol 2013; 10: 46.
[http://dx.doi.org/10.1186/1743-8977-10-46] [PMID: 24047432]

[61] Halamoda Kenzaoui B, Chapuis Bernasconi C, Guney-Ayra S, Juillerat-Jeanneret L. Induction of oxidative stress, lysosome activation and autophagy by nanoparticles in human brain-derived endothelial cells. Biochem J 2012; 441(3): 813-21.
[http://dx.doi.org/10.1042/BJ20111252] [PMID: 22026563]

[62] Zucca FA, Basso E, Cupaioli FA, *et al.* Neuromelanin of the human substantia nigra: an update. Neurotox Res 2014; 25(1): 13-23.
[http://dx.doi.org/10.1007/s12640-013-9435-y] [PMID: 24155156]

[63] Sulzer D, Mosharov E, Talloczy Z, Zucca FA, Simon JD, Zecca L. Neuronal pigmented autophagic vacuoles: lipofuscin, neuromelanin, and ceroid as macroautophagic responses during aging and disease. J Neurochem 2008; 106(1): 24-36.
[http://dx.doi.org/10.1111/j.1471-4159.2008.05385.x] [PMID: 18384642]

[64] Wischik CM, Harrington CR, Storey JM. Tau-aggregation inhibitor therapy for Alzheimers disease. Biochem Pharmacol 2014; 88(4): 529-39.
[http://dx.doi.org/10.1016/j.bcp.2013.12.008] [PMID: 24361915]

[65]   Gheshlaghi ZN, Riazi GH, Ahmadian S, Ghafari M, Mahinpour R. Toxicity and interaction of titanium dioxide nanoparticles with microtubule protein. Acta Biochim Biophys Sin (Shanghai) 2008; 40(9): 777-82.
        [http://dx.doi.org/10.1093/abbs/40.9.777] [PMID: 18776989]

[66]   Mao Z, Xu B, Ji X, *et al.* Titanium dioxide nanoparticles alter cellular morphology *via* disturbing the microtubule dynamics. Nanoscale 2015; 7(18): 8466-75.
        [http://dx.doi.org/10.1039/C5NR01448D] [PMID: 25891938]

[67]   Moghimi SM, Peer D, Langer R. Reshaping the future of nanopharmaceuticals: ad iudicium. ACS Nano 2011; 5(11): 8454-8.
        [http://dx.doi.org/10.1021/nn2038252] [PMID: 21992178]

[68]   Andrieux K, Couvreur P. Polyalkylcyanoacrylate nanoparticles for delivery of drugs across the blood-brain barrier. Wiley Interdiscip Rev Nanomed Nanobiotechnol 2009; 1(5): 463-74.
        [http://dx.doi.org/10.1002/wnan.5] [PMID: 20049811]

[69]   Torchilin VP. Micellar nanocarriers: pharmaceutical perspectives. Pharm Res 2007; 24(1): 1-16.
        [http://dx.doi.org/10.1007/s11095-006-9132-0] [PMID: 17109211]

[70]   Adams ML, Lavasanifar A, Kwon GS. Amphiphilic block copolymers for drug delivery. J Pharm Sci 2003; 92(7): 1343-55.
        [http://dx.doi.org/10.1002/jps.10397] [PMID: 12820139]

[71]   Hutter E, Boridy S, Labrecque S, *et al.* Microglial response to gold nanoparticles. ACS Nano 2010; 4(5): 2595-606.
        [http://dx.doi.org/10.1021/nn901869f] [PMID: 20329742]

[72]   Sofi F, Macchi C, Abbate R, Gensini GF, Casini A. Effectiveness of the Mediterranean diet: can it help delay or prevent Alzheimers disease? J Alzheimers Dis 2010; 20(3): 795-801.
        [PMID: 20182044]

[73]   Tabrez S, Priyadarshini M, Urooj M, *et al.* Cancer chemoprevention by polyphenols and their potential application as nanomedicine. J Environ Sci Health C Environ Carcinog Ecotoxicol Rev 2013; 31(1): 67-98.
        [http://dx.doi.org/10.1080/10590501.2013.763577] [PMID: 23534395]

[74]   Beard JL, Connor JR, Jones BC. Iron in the brain. Nutr Rev 1993; 51(6): 157-70.
        [http://dx.doi.org/10.1111/j.1753-4887.1993.tb03096.x] [PMID: 8371846]

[75]   Cui Z, Lockman PR, Atwood CS, *et al.* Novel D-penicillamine carrying nanoparticles for metal chelation therapy in Alzheimers and other CNS diseases. Eur J Pharm Biopharm 2005; 59(2): 263-72.
        [http://dx.doi.org/10.1016/j.ejpb.2004.07.009] [PMID: 15661498]

[76]   Liu G, Men P, Perry G, Smith MA. Chapter 5 - Development of iron chelator-nanoparticle conjugates as potential therapeutic agents for Alzheimer disease. Prog Brain Res 2009; 180: 97-108.
        [http://dx.doi.org/10.1016/S0079-6123(08)80005-2] [PMID: 20302830]

[77]   Wong HL, Wu XY, Bendayan R. Nanotechnological advances for the delivery of CNS therapeutics. Adv Drug Deliv Rev 2012; 64(7): 686-700.
        [http://dx.doi.org/10.1016/j.addr.2011.10.007] [PMID: 22100125]

[78]   Lin MT, Beal MF. Alzheimers APP mangles mitochondria. Nat Med 2006; 12(11): 1241-3.
        [http://dx.doi.org/10.1038/nm1106-1241] [PMID: 17088888]

[79]   Kamat PK, Kalani A, Kyles P, Tyagi SC, Tyagi N. Autophagy of mitochondria: a promising therapeutic target for neurodegenerative disease. Cell Biochem Biophys 2014; 70(2): 707-19.
        [http://dx.doi.org/10.1007/s12013-014-0006-5] [PMID: 24807843]

[80]   Navarro-Yepes J, Zavala-Flores L, Anandhan A, *et al.* Antioxidant gene therapy against neuronal cell death. Pharmacol Ther 2014; 142(2): 206-30.
        [http://dx.doi.org/10.1016/j.pharmthera.2013.12.007] [PMID: 24333264]

[81]   Leuner K, Schütt T, Kurz C, *et al.* Mitochondrion-derived reactive oxygen species lead to enhanced amyloid beta formation. Antioxid Redox Signal 2012; 16(12): 1421-33.
[http://dx.doi.org/10.1089/ars.2011.4173] [PMID: 22229260]

[82]   Yao J, Du H, Yan S, *et al.* Inhibition of amyloid-beta (Abeta) peptide-binding alcohol dehydrogenase-Abeta interaction reduces Abeta accumulation and improves mitochondrial function in a mouse model of Alzheimers disease. J Neurosci 2011; 31(6): 2313-20.
[http://dx.doi.org/10.1523/JNEUROSCI.4717-10.2011] [PMID: 21307267]

[83]   Götz J. Tau and transgenic animal models. Brain Res Brain Res Rev 2001; 35(3): 266-86.
[http://dx.doi.org/10.1016/S0165-0173(01)00055-8] [PMID: 11423157]

[84]   Masserini M. Nanoparticles for brain drug delivery ISRN Biochem 2013;2013:238428.
[http://dx.doi.org/10.1155/2013/238428]

[85]   Hroudova J, Singh N, Fisar Z. Mitochondrial dysfunctions in neurodegenerative diseases: relevance to Alzheimer's disease Biomed Res Int 2014;2014:175062.
[http://dx.doi.org/10.1155/2014/175062]

[86]   Su B, Wang X, Zheng L, Perry G, Smith MA, Zhu X. Abnormal mitochondrial dynamics and neurodegenerative diseases. Biochim Biophys Acta 2010; 1802(1): 135-42.
[http://dx.doi.org/10.1016/j.bbadis.2009.09.013] [PMID: 19799998]

[87]   Wang X, Su B, Lee HG, *et al.* Impaired balance of mitochondrial fission and fusion in Alzheimers disease. J Neurosci 2009; 29(28): 9090-103.
[http://dx.doi.org/10.1523/JNEUROSCI.1357-09.2009] [PMID: 19605646]

[88]   Rui Y, Tiwari P, Xie Z, Zheng JQ. Acute impairment of mitochondrial trafficking by beta-amyloid peptides in hippocampal neurons. J Neurosci 2006; 26(41): 10480-7.
[http://dx.doi.org/10.1523/JNEUROSCI.3231-06.2006] [PMID: 17035532]

[89]   Barsoum MJ, Yuan H, Gerencser AA, *et al.* Nitric oxide-induced mitochondrial fission is regulated by dynamin-related GTPases in neurons. EMBO J 2006; 25(16): 3900-11.
[http://dx.doi.org/10.1038/sj.emboj.7601253] [PMID: 16874299]

[90]   Cho DH, Nakamura T, Fang J, *et al.* S-nitrosylation of Drp1 mediates beta-amyloid-related mitochondrial fission and neuronal injury. Science 2009; 324(5923): 102-5.
[http://dx.doi.org/10.1126/science.1171091] [PMID: 19342591]

[91]   Wang X, Su B, Siedlak SL, *et al.* Amyloid-beta overproduction causes abnormal mitochondrial dynamics *via* differential modulation of mitochondrial fission/fusion proteins. Proc Natl Acad Sci USA 2008; 105(49): 19318-23.
[http://dx.doi.org/10.1073/pnas.0804871105] [PMID: 19050078]

[92]   Wang X, Su B, Liu W, *et al.* DLP1-dependent mitochondrial fragmentation mediates 1-methyl-4-phenylpyridinium toxicity in neurons: implications for Parkinsons disease. Aging Cell 2011; 10(5): 807-23.
[http://dx.doi.org/10.1111/j.1474-9726.2011.00721.x] [PMID: 21615675]

[93]   Kamp F, Exner N, Lutz AK, *et al.* Inhibition of mitochondrial fusion by α-synuclein is rescued by PINK1, Parkin and DJ-1. EMBO J 2010; 29(20): 3571-89.
[http://dx.doi.org/10.1038/emboj.2010.223] [PMID: 20842103]

[94]   Xie W, Chung KK. Alpha-synuclein impairs normal dynamics of mitochondria in cell and animal models of Parkinsons disease. J Neurochem 2012; 122(2): 404-14.
[http://dx.doi.org/10.1111/j.1471-4159.2012.07769.x] [PMID: 22537068]

[95]   Yang Y, Ouyang Y, Yang L, *et al.* Pink1 regulates mitochondrial dynamics through interaction with the fission/fusion machinery. Proc Natl Acad Sci USA 2008; 105(19): 7070-5.
[http://dx.doi.org/10.1073/pnas.0711845105] [PMID: 18443288]

[96]   Yu W, Sun Y, Guo S, Lu B. The PINK1/Parkin pathway regulates mitochondrial dynamics and function in mammalian hippocampal and dopaminergic neurons. Hum Mol Genet 2011; 20(16): 3227-40.
[http://dx.doi.org/10.1093/hmg/ddr235] [PMID: 21613270]

[97]   Choubey V, Cagalinec M, Liiv J, *et al.* BECN1 is involved in the initiation of mitophagy: it facilitates PARK2 translocation to mitochondria. Autophagy 2014; 10(6): 1105-19.
[http://dx.doi.org/10.4161/auto.28615] [PMID: 24879156]

[98]   Exner N, Treske B, Paquet D, *et al.* Loss-of-function of human PINK1 results in mitochondrial pathology and can be rescued by parkin. J Neurosci 2007; 27(45): 12413-8.
[http://dx.doi.org/10.1523/JNEUROSCI.0719-07.2007] [PMID: 17989306]

[99]   Lutz AK, Exner N, Fett ME, *et al.* Loss of parkin or PINK1 function increases Drp1-dependent mitochondrial fragmentation. J Biol Chem 2009; 284(34): 22938-51.
[http://dx.doi.org/10.1074/jbc.M109.035774] [PMID: 19546216]

[100]   Sandebring A, Thomas KJ, Beilina A, *et al.* Mitochondrial alterations in PINK1 deficient cells are influenced by calcineurin-dependent dephosphorylation of dynamin-related protein 1. PLoS One 2009; 4(5): e5701.
[http://dx.doi.org/10.1371/journal.pone.0005701] [PMID: 19492085]

[101]   Wang X, Yan MH, Fujioka H, *et al.* LRRK2 regulates mitochondrial dynamics and function through direct interaction with DLP1. Hum Mol Genet 2012; 21(9): 1931-44.
[http://dx.doi.org/10.1093/hmg/dds003] [PMID: 22228096]

[102]   Busquets MA, Sabaté R, Estelrich J. Potential applications of magnetic particles to detect and treat Alzheimers disease. Nanoscale Res Lett 2014; 9(1): 538.
[http://dx.doi.org/10.1186/1556-276X-9-538] [PMID: 25288921]

[103]   Mahmoudi M, Quinlan-Pluck F, Monopoli MP, *et al.* Influence of the physiochemical properties of superparamagnetic iron oxide nanoparticles on amyloid β protein fibrillation in solution. ACS Chem Neurosci 2013; 4(3): 475-85.
[http://dx.doi.org/10.1021/cn300196n] [PMID: 23509983]

[104]   Spuch C, Saida O, Navarro C. Advances in the treatment of neurodegenerative disorders employing nanoparticles. Recent Pat Drug Deliv Formul 2012; 6(1): 2-18.
[http://dx.doi.org/10.2174/187221112799219125] [PMID: 22272933]

[105]   Cheng KK, Yeung CF, Ho SW, Chow SF, Chow AH, Baum L. Highly stabilized curcumin nanoparticles tested in an *in vitro* blood-brain barrier model and in Alzheimers disease Tg2576 mice. AAPS J 2013; 15(2): 324-36.
[http://dx.doi.org/10.1208/s12248-012-9444-4] [PMID: 23229335]

[106]   Dhawan S, Kapil R, Singh B. Formulation development and systematic optimization of solid lipid nanoparticles of quercetin for improved brain delivery. J Pharm Pharmacol 2011; 63(3): 342-51.
[http://dx.doi.org/10.1111/j.2042-7158.2010.01225.x] [PMID: 21749381]

[107]   Wang ZH, Wang ZY, Sun CS, Wang CY, Jiang TY, Wang SL. Trimethylated chitosan-conjugated PLGA nanoparticles for the delivery of drugs to the brain. Biomaterials 2010; 31(5): 908-15.
[http://dx.doi.org/10.1016/j.biomaterials.2009.09.104] [PMID: 19853292]

[108]   Chonpathompikunlert P, Yoshitomi T, Vong LB, Imaizumi N, Ozaki Y, Nagasaki Y. Recovery of Cognitive Dysfunction *via* Orally Administered Redox-Polymer Nanotherapeutics in SAMP8 Mice. PLoS One 2015; 10(5): e0126013.
[http://dx.doi.org/10.1371/journal.pone.0126013] [PMID: 25955022]

[109]   Yu Y, Pang Z, Lu W, Yin Q, Gao H, Jiang X. Self-assembled polymersomes conjugated with lactoferrin as novel drug carrier for brain delivery. Pharm Res 2012; 29(1): 83-96.
[http://dx.doi.org/10.1007/s11095-011-0513-7] [PMID: 21979908]

[110] Liu Z, Gao X, Kang T, *et al.* B6 peptide-modified PEG-PLA nanoparticles for enhanced brain delivery of neuroprotective peptide. Bioconjug Chem 2013; 24(6): 997-1007.
[http://dx.doi.org/10.1021/bc400055h] [PMID: 23718945]

[111] Garbayo E, Estella-Hermoso de Mendoza A, Blanco-Prieto MJ. Diagnostic and therapeutic uses of nanomaterials in the brain. Curr Med Chem 2014; 21(36): 4100-31.
[http://dx.doi.org/10.2174/0929867321666140815124246] [PMID: 25139519]

[112] Huang R, Ke W, Liu Y, Jiang C, Pei Y. The use of lactoferrin as a ligand for targeting the polyamidoamine-based gene delivery system to the brain. Biomaterials 2008; 29(2): 238-46.
[http://dx.doi.org/10.1016/j.biomaterials.2007.09.024] [PMID: 17935779]

[113] Huang R, Han L, Li J, *et al.* Neuroprotection in a 6-hydroxydopamine-lesioned Parkinson model using lactoferrin-modified nanoparticles. J Gene Med 2009; 11(9): 754-63.
[http://dx.doi.org/10.1002/jgm.1361] [PMID: 19554623]

[114] Hu K, Shi Y, Jiang W, Han J, Huang S, Jiang X. Lactoferrin conjugated PEG-PLGA nanoparticles for brain delivery: preparation, characterization and efficacy in Parkinsons disease. Int J Pharm 2011; 415(1-2): 273-83.
[http://dx.doi.org/10.1016/j.ijpharm.2011.05.062] [PMID: 21651967]

[115] Joshi N, Basak S, Kundu S, De G, Mukhopadhyay A, Chattopadhyay K. Attenuation of the early events of α-synuclein aggregation: a fluorescence correlation spectroscopy and laser scanning microscopy study in the presence of surface-coated Fe3O4 nanoparticles. Langmuir 2015; 31(4): 1469-78.
[http://dx.doi.org/10.1021/la503749e] [PMID: 25561279]

[116] Milowska K, Grochowina J, Katir N, *et al.* Viologen-Phosphorus Dendrimers Inhibit α-Synuclein Fibrillation. Mol Pharm 2013; 10(3): 1131-7.
[http://dx.doi.org/10.1021/mp300636h] [PMID: 23379345]

[117] Weissig V, DSouza GG, Torchilin VP. DQAsome/DNA complexes release DNA upon contact with isolated mouse liver mitochondria. J Control Release 2001; 75(3): 401-8.
[http://dx.doi.org/10.1016/S0168-3659(01)00392-3] [PMID: 11489326]

[118] DSouza GG, Cheng SM, Boddapati SV, Horobin RW, Weissig V. Nanocarrier-assisted sub-cellular targeting to the site of mitochondria improves the pro-apoptotic activity of paclitaxel. J Drug Target 2008; 16(7): 578-85.
[http://dx.doi.org/10.1080/10611860802228855] [PMID: 18686127]

[119] Yamada Y, Harashima H. Mitochondrial drug delivery systems for macromolecule and their therapeutic application to mitochondrial diseases. Adv Drug Deliv Rev 2008; 60(13-14): 1439-62.
[http://dx.doi.org/10.1016/j.addr.2008.04.016] [PMID: 18655816]

[120] Suzuki R, Yamada Y, Harashima H. Efficient cytoplasmic protein delivery by means of a multifunctional envelope-type nano device. Biol Pharm Bull 2007; 30(4): 758-62.
[http://dx.doi.org/10.1248/bpb.30.758] [PMID: 17409516]

[121] Furukawa R, Yamada Y, Takenaga M, Igarashi R, Harashima H. Octaarginine-modified liposomes enhance the anti-oxidant effect of Lecithinized superoxide dismutase by increasing its cellular uptake. Biochem Biophys Res Commun 2011; 404(3): 796-801.
[http://dx.doi.org/10.1016/j.bbrc.2010.12.062] [PMID: 21168389]

[122] Paunesku T, Rajh T, Wiederrecht G, *et al.* Biology of TiO2-oligonucleotide nanocomposites. Nat Mater 2003; 2(5): 343-6.
[http://dx.doi.org/10.1038/nmat875] [PMID: 12692534]

[123] Salnikov V, Lukyánenko YO, Frederick CA, Lederer WJ, Lukyánenko V. Probing the outer mitochondrial membrane in cardiac mitochondria with nanoparticles. Biophys J 2007; 92(3): 1058-71.
[http://dx.doi.org/10.1529/biophysj.106.094318] [PMID: 17098804]

[124] Yoong SL, Wong BS, Zhou QL, *et al.* Enhanced cytotoxicity to cancer cells by mitochondria-targeting MWCNTs containing platinum(IV) prodrug of cisplatin. Biomaterials 2014; 35(2): 748-59.
[http://dx.doi.org/10.1016/j.biomaterials.2013.09.036] [PMID: 24140044]

[125] Marrache S, Dhar S. Engineering of blended nanoparticle platform for delivery of mitochondria-acting therapeutics. Proc Natl Acad Sci USA 2012; 109(40): 16288-93.
[http://dx.doi.org/10.1073/pnas.1210096109] [PMID: 22991470]

[126] Wasiak T, Ionov M, Nieznanski K, *et al.* Phosphorus dendrimers affect Alzheimers (Aβ128) peptide and MAP-Tau protein aggregation. Mol Pharm 2012; 9(3): 458-69.
[http://dx.doi.org/10.1021/mp2005627] [PMID: 22206488]

[127] Dadras A, Riazi GH, Afrasiabi A, Naghshineh A, Ghalandari B, Mokhtari F. *In vitro* study on the alterations of brain tubulin structure and assembly affected by magnetite nanoparticles. J Biol Inorg Chem 2013; 18(3): 357-69.
[http://dx.doi.org/10.1007/s00775-013-0980-x] [PMID: 23397429]

[128] Tan YJ, Sui D, Wang WH, Kuo MH, Reid GE, Bruening ML. Phosphopeptide enrichment with TiO2-modified membranes and investigation of tau protein phosphorylation. Anal Chem 2013; 85(12): 5699-706.
[http://dx.doi.org/10.1021/ac400198n] [PMID: 23638980]

[129] Hasadsri L, Kreuter J, Hattori H, Iwasaki T, George JM. Functional protein delivery into neurons using polymeric nanoparticles. J Biol Chem 2009; 284(11): 6972-81.
[http://dx.doi.org/10.1074/jbc.M805956200] [PMID: 19129199]

[130] Aspenström P, Fransson A, Saras J. Rho GTPases have diverse effects on the organization of the actin filament system. Biochem J 2004; 377(Pt 2): 327-37.
[http://dx.doi.org/10.1042/bj20031041] [PMID: 14521508]

[131] Chin-Chan M, Navarro-Yepes J, Quintanilla-Vega B. Environmental pollutants as risk factors for neurodegenerative disorders: Alzheimer and Parkinson diseases. Front Cell Neurosci 2015; 9: 124.
[http://dx.doi.org/10.3389/fncel.2015.00124] [PMID: 25914621]

CHAPTER 3

Neuronal Mechanisms for Nanotopography Sensing

Ilaria Tonazzini[1,2] and **Marco Cecchini**[1,*]

[1] *NEST, Istituto Nanoscienze-CNR and Scuola Normale Superiore, Piazza San Silvestro 12, 56127 Pisa, Italy*

[2] *Fondazione Umberto Veronesi, Piazza Velasca 5, 20122 Milano, Italy*

Abstract: Cell contact interaction with extracellular environment cooperates in coordinating several physio-pathological processes *in vivo*, and can be exploited to manipulate cell responses *in vitro*. Thanks to recent developments in micro/nano-engineering techniques, nano/micro-structured surfaces have been introduced capable of controlling neuronal cell adhesion, differentiation, migration, and neurite orientation by interfering with the cell adhesion machinery. In particular, this process is mediated by focal adhesion (FA) establishment and maturation. FAs cross-talk with the actin fibers and act as topographical sensors, by integrating signals from the extracellular environment. Here, we describe the mechanisms of nanotopography sensing in neuronal cells. In particular, experiments addressing the role of FAs, myosinII-dependent cell contractility, and actin dynamics in neuronal contact guidance along directional nanostructured surfaces are reviewed and discussed.

Keywords: Actin contractility, Contact guidance, Cytoskeleton, Focal adhesions, Mechano-transduction, Neuron, Nanostructured substrates, Neurite, Nanograting, Nanogroove.

INTRODUCTION

Nowadays it is accepted that cells respond to the morphology of the extracellular environment at the nanoscale level. The contact interaction of cells with extracellular physical features cooperates in regulating physiological (*e.g.* embryogenesis, cell migration) [1] and pathological processes [2] *in vivo*, and can be exploited to modulate cell responses *in vitro* [3 - 5]. In the central nervous system (CNS), the sensing of the extracellular environment combines with intracellular signaling patterns that is integrated by cells to establish the final neuronal polarity, differentiation, migration, neurite path-finding and the final

---

* **Corresponding author Marco Cecchini:** NEST, Istituto Nanoscienze-CNR and Scuola Normale Superiore, Piazza San Silvestro 12, 56127 Pisa, Italy; Tel/Fax: +39050509459; E-mail: marco.cecchini@nano.cnr.it

architecture of the functional network of neuronal connections [6 - 8]. The mammalian neuronal network is an example of highly polarized tissue, where cell development is driven by molecular stimuli, acting over long distance, and by physical signals that act locally through direct contact sensing [9, 10]. Neurons polarize and produce long cellular extensions (the neurites) whose development is governed by the formation and maturation of focal adhesions (FAs), the integrin-based cellular structures anchoring the cell to the external environment [6]. During neurite development, focal adhesions act as *topographical sensors* and integrate both physical and chemical signals from the extracellular matrix [11]. The maturation of FAs is in fact finely tuned by multiple information regarding the extracellular matrix properties such as mechanical stiffness, density of adhesion points, their chemical identity and geometry, and surface topography [12]. Importantly, through the modulation of FA maturation, a specific extracellular configuration can regulate the cell fate and in particular neuronal polarization, migration [13] and function. For example, synaptic plasticity has been recently suggested to involve the surrounding extracellular matrix signaling [14,15]. These processes involve coordinated interactions between FAs and the cell cytoskeleton. In order to build a correct brain architecture, a coordinated rearrangement of the cytoskeleton in response to extracellular cues is essential, and focal adhesion kinase (FAK) was recognized as a key neuronal enzyme [16]. For sake of example, the activation of the FA effector FAK is dispensable for glial-independent migration of interneurons but is required for the normal interaction of pyramidal neurons with radial glial fibers during cortical migration [13]. FAK is also required by both attractive and repulsive stimuli to control cytoskeletal dynamics and axon outgrowth and disassembly, working as a versatile molecular integrator that can switch to different functions depending on its activation site [17].

All cells grow and live while embedded in a dense and complex environment, the extracellular matrix (ECM), which contains an array of structural and directional cues. In particular, beyond chemical recognition, three independent ECM parameters play a major role in governing cell behaviour: topography, stiffness, and density of adhesion points. Topographical features in the micron and submicron range act as physical boundaries providing a local constraint to the formation and maturation of FAs. Cells apply force to the developing adhesions through acto-myosin contractility and the mechanical response of the matrix controls the further maturation of the adhesion points in a molecularly regulated feedback loop. Thanks to recent developments in micro/nano-engineering tools, the processes that control cell and, in particular, neuronal guidance and polarization can now be investigated *in vitro* using nano/micro-structured surfaces [18 - 21]. Nano/micro-textured substrates were demonstrated to be capable of tuning neuronal and glial cell adhesion, differentiation, polarization, migration, neurite orientation and even stem-cell fate [3, 22 - 25]. In particular, non-conventional lithographic technologies such as nanoimprint lithography (Fig. **1a**), electrospinning, or soft replica moulding to name but a few, allow the production of biocompatible substrates with customizable topography in the critical ranges affecting cellular functions. The application of these methods yielded patterned surfaces with lateral resolution ranging from few microns (*e.g.*, photolithography) down to tens of nanometers (*e.g.*, electron beam lithography [26]) and showed good versatility and reliability coupled with the use of polymeric materials [27 - 29]. The combination of these techniques with optically transparent polymers such as tissue culture polystyrene (TCPS), polyethylene terephthalate (PET) and cyclic olefin copolymer (COC) yielded patterned substrates where contact guidance could be observed by means of high resolution microscopy in living cells [28, 30, 31].

Here we focus on reviewing the influence of topography on the responses of neuronal cells. In order to control the assembly and maturation of FAs, and thus to induce specific cellular guidance, biomimetic scaffolds can provide modulation of the topographical parameters within the physiological ranges that are resolved by cells. The physical parameters in the substrate topography, which were reported to modulate contact guidance and formation of FAs in mammalian cells are the size, aspect ratio, and lateral spacing of the topographical features together with the isotropy and degree of disorder of the pattern.

In this framework, governing neuronal cell adhesion, migration, and axonal outgrowth are critical elements for regenerative medicine applications and for developing artificial neuronal interfaces, but at the same time these substrates open new experimental perspectives for the study of the molecular mechanisms at the base of neuronal environmental sensing. The molecular/cellular processes regulating synaptic plasticity, and learning are in fact an adaptation of the mechanisms used by all cells to regulate cell motility and shaping [32] as they involve the same complex machinery (*i.e.* actin fiber regulation and cell-cell/-ECM interaction signalling) where also the activation of FAs is emerging as pivotal [33]. Therefore, new knowledge about the sensing mechanisms of neuronal cells might impact also our understanding of several CNS disorders and providing new insight into the mechanisms leading to several neurological and neuropsychiatric disorders associated with connectivity and cognitive impairments.

Focal Adhesions and Cytoskeleton during Neuronal Cell-nanograting Interaction

In the last years nanogratings (NGs), anisotropic topographies composed by alternating lines of grooves and ridges with sub-micrometer lateral dimension (Fig. **1**), have been intensively studied and turned up as one of the most compelling systems for inducing neuronal alignment *via* pure cell mechanotransduction. In detail, nano/microgratings induced qualitatively similar effects on single cell polarization, and neurite network development in several different neuronal models, such as differentiating PC12 and neuroblastoma cells [20, 27, 31], invertebrate leech neurons [34], neuro-differentiated murine embryonic stem cells and primary murine hippocampal neurons [24] (Fig. **2**). Overall, NGs with linewidths in the 500 nm – 2 µm range can induce neuronal polarization and neurite alignment along the substrate direction, with different efficiency according to the cell morphological characteristics.

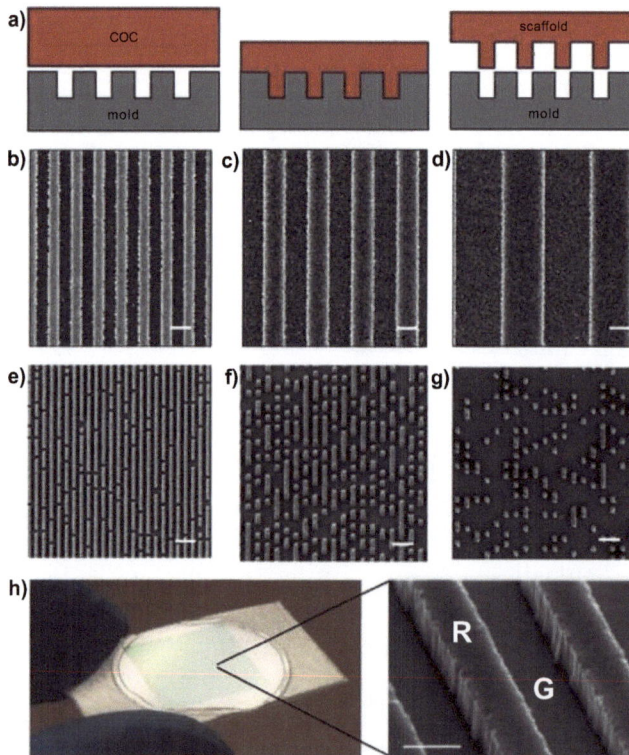

Fig. (1). Anisotropic nanogratings (NGs). **a)** Scheme of the fabrication process: the silicon molds are fabricated by electron beam lithography (EBL) and reactive-ion etching (RIE); then thermal nano-imprint lithography (NIL) is exploited to transfer the nanopattern from the molds to thermoplastic materials as the copolymer 2-norbornene ethylene (COC). **b-d)** SEM images of different COC NGs: NGs with ridge width (= groove width) of 500 nm (**b**), 1 µm (**c**), 2 µm (**d**), and depth of 350 nm; scale bars = 1 µm. **e-g)** SEM images of COC *noisy* NGs: an increasing amount of random nano-modifications is inserted in NGs with ridge width (= groove width) of 500 nm, varying from 10% (**e**), to 40% (**f**), and 80% (**g**). The presence of noise leads to the modulation of substrate directionality; scale bars = 2 µm. **h)** Image of a thermoplastic substrate with a squared NG patterned area; *inset*= detailed SEM image of the NG pattern, with ridges (R) and grooves (G) of 1 µm width; scale bar = 1 µm.

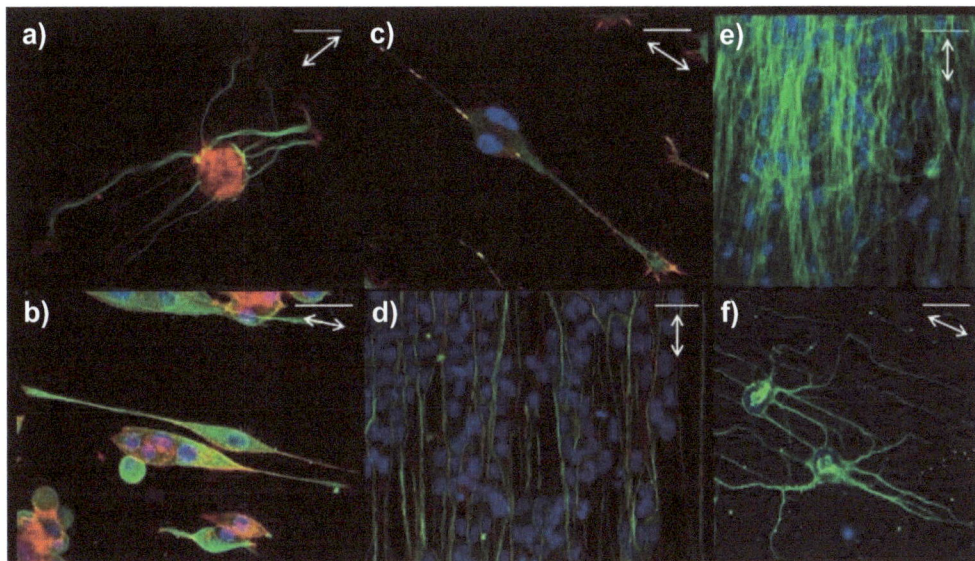

Fig. (2). Confocal images of different neuronal cells cultured on thermoplastic nanogratings. On 1-μm-line width NGs: **a)** A leech neuron grown on T2 and immunostained for Tubulin-α (*green*) and actin (*red*); **b)** SH-SY5Y differentiated cells on T2 and immunostained for Tubulin-βIII (*green*) and actin (*red*). Smaller neuronal cells grown on 500 nm-line width NGs: **c)** F11 cells immunostained for vinculin (*green*) and actin (*red*); **d)** PC12 cells immunostained for Tubulin-βIII (*green*) and actin (*red*); **e)** Mouse embryonic stem cells differentiated in neurons and immunostained for Tubulin-βIII (*green*); **f)** murine hippocampal neurons (day 14 *in vitro*) immunostained for Tubulin-α (*green*). All samples have been stained with DAPI to visualize nuclei (*blue*). Scale bars = 30 μm; inset = NG direction. Modified from [34].

The interaction of differentiating PC12 cells with NGs with groove depth between 250 and 500 nm results in filopodium dynamical sensing and neurite spatially selective consolidation, leading to a highly specific bipolar cell morphotype, and neurite alignment to the NG tracks [27, 35]. The result of the interaction between differentiated PC12 cells and the NG is an angular modulation of neurite persistence: the aligned neurites receive a positive feedback and persist longer than misaligned neurites and even longer than neurites growing on flat substrates [36]. Importantly, the shift from a multipolar to a bipolar morphology and the development of two aligned neurites on NGs are based on an active reading of the substrate topography.

This bipolarity selection is obtained by a spatial modulation of FA reinforcement. After the initial contact with the extracellular matrix (ECM), integrin receptors cluster and begin the enrollment of several cellular components thus inducing the formation of a cytoplasmic protein complex, the adhesion plaque. Adhesion plaques grow and enlarge in a hierarchical fashion increasing in size and enriching in characteristic proteins such as vinculin, focal adhesion kinase, talin, and paxillin. In this way, initial small contacts (<1 μm^2) maturate into larger FAs, whose maturation is regulated by a local balance between the force generated by cell contractility and ECM [12]. Mature FAs finally establish connections with actin fibers and can thus set up the cell shape. The maturation of FAs, regulation of actin filament interactions, and cell contractility cooperate to translate ECM guidance cues into specific cell polarity states [37].

NGs designed for high-resolution live-cell microscopy have been recently exploited to study the dynamics of focal adhesions, cell contractility, and actin fibers in neuronal contact guidance [36]. By taking advantage from a fluorescently-tagged paxillin vector, it was found that FAs are generally formed at the neurite tips and develop exclusively on top of NG ridges (Fig. **3a**). Neuronal polarization and contact guidance are based on a geometrical constraint of FAs resulting in an angular modulation of their maturation and endurance. In fact, cell contractility induces the maturation of nascent adhesions at the tip of aligned neurites (within 15 ° to the direction of the NG), whereas those produced at the tip of misaligned processes remain immature and are eventually lost. This allows the consolidation of aligned neurites and drives the collapse of the misaligned ones. Nanotopographies able to interfere with focal adhesion development during neuronal differentiation were then designed [3]. By varying a single topographical parameter (*i.e.* ridge width) the orientation and maturation of FAs was finely modulated producing an independent control over the number and the growth direction of neurites in neurodifferentiated PC12 cells (Fig. **3b**). In this study, it was possible to establish that neurite pathfinding is responsive to ridge-width variations between 500 and 1000 nm, while the mechanism controlling neuronal polarity establishment is sensitive for variations in ridge width within 1000 and 1500 nm [3].

In this scenario, ROCK-mediated contractility contributes to polarity selection during neuronal differentiation. In fact, in the neurites developing on NGs ROCK-mediated myosin-II contractility cross-talks with the adhesions, with the outcome that only aligned adhesions are allowed to grow, while misaligned ones, blocked by the topographical constraint, cannot develop into mature FAs [36]. Consistently, when the FA maturation is inhibited by the block of ROCK or of myosin-II activities during the interaction between cells and topographical features, the mechanism selecting bipolar PC12 cells with two aligned neurites fails to transfer the substrate symmetry to the cells [36], thus confirming that FAs directly interact with the actin cytoskeleton and thus tune actin-dependent contractility during neuronal polarization.

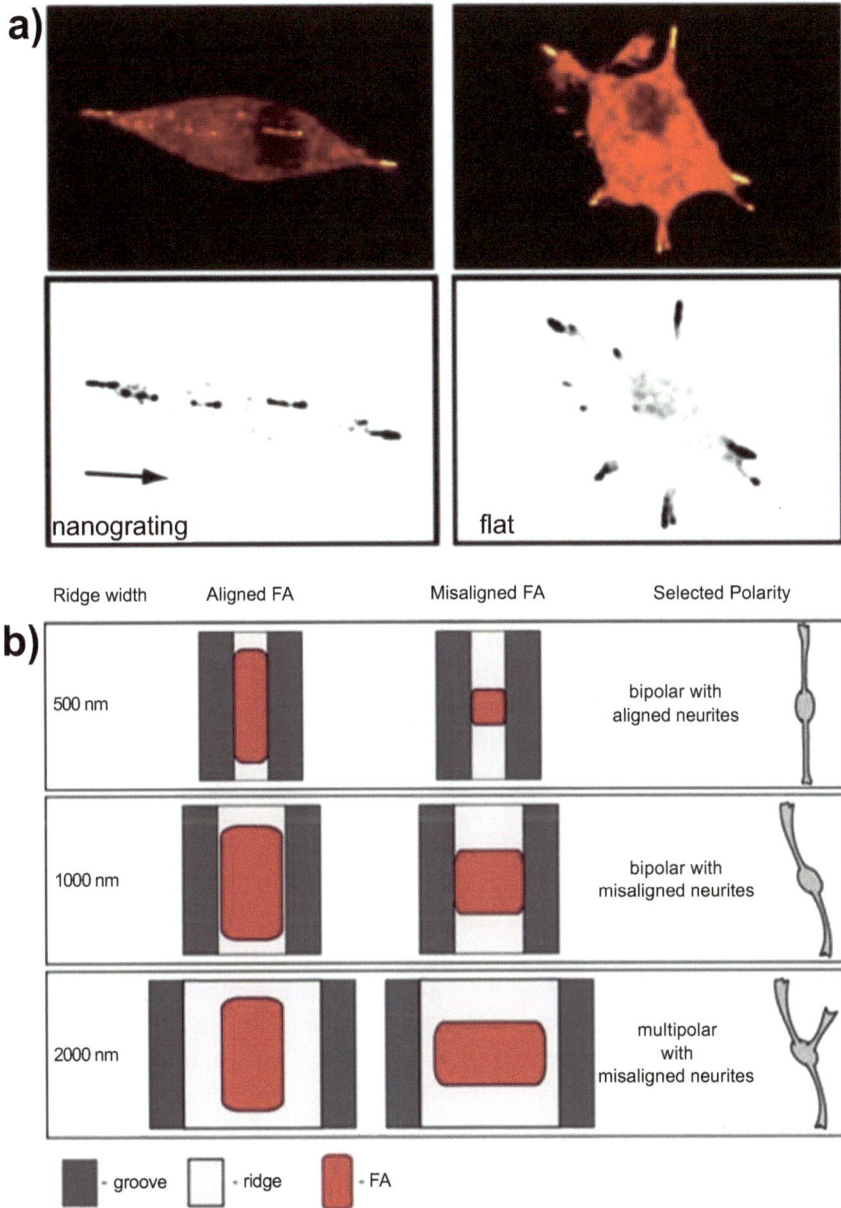

Fig. (3). Angular modulation of focal adhesion size. **a)** Distribution of paxillin-EGFP fluorescent signal in NGF-differentiating PC12 cells on NGs (*left*) and on flat substrate (*right*). In the lower row, inverted fluorescent signal at the cell-substrate interface shows the paxillin-EGFP accumulation in FAs (*black spots*). Black arrows = NG direction. **b)** Focal adhesions constrain on NGs with increasing ridge width. For each nanograting the corresponding ridge width is reported. The cartoons depict a schematic model of the relative size of aligned and misaligned focal adhesions (*red rectangles*) growing on top of ridges (*light grey lanes*).

The importance of actin fiber contractility also emerged at growth cone level. In fact, growth cones (GCs) - dynamic structures rich in actin filaments at the tips of neurites - probe external information by filopodia and integrate multiple physico-chemical cues. Filopodia sensing governed cell contact guidance, in leech neurons. Indeed, neurite topographical reading was mediated by an anisotropic filopodium organization in the growth cones on NGs, thus presenting the aligned filopodia longer and richer in fibrous actin [34].

Neuronal Differentiation Drives Nanotopography Sensing

In the PC12 neuronal model the substrate reading process occurs upon induction of neuronal differentiation. Importantly, the topographical guidance and neurite pathfinding of PC12 cells requires the stimulation of the neurotrophic growth factor (NGF)-induced neuronal differentiation signaling [30]. In addition, the behavior of the human neuroblastoma cell line SHSY5Y on anisotropic nanotopographies was probed, in proliferative or differentiating conditions, and a similar response to that of PC12s was found upon differentiation with retinoic acid (RA) and brain-derived neurotrophic factor (BDNF) [24]. Specifically, RA/BDNF differentiation ameliorated neuronal alignment, resulting in an approximately fourfold improvement of the tubulin cytoskeleton orientation along the substrate topography. This implies that common molecular mechanisms are involved for nanotopography sensing and that a different stimulation of different neuronal cells can modulate their response to similar topographical patterns. Alternative neuro-differentiation protocols were also tested in PC12s, by using activators of the cAMP pathway [*i.e.* Forskolin, an adenylyl cyclase activator resulting in increased levels of cyclic adenosine monophosphate (cAMP); 8CPT-2Me-cAMP, an agonist activating a protein in cAMP signalling pathway downstream adenylyl cyclase and cAMP (EPAC) and PKA independent] [30]. Several differentiation protocols based on forskolin led to neuronal-like differentiation but less efficient neurite alignment along the NGs [30]. In PC12s the topographical guidance was reduced by forskolin with respect to what measured for NGF, and this effect was correlated to an interference with FA maturation. These results notify that a fine regulation exists, during neuronal development, which controls the response of growing neurites to the ECM signals. In line with these findings, recently it has also been demonstrated that several neuronal sub-populations, or even the same neuronal types but at different development stage, can react differently to the same guidance cue during migration and differentiation [13, 14, 38].

Reading of the topographical guidance cues is not a univocal process but a function of the molecular differentiation pathway that is active in the cell. The complex interaction between substrate features, neurotrophic molecules, and cellular adhesion system cooperate to define the neuronal cell shaping during topographical guidance. Thus, molecular modulation of contact sensing appears to have a deep impact on the neuronal response and it must be taken in account for the development of nanostructures as neuro-regenerative scaffolds.

Noise Tolerance in Neuronal Contact Guidance

Since NGs were shown to be powerful tools to investigate topographical reading, novel NG geometries were introduced with the aim to study neurite contact guidance in the presence of topographical noise. It is well established that cells in tissues are exposed to nanotopographical stimuli determined by the detail of the local micro/nano-environment [8] and that directional textures typically necessarily coexists with bio-topographical noise. Examples are protein misfolding, sclerotic plaques, defects of ECM proteins or of their proteolysis, mutations in collagen fiber banding, scar-tissue invasion, formation of gaps and/or neuroma after nerve injury. The impact of noise on neuronal topographical sensing was investigated *in vitro* exploiting NG scaffolds with a controlled level of anisotropy (Fig. **1e-g**). In detail, neuron feedback on imperfect directional stimuli was studied by using NGs with a controlled degree of random nanotopographical noise and overall substrate directionality (*noisy NGs*) [39]. Reduced substrate directionality induced a sigmoidal loss of neurite alignment with a clear threshold response, showing that neurite path-finding is tolerant to a certain degree of noise and that it can successfully read hidden topographical information. Nanotopographical noise impacted FA maturation and alignment, which exhibited a non-linear behavior correlating with that of neurite alignment [40].

Given that myosin-II contractility signaling was demonstrated to regulate neuronal mechanotransduction [36], the effect of pharmacological treatments targeting this pathway (*e.g.* blebbistatin and nocodazole) on FA maturation and neurite navigation along noisy NGs was also investigated. Importantly, the noise tolerance of neurite guidance resulted improved by increased cell contractility [40]. Once the neurite NGF-stimulated extension started, nocodazole (*i.e.* a potent microtubule depolymerizing agent that can increase cell contractility *via* Rho-A pathway activation if administrated at low nM concentration) improved neurite alignment on noisy NGs, shifting the alignment threshold to higher noise levels; the opposite effect was instead caused by blebbistatin, a drug that inhibits cell contractility (Fig. **4a**). Neurite path-finding is thus capable through FAs of read partially ordered topographies, as expected given the presence of physiological structural noise in the extracellular environment *in vivo*.

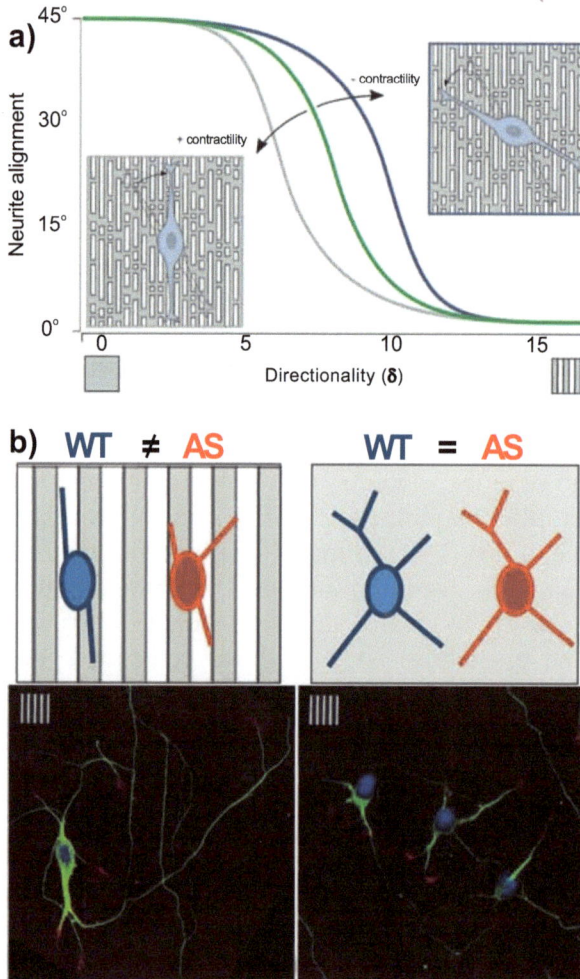

Fig. (4). Noise tolerance in neuronal contact guidance. **a)** Scheme of neurite guidance to noisy directional stimuli. Neurite guidance does not decrease linearly with substrate directionality, but it is a threshold process with a sigmoidal trend. Nocodazole increases cell contractility and improves neurite alignment (*grey line*). Conversely, the administration of blebbistatin, myosin-II inhibitor, affects the neurite guidance and leads to an overall reduction of alignment (*blue line*). **b)** Scheme of WT (*blue*) and Ube3a-deficient (AS) (*red*) hippocampal neurons development on anisotropic NGs (where they show different contact guidance responses) *versus* flat isotropic standard surfaces (where they behave the same), at early developmental stage *in vitro*. Representative confocal images of WT and AS neurons on 1 μm-width NGs (*insets*) grown for one week and immunostained for actin (*red*), tubulin (*green*) and nuclei (*blue*).

Neuronal contact guidance was also studied in "noisy" neurons on perfect nanogratings. A "noisy" neuron can be interpreted as a neuron with a molecular impairment, such as a neuronal model of a pathological disorder (*in vitro*). Although deficits in brain micro-connectivity have been recently implicated in the pathogenesis of cognitive disorders (*e.g.* autism, schizophrenia) [41, 42], the role of neuronal contact sensing mechanisms in these pathological conditions is under-investigated. Recent findings indicate that during neuronal development the pathways that regulate extracellular sensing (*e.g.* adhesion, cytoskeleton) are regulated by ubiquitination, in particular through E3 ubiquitin ligases, which function by ligating ubiquitin to specific target proteins [43]. In fact, many E3 ubiquitin ligases emerged as key regulators of several aspects of neuronal morphogenesis and connectivity at distinct temporal stages [43, 44]. Among these, Ubiquitin E3a ligase (UBE3A) is critical for neurodevelopment, as its loss leads to the neurocognitive disorder Angelman Syndrome (AS; OMIN 105830) [45]. Recently, defects have been reported in UBE3A-deficient models at the level of dendritic arborization in pyramidal neurons and of spine actin reorganization [46, 47], but how UBE3A loss results in neurocognitive impairments is still unclear. The contact sensing of Wild-Type (WT) and UBE3A-deficient (AS model) hippocampal neurons (HNs) was investigated by taking advantage of gratings with different topographical characteristics, with the aim to match their abilities to read physical directional stimuli [48]. Neurite growth, alignment to the topography and branching were studied by live and confocal imaging at different temporal phases. Importantly no differences in the neurite morphology and polarization emerged on standard isotropic substrates, where no topographical information were present. Conversely, on NGs at early stages *in vitro* (within 3 days *in vitro*) AS neurites were less efficient in retrieving substrate directional information than WT neurites, especially on NGs with ridge width of 1 µm, and their extension along NGs was reduced in comparison to WT HNs (Fig. **4b**) [48]. By using NGs, it was possible to demonstrate defective behavior in the contact guidance and neurite polarization of AS HNs in response to pure topographical signals, which was linked to the impaired activation of focal adhesion pathway, in particular at the level of FAK effector. These results demonstrated the use of NGs as advanced tools for testing the molecular mechanisms of cell contact guidance and shaping/polarization dynamics in neuro-pathological models and finally suggested that the neuronal contact sensing machinery might be affected in Angelman Syndrome.

CONCLUSIONS

Nano-microgrooved substrates are a useful tool to investigate the neuronal mechanisms for nanotopography sensing *in vitro*. It has been shown that: *i)* neuronal polarization is induced by a geometrical constraint of the FAs and requires ROCK-mediated contractility; *ii)* the substrate reading process occurs after induction of neuronal differentiation; *iii)* the loss of neurite guidance does not increase linearly with nanotopographical noise, but it is a threshold process; *iv)* pharmacological tuning of cell contractility can be a valid strategy to improve neuronal guidance; *v)* neuronal contact guidance can be investigated by nanostructured substrates in neuro-pathological models. The development of efficient regenerative scaffolds would improve thanks to the comprehension of the molecular mechanisms regulating neuronal interaction with the local physical properties of the extracellular environment. This knowledge might benefit the rational engineering of neuro-regenerative interfaces to enhance peripheral nerve wound healing. Moreover, further insight into the basic mechanisms of neuronal wiring and guidance mechanisms will improve our understanding of the nervous system and might help to identify novel mechanisms and possible targets for several brain disorders.

CONFLICT OF INTEREST

The authors confirm that they have no conflict of interest to declare for this publication.

ACKNOWLEDGMENTS

Declared none.

REFERENCES

[1] Ingber DE. Mechanical control of tissue morphogenesis during embryological development. Pathology 2006; 266: 255-66.

[2] Jaalouk DE, Lammerding J. Mechanotransduction gone awry. Nat Rev Mol Cell Biol 2009; 10(1): 63-73.

[3] Ferrari A, Cecchini M, Dhawan A, *et al.* Nanotopographic control of neuronal polarity. Nano Lett 2011; 11(2): 505-11.

[4] Thiery JP. Cell adhesion in development: a complex signaling network. Curr Opin Genet Dev 2006; 13(4): 365-71.

[5] Di Rienzo C, Jacchetti E, Cardarelli F, Bizzarri R, Beltram F, Cecchini M. Unveiling LOX-1 receptor interplay with nanotopography: mechanotransduction and atherosclerosis onset. Sci Rep 2013; 3: 1141.

[6] Myers JP, Santiago-Medina M, Gomez TM. Regulation of axonal outgrowth and pathfinding by integrin-ECM interactions. Dev Neurobiol 2011; 71(11): 901-23.

[7] Woo S, Gomez TM. Rac1 and RhoA promote neurite outgrowth through formation and stabilization of growth cone point contacts. J Neurosci 2006; 26: 1418-28.

[8] Yokota Y, Gashghaei HT, Han C, Watson H, Campbell KJ, Anton ES. Radial glial dependent and independent dynamics of interneuronal migration in the developing cerebral cortex. PLoS One 2007; 2(8): e794.

[9] Tessier Lavigne M, Goodman CS. The molecular biology of axon guidance. Science 1996; 274(5290): 1123-33.

[10] Vitriol EA, Zheng JQ. Growth cone travel in space and time: the cellular ensemble of cytoskeleton, adhesion, and membrane. Neuron 2012; 73: 1068-81.

[11] Geiger B, Spatz JP, Bershadsky AD. Environmental sensing through focal adhesions. Nat Rev Mol Cell Biol 2009; 10(1): 21-33.

[12]   Geiger B, Yamada KM. Molecular Architecture and Function of cell-matrix adhesions. J Cell Sci 2011; 114: 3583-90.

[13]   Valiente M, Ciceri G, Rico B, Marín O. Focal adhesion kinase modulates radial glia-dependent neuronal migration through connexin-26. J Neurosci 2011; 31: 11678-91.

[14]   Di Cristo G, Chattopadhyaya B, Kuhlman SJ, *et al.* Activity-dependent PSA expression regulates inhibitory maturation and onset of critical period plasticity. Nat Neurosci 2007; 10: 1569-77.

[15]   Dityatev A, Rusakov D. Molecular signals of plasticity at the tetrapartite synapse. Curr Opin Neurobiol 2011; 21: 353-9.

[16]   Navarro AI, Rico B. Focal adhesion kinase function in neuronal development. Curr Opin Neurobiol 2014; 27: 89-95.

[17]   Chacón MR, Fazzari P. FAK: Dynamic integration of guidance signals at the growth cone. Cell Adhes Migr 2011; 5(1): 52-5.

[18]   Eshghi S, Schaffer DV. Engineering microenvironments to control stem cell fate and function. StemBook. Cambridge, MA: Harvard Stem Cell Institute 2008.

[19]   Johansson F, Carlberg P, Danielsen N, Montelius L, Kanje M. Axonal outgrowth on nanoimprinted patterns. Biomaterials 2006; 27: 1251-8.

[20]   Wieringa P, Tonazzini I, Micera S, Cecchini M. Nanotopography induced contact guidance of the F11 cell line during neuronal differentiation: a neuronal model cell line for tissue scaffold development. Nanotechnology 2012; 23(27): 275102.

[21]   Antonini S, Meucci S, Jacchetti E, *et al.* Sub-micron lateral topography affects endothelial migration by modulation of focal adhesion dynamics. Biomed Mater 2015; 10(3): 035010.

[22]   Tonazzini I, Bystrenova E, Chelli B, *et al.* Multiscale morphology of organic semiconductor thin films controls the adhesion and viability of human neural cells. Biophys J 2010; 98(12): 2804-12.

[23]   Gomez N, Chen S, Schmidt CE. Polarization of hippocampal neurons with competitive surface stimuli: contact guidance cues are preferred over chemical ligands. J R Soc Interface 2007; 4: 223-33.

[24]   Tonazzini I, Cecchini A, Elgersma Y, Cecchini M. Interaction of SH-SY5Y Cells with Nanogratings During Neuronal Differentiation: Comparison with Primary Neurons. Adv Healthc Mater 2014; 3(4): 581-7.

[25]   Tonazzini I, Jacchetti E, Meucci S, Beltram F, Cecchini M. Schwann Cell Contact Guidance *versus* Boundary -Interaction in Functional Wound Healing along Nano and Microstructured Membranes. Adv Healthc Mater 2015; 4(12): 1849-60.

[26]   Shin H. Fabrication methods of an engineered microenvironment for analysis of cell-biomaterial interactions. Biomaterials 2007; 28: 126-33.

[27]   Cecchini M, Bumma G, Serresi M, Beltram F. PC12 differentiation on biopolymer nanostructures. Nanotechnology 2007; 18(50): 505103.

[28]   Cecchini M, Ferrari A, Beltram F. PC12 polarity on biopolymer nanogratings. J Phys Conf Ser 2008; 100: 012003.

[29]   Liliensiek SJ, Wood JA, Yong J, Auerbach R, Nealey PF, Murphy CJ. Modulation of human vascular endothelial cell behaviors by nanotopographic cues. Biomaterials 2010; 31(20): 5418-26.

[30]   Ferrari A, Faraci P, Cecchini M, Beltram F. The effect of alternative neuronal differentiation pathways on PC12 cell adhesion and neurite alignment to nanogratings. Biomaterials 2010; 31(9): 2565-73.

[31]   Ferrari A, Cecchini M, Degl Innocenti R, Beltram F. Directional PC12 cell migration along plastic nanotracks. IEEE Trans Biomed Eng 2009; 56(11): 2692-6.

[32]   Baudry M, Bi X. Learning and memory: an emergent property of cell motility. Neurobiol Learn Mem 2013; 104: 64-72.

[33] Monje FJ, Kim E-J, Pollak DD, *et al.* Focal Adhesion Kinase Regulates Neuronal Growth, Synaptic Plasticity and Hippocampus-Dependent Spatial Learning and Memory. Neurosignals 2011; 20(1): 1-14.

[34] Tonazzini I, Pellegrini M, Pellegrino M, Cecchini M. Interaction of leech neurons with topographical gratings : comparison with rodent and human neuronal lines and primary cells. Interface Focus 2014; 4(1): 20130047.

[35] Cecchini M, Signori F, Pingue P, Bronco S, Ciardelli F, Beltram F. High-resolution poly(ethylene terephthalate) (PET) hot embossing at low temperature: thermal, mechanical, and optical analysis of nanopatterned films. Langmuir 2008; 24(21): 12581-6.

[36] Ferrari A, Cecchini M, Serresi M, Faraci P, Pisignano D, Beltram F. Neuronal polarity selection by topography-induced focal adhesion control. Biomaterials 2010; 31(17): 4682-94.

[37] Bershadsky AD, Ballestrem C, Carramusa L, *et al.* Assembly and mechanosensory function of focal adhesions: experiments and models Eur J Cell Biol 2006; 85: 165e73

[38] Smeal RM, Tresco P. The influence of substrate curvature on neurite outgrowth is cell type dependent. Exp Neurol 2008; 213: 281-92.

[39] Meucci S, Tonazzini I, Beltram F, Cecchini M. Biocompatible noisy nanotopographies with specific directionality for controlled anisotropic cell cultures. Soft Matter 2012; 8(4): 1109.

[40] Tonazzini I, Meucci S, Faraci P, Beltram F, Cecchini M. Neuronal differentiation on anisotropic substrates and the influence of nanotopographical noise on neurite contact guidance. Biomaterials 2013; 34(25): 6027-36.

[41] Fan Y, Abrahamsen G, Mills R, *et al.* Focal adhesion dynamics are altered in schizophrenia. Biol Psychiatry 2013; 74(6): 418-26.

[42] McFadden K, Minshew NJ. Evidence for dysregulation of axonal growth and guidance in the etiology of ASD. Front Hum Neurosci 2013; 7: 671.

[43] Kawabe H, Brose N. The role of ubiquitylation in nerve cell development. Nat Rev Neurosci 2011; 12(5): 251-68.

[44] Yamada T, Yang Y, Bonni A. Spatial organization of ubiquitin ligase pathways orchestrates neuronal connectivity. Trends Neurosci 2013; 36(4): 218-26.

[45] Jiang YH, Armstrong D, Albrecht U, *et al.* Mutation of the Angelman ubiquitin ligase in mice causes increased cytoplasmic p53 and deficits of contextual learning and long-term potentiation. Neuron 1998; 21(4): 799-811.

[46] Miao S, Chen R, Ye J, *et al.* The angelman syndrome protein ube3a is required for polarized dendrite morphogenesis in pyramidal neurons. J Neurosci 2013; 33(1): 327-33.

[47] Baudry M, Kramar E, Xu X, *et al.* Ampakines promote spine actin polymerization, long-term potentiation, and learning in a mouse model of Angelman syndrome. Neurobiol Dis 2012; 47(2): 210-5.

[48] Tonazzini I, Meucci S, Van Woerden GM, Elgersma Y, Cecchini M. Impaired Neurite Contact Guidance in Ubiquitin Ligase E3a (Ube3a)-Deficient Hippocampal Neurons on Nanostructured Substrates. Adv Healthc Mater 2016; 5(7): 850-62.

CHAPTER 4

Drug Delivery to the Brain by Liposomal Carrier Systems

Anne Mahringer and **Gert Fricker**[*]

Ruprecht-Karls University, Institute of Pharmacy and Molecular Biotechnology, Im Neuenheimer Feld 329, 69120 Heidelberg, Germany

Abstract: Endothelial cells of brain microvessels limit the entry into the brain for xenobiotics and many drugs, which otherwise may be therapeutically active in the central nervous system. The ABC transporters, P-glycoprotein and Breast cancer resistance protein, which are predominantly located in the luminal surface of capillary endothelial cells, are key players for this barrier function. Thus, particular efforts have been made to overcome the blood-brain barrier or to circumvent these efflux pumps. The various options for drug transport into the brain include encapsulation of active compounds into delivery systems, *e.g.* liposomes, which are able to by-pass the export pumps and to convey their payload across the endothelial barrier. The applied systems target receptors at the luminal surface of the blood-brain barrier by using antibody-coupled immunoliposomes, liposomes conjugated to receptor-targeting vectors such as insulin, transferrin and apolipoproteins or cationized albumin-coupled liposomes.

Keywords: Albumin, ApoE, Blood-brain barrier, Immunoliposomes, Insulin, P-glycoprotein, Transferrin.

INTRODUCTION

Delivery of drugs to the central nervous system (CNS) remains to be a major challenge in modern pharmacotherapy. In 2007, William Pardridge gave an excellent summary on drug distribution to the CNS [1]. He stated that out of >7000 drugs in the Comprehensive Medicinal Chemistry database only 5% are used for CNS treatment, and that these compounds are limited to depression, schizophrenia and insomnia [2]. Most drugs do not reach the brain because they are not able to cross the blood-brain barrier (BBB), which is formed by endothelial cells of brain microvessels. In principle, drugs may reach the brain by several paracellular and transcellular routes across the BBB (Fig.1), from which

[*] **Corresponding author Gert Fricker:** Ruprecht-Karls University, Institute of Pharmacy and Molecular Biotechnology, Im Neuenheimer Feld 329, 69120 Heidelberg, Germany; Tel/Fax: +49 6221 54 8336; E-mail: gert.fricker@uni-hd.de

Giovanni Tosi (Ed)

the paracellular pathway is neglectable due to the very tight junctions between adjacent cells of brain capillaries. In addition, a key problem is the expression of export proteins in the luminal membranes of the endothelial cells, mainly P-glycoprotein (P-gp, ABCB1), Breast cancer resistance protein (Bcrp, ABCG2) and several Multidrug resistance-related proteins (Mrps, ABCCs) recognizing a huge variety of different substrates including most drugs and drug candidates. One possibility to overcome this obstacle is the use of colloidal carriers, which are able to by-pass these export proteins and to be transferred across the barrier by cytotic transport processes. Such carriers may include nanocrystals, micelles, polymeric nanoparticles, solid lipid nanoparticles and liposomes, respectively. The ultimate goal of these nanocarriers would be an exclusive delivery of drugs to the CNS. This idea of an unfailing "magic bullet" was originally developed by Paul Ehrlich at the beginning of the 20th century and it is more acute in medicine and pharmacy than ever. However, it is difficult to get a drug to the brain without coming into contact with other parts of the body. All hitherto existing efforts clearly show the main problems of drug targeting. Drug delivery systems must feature a satisfactory loading capacity and signal molecules or vectors, that recognize their targets with sufficient selectivity, have to be attached to the surface of the nanocarriers. Finally, they have to be inert and biocompatible, implying no own pharmacological effects by the inserted materials, no immune response of the body and biodegradability. Amongst the available nanocarriers liposomes fulfil these requirements quite well and are, therefore, suitable for CNS drug delivery. In the present article we describe features of such liposomes.

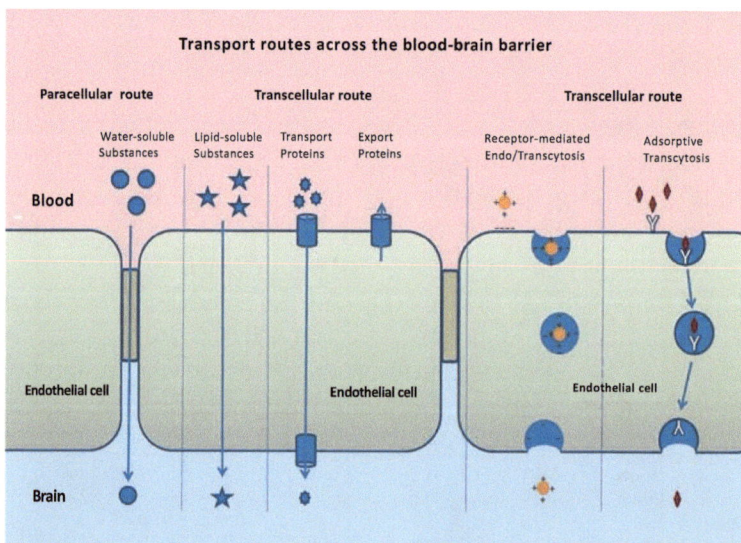

Fig. (1). Para- and transcellular as well as carrier-, receptor and adsorption-mediated transport pathways across the blood-brain barrier (modified from Abbott *et al.* [39]).

LIPOSOMES

Generally, liposomes are self-assembling phospholipid vesicles with an aqueous inner space, surrounded by a bilayer of naturally occurring phospholipids. Intravenously (i.v.) administered they may bind unspecifically to blood components, the opsonins, and are subsequently trapped by the reticuloendothelial system (RES) of liver, spleen, lung or bone marrow. A significant improvement allowing to overcome this phenomenon is the hydrophilisation of the liposomal surface by attachment of sterically hindering polyethylene glycol (PEG) chains (stealth liposomes). Circumvention of the RES can be accomplished by coupling signal molecules to the ends of PEG residues, thus avoiding internalization by the RES and improving the targeting effect. Over the past 2 decades interesting developments within molecular biology have offered new dimensions for the identification of ligand-receptor interactions and for the manufacture of appropriate signal molecules. *E.g.* "immunoliposomes", have successfully been tested in animal studies, in which selective antibodies targeting epitopes of receptors at the surface of brain microvessels have been coupled to PEG residues. These receptors include the insulin receptor, which transports insulin of peripheral origin into the brain [3, 4], insulin-like growth factor I and II receptors (IGFIR, IGFIIR) [1, 5], LDL receptor, leptin receptor (OBR) [6], or the receptor of advanced glycation endproducts (RAGE) [7]. Further receptors at the BBB are the low-density lipoprotein-related receptors 1 (LRP1) and 2 (LRP2, megalin) [8] and the transferrin receptor (TFR), which is expressed at the luminal as well as at the abluminal membrane of endothelial cells in the BBB, thus acting in a bidirectional way [9, 10].

TRANSFERRIN RECEPTOR

The transferrin receptor (TFR) moves apotransferrin fast in the brain-to-blood direction and holotransferrin from blood-to-brain. It is one of the most promising receptors for effective brain-targeting as it is highly expressed at the BBB. Here, some interesting examples are mentioned to demonstrate its capabilities: One of the first studies focussing this receptor used immunoliposomes for the delivery of the antineoplastic agent daunomycin to the rat brain [11]. Thiolated antibodies (OX26) were coupled to maleimide-grafted 85nm liposomes sterically stabilized with polyethylene glycol (PEG). Whereas no brain uptake of PEG-conjugated liposomes carrying [^{3}H]daunomycin was observed, coupling of thirty OX26 antibodies per liposome resulted in optimal brain delivery, which showed saturation at higher antibody densities. Brain targeting was not seen in immuno-liposomes conjugated with a mouse IgG2a isotype control. Further on, coinjection of free OX26 saturated the plasma clearance of the immunoliposomes. The study impressively demonstrated that these PEG-conjugated immunoliposomes might harbor ≥10.000 drug molecules [11].

The concept of using OX26-decorated immunoliposomes was also successfully extended to the CNS delivery of a biological [12] in an experimental 6-hydroxydopamine model of Parkinson's disease in order to normalize tyrosine hydroxylase activity in the striatum of adult rats. The tyrosine hydroxylase expression plasmid was incorporated into liposomes targeting the rat TFR to undergo both receptor-mediated transcytosis across the BBB and subsequent receptor-dependent endocytosis into neurons beyond the BBB by accessing the TFR. Thus, the striatal tyrosine hydroxylase activity ipsilateral to the intracerebral injection of the neurotoxin could be normalized.

Comparable to OX26, the MAb *versus* the rat TFR, the 8D3 antibody is directed against the mouse TFR. Significant transport to the brain of mice was also achieved by biotinylated PEG-stabilized liposomes, which had been coupled to anti-mouse TFR antibody 8D3 [13] to deliver double-stranded oligodeoxy-nucleotide-polyethylenimine complexes. PEGylated liposomes without the vector showed virtually no brain uptake after correction for organ plasma volume. In contrast, 8D3-targeted liposomes exhibited increased brain uptake of 0.33% ID/g at 1 h after i.v. bolus injection. The authors explicitly emphasize the delivery of intact double-stranded native, phosphodiester-based oligodeoxynucleotide, which is due to the protective shield of the shuttle system in the systemic circulation during the delivery process. 8D3-coupled liposomes were also useful to deliver bacterial ß-galactosidase linked to a brain-specific promoter taken from the 5' flanking sequence of the human glial fibrillary acidic protein (GFAP) gene [14]. Confocal microscopy colocalized immunoreactive bacterial ß-galactosidase together with GFAP in brain astrocytes.

An interesting brain drug delivery system for chemotherapy of glioblastoma consists of transferrin (TF) and cell-penetrating peptide dual-functionalized liposomes, TF/TAT-lip (transactivator of transcription protein of the human immunodeficiency virus type 1) [15], which had been loaded with doxorubicin (DOX) as a model drug. Cellular uptake was studied in both brain capillary endothelial cells (BCECs) of rats and U87 glioblastoma cells. The TF/TAT-lip-DOX presents the best anti-proliferative activity against U87 cells. *In vivo* experiments in an orthotopic rat brain tumor model revealed significantly prolonged survival times after administration of these liposomes compared to treatment with non-surface-decorated liposomes.

Dual-targeting daunorubicin liposomes modified with 4-aminophenyl-α-D-mannopyranoside, a mannose derivative, and transferrin molecules have recently been developed to effectively treat brain tumors [16]. The plasma drug concentration-time profile of the daunorubicin-carrying liposomes in mice decreased more slowly than free daunorubicin in the initial phase and maintained higher drug levels in the terminal phase, resulting in longer blood exposure compared with the free drug. An important observation was the fact that daunorubicin levels were lower in heart tissue and significantly higher in brain tissue after administration of the dual-targeting vectors if compared to the free drug. Moreover, the daunorubicin liposomes, that are assumed to be selectively endocytosed *via* glucose transporters in addition to the TFR, showed a significantly enhanced drug transport potential across the BBB.

In another dual-targeting study [17] a cyclic arginine-glycine-aspartic acid (RGD) peptide and TF were utilized as directing ligands for paclitaxel-loaded liposomes against glioblastoma. Cyclic RGD peptides are specific target-seeking ligands against the integrin receptor $\alpha v\beta 3$ expressed on the tumor neovasculature. *In vitro* cell uptake and three-dimensional tumor spheroid penetration studies demonstrated that the system could aim at endothelial and tumor cells, as well as penetrate the tumor cells to reach the core of the tumor spheroids. *In vivo* imaging demonstrated brain distribution of these RGD/TF-liposomes.

A novel approach uses a bispecific mechanism, receptor mediation across endothelial cells *via* TFR combined with external non-invasive magnetic force, by ferrous magnet-based liposomes for BBB transmigration enhancement [18]. Homogenous magnetic nanoparticles with a size of ~10nm were embedded into transferrin-decorated PEGylated liposomes. These magneto-liposomes (MLs) showed a single mono-dispersion, ~130 ± 10nm diameter and a remarkable magnetite encapsulation efficiency of nearly 60%. The hydrodynamic size of the MLs was stable for over two months at 4°C. Compared to magnetic force- or transferrin receptor-mediated transportation alone their synergy resulted in 50-100% increased transmigration without affecting BBB integrity. Consequently, confocal microscopy and iron concentration in BBB-composing cells confirmed high cellular uptake of ML particles due to the concerted effect.

Double-labelling of liposomes with TFR antibodies and apolipoprotein E (ApoE) being recognized by the LDL/LRP-receptors was used to study an additive and synergistic effect of both ligands [19]. Mono- and dual-decorated liposomes were prepared by immobilization of monoclonal antibodies (MAb) against TFR (OX26 or RI7217) and/or a peptide analogue of ApoE3 to target low-density lipoprotein-related receptor (LRP). Liposomal uptake and transport across immortalized hCMEC/D3 endothelial cell monolayers were significantly affected by the double decoration, but only in absence of apolipoproteins in the medium. Intact vesicle-mediated transcytosis was confirmed by equal transport extent of hydrophilic and lipophilic labels. However, *in vivo* the targeting ability of dual-decorated liposomes was not higher than that of MAb-decorated liposomes, which was explained by the presence of apolipoproteins in the blood circulation.

A comparative study investigated the transfection ability and efficiency of 4 types of liposomes and immunoliposomes for exogenous gene delivery into the brain in rats [20]. Liposomes encapsulating pCMV (human cytomegalovirus promoter)-LacZ plasmid at 80μg or 300μg, immunoliposomes encapsulating 80μg transferrin receptor antibodies (OX26)-pCMV-LacZ plasmid or brain-specific immunoliposomes encapsulating 80μg OX26-pGFAP (glial fibrillary acidic protein promoter)-LacZ plasmid. A control group received no injected agent. The LacZ mRNA levels (1 h post-injection) and beta-galactosidase activity (48 h post-injection) in the brain and peripheral organs were assayed using real-time reverse transcription-polymerase chain reaction and histochemical staining, respectively. Both immunoliposomes delivered exogenous DNA containing the LacZ gene into the brain after venous injection, resulting in extensive LacZ expression in the brain. Furthermore, the brain-specific OX26-pGFAP-LacZ immunoliposome decreased the non-specific expression of LacZ in peripheral organs without affecting transfection efficiency in the brain.

INSULIN RECEPTOR

Insulin receptors were among the first to be proposed as targets for colloidal carrier transport across the BBB. Zhang *et al.* [21] demonstrated successful non-viral gene transfer (plasmids encoding either luciferase or ß-galactosidase) to primate brain after encapsulation into PEGylated immunoliposome, which had been coupled to a monoclonal antibody to the human insulin receptor. Neuronal expression of the ß-galactosidase gene in brain was demonstrated by histochemistry and confocal microscopy.

LDL RECEPTOR

The presence of a low-density lipoprotein receptor (LDLR) was first described in bovine brain capillaries by Méresse *et al.* [22]. Apparently, the expression of this receptor is modulated by soluble factors released from astrocytes [23]. The receptor recognizes apolipoproteins circulating in the blood stream, mainly apolipoprotein E (ApoE). *In vitro* studies with rodent brain capillary endothelial cells (BCEC) and the cell line RBE4 as well as *ex vivo* studies with functionally intact brain capillaries from porcine brain showed that these liposomes coupled to a peptide of 26 amino acids length, derived from the binding site of human apolipoprotein E4 (ApoE4) were effectively endocytosed by the endothelial cells (Fig. **2**) [24]. In contrast to random peptide-coupled liposomes, the ApoE4-fragment-linked liposomes were rapidly taken up by cultured BCECs and RBE4 cells. Uptake could be inhibited by ApoE4, free peptide, and antibodies against the LDLR in a concentration-dependent manner.

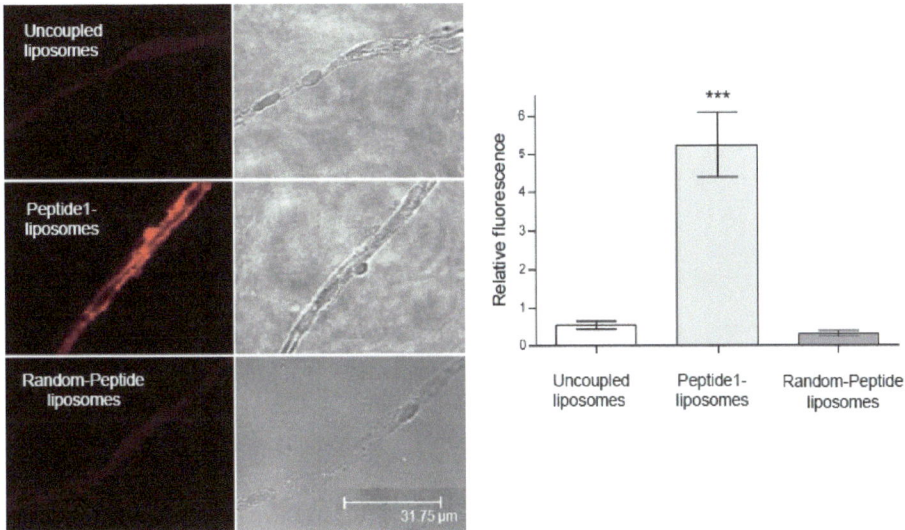

Fig. (2). PEGylated and Rhodamine PE-labelled liposomes attached to Peptide-1 carrying the amino acid sequence of the ApoE4-binding domain showed an increased accumulation in the endothelium of mouse brain capillary endothelial cells when compared to liposomes that remained either uncoupled or were carrying a random amino acid sequence (Random-Peptide). The enhanced uptake into brain capillary endothelial cells from the blood compartment suggest an LDL receptor-mediated mechanism (adapted from Hülsermann *et al.* [25]).

An interesting study showed that statins reduced the efflux activity of P-gp and Bcrp in a human brain microvessel cell line (hCMEC/D3) by increasing the synthesis of NO, which elicits the nitration of critical tyrosine residues on these transporters. Statins also increased the number of low-density lipoprotein (LDL) receptors in BBB cells as well as in tumor cells like human glioblastoma. The association of statins plus drug-loaded liposomes engineered as LDLs was effective as a vehicle for doxorubicin to cross the BBB [25].

LEPTIN RECEPTOR

Leptin, a peptide hormone consisting of 167 amino acids, is a major regulator of body weight [6], and controls together with the peptide hormone ghrelin hunger and satiety. It was suggested that a negative feedback loop exists between leptin and body weight [26]. Fat induced leptin crosses the BBB by transcytosis and interacts with leptin receptors (OBR) in the arcuate nucleus to inhibit feeding and to increase thermogenesis, which decreases fat mass. Absent or impaired OBRs are discussed to be a cause of obesity in humans.

In a recent *in vitro* study [27] the function of a leptin-derived peptide (Lep(70-89)) as a ligand for mouse brain-derived endothelial cells (MBEC4) was investigated. Lep(70-89)-modified liposomes, prepared with a PEG spacer (Lep(70-89)-PEG-LPs) exhibited a significantly higher cellular uptake than peptide-unmodified liposomes (PEG-LPs). In this study it became not clear whether the OBR was directly involved in liposomal transfer across the cells, but cellular uptake was inhibited by amiloride, while no significant inhibitory effect was determined by the presence of chlorpromazine and filipin III, proposing that macropinocytosis is largely involved in the cellular uptake of Lep(70-89)- PEG-LPs. Imaging studies revealed that Lep(70-89)-PEG-LPs were not co-localized with endosomes/lysosomes. The authors concluded that Lep(70-89)- SPEG-liposomes were taken up *via* macropinocytosis and were subject to non-classical intracellular trafficking, resulting in the circumvention of lysosomal degradation in endothelial cells.

RECEPTOR OF ADVANCED GLYCATION END PRODUCTS

The receptor of advanced glycation end products (RAGE), for advanced glycation end products, which was first characterized by Neeper *et al.* [7], is a 35kD transmembrane receptor of the immunoglobulin E super-family. This multi-ligand receptor is involved in inflammatory disorders, diabetic complications, tumor growth, and Alzheimer's disease [28]. RAGE appears to interact with Amyloid ß (Aß) resulting in transport of Aß across the BBB [29, 30]. Anti-Aβ-MAb (Aβ-MAb)-decorated immunoliposomes (ILP) had been used to target RAGE [31]. Uptake of these liposomes into human brain endothelial hCMEC/D3 cells was measured. Internalization efficiency was higher than for control PEGylated liposomes and increased significantly when cells were pre-incubated with Aβ1-42 peptides. Transcytosis of Aβ-MAb-ILP through monolayers was 2.5 times higher when monolayers were exposed to Aβ1-42. The Aβ peptide-induced increase in binding (and transport) was regulated by RAGE, as proven after blocking the receptor by a specific MAb. Aβ1-42 peptides did not modify the barrier tightness and integrity, as analysed by transendothelial resistance and Lucifer Yellow permeability. Additionally, hCMEC/D3 cell viability was not affected by Aβ-MAb-liposomes.

ABSORPTIVE-MEDIATED TRANSCYTOSIS ACROSS THE BBB

Not only linkage to receptor-directed antibodies can be used for brain-targeting drug delivery but also coupling liposomes to plasma proteins, such as cationized albumin, which undergoes absorptive-mediated endocytosis at the BBB [31, 32]. The suitability of albumin-coupled liposomes as drug carriers for brain-specific targeting has previously been investigated using bovine serum albumin (BSA) and cationized albumin (cBSA), respectively, as targeting vectors [33]. Sterically stabilized PEG-liposomes were covalently coupled to either thiolated BSA or cBSA. Liposomes were loaded with carboxy-fluorescein and rhodamine-labelled dipalmitoyl-phosphatidyl-ethanolamine as hydrophilic and lipophilic marker compounds, respectively. In contrast to BSA-coupled liposomes, cBSA-linked liposomes were rapidly taken up by cultured porcine brain capillary endothelial cells. Cellular uptake could be inhibited by free cationized albumin, phenylarsineoxide, nocodazole, and filipin, but not by depolymerization, suggesting a caveolae-mediated incorporation process. Immunostaining demonstrated a high expression of caveolin in the capillary endothelium. *In vivo* experiments in rats [34] revealed that after i.v. application of fluorescent-labelled cBSA-liposomes with a size of 120-150nm, fluorescence-associated with liposomes could be detected in brain capillary surrounding tissue after 3, 6 and 24h, suggesting successful brain delivery.

ULTRASOUND AND LIPOSOMES TO OVERCOME THE BBB

Among the different strategies to deliver therapeutics into the CNS the focussed ultrasound sonication (FUS) together with microbubbles (MBs) or liposomes gains increasingly interest for the delivery of small molecules and biologics. This approach has been shown to be effective in transiently disrupting the BBB for non-invasive drug delivery [35, 36]. Several studies prove the value of this strategy. To name some of them, liposomal doxorubicin was administered in a rat orthotopic glioma model prior to focussed ultrasound [37]. Tumor doxorubicin concentrations increased monotonically in control tumors at 9, 14 and 17 d, however, with FUS-induced disruption, the doxorubicin concentrations were enhanced significantly regardless of the stage of tumor growth. These results suggest that even large/late-stage tumors can benefit from FUS-induced drug delivery enhancement.

Another CNS-malignancy – CNS lymphoma – was treated with methotrexate (MTX) transported in microbubble-associated liposomes [38]. To date, MTX is delivered to the CNS by high systemic doses causing long-term neurotoxicity, or by intrathecal administration, which may lead to infections or hemorrhagic complications. Here, microbubbles were manufactured and coupled to MTX-loaded liposomes. These liposome-coupled microbubbles had a high drug-loading capacity of app. 9% and a diameter of app. 2.64 µm. When rats were subjected to sonication, MTX-liposome-coupled microbubbles led to targeted disruption of the BBB without noticeable tissue or capillary damage. Consequently, treatment with MTX-liposome-coupled microbubbles and ultrasound resulted in a more significantly elevated brain MTX concentration than all other treatments, suggesting that the combination of liposomes and ultrasound may hold great promise as novel therapy for primary CNS lymphoma and other CNS malignancies.

CONCLUSION

Drug targeting to the CNS has recently gained remarkable interest, whereas application of normal liposomes yielded more or less disillusioning results in clinical trials within the last 25 years, the above discussed vector technologies offer promising tools for new therapeutic areas. Modified liposomes loaded with small molecules as well as biologicals including proteins or nucleic acid molecules show first positive clinical results. Even if broader use of the mentioned systems and their validation in *in vivo* models will require some time, the continuously ongoing optimisation of carrier systems by use of enzymatically cleavable bonds and improved signal structures will lead to new biodegradable systems, offering new possibilities of drug delivery to the CNS.

CONFLICT OF INTEREST

The authors confirm that they have no conflict of interest to declare for this publication.

ACKNOWLEDGMENTS

Declared none.

REFERENCES

[1]   Pardridge WM. Blood-brain barrier delivery. Drug Discov Today 2007; 12(1-2): 54-61. [http://dx.doi.org/10.1016/j.drudis.2006.10.013] [PMID: 17198973]

[2]   Ghose AK, Viswanadhan VN, Wendoloski JJ. A knowledge-based approach in designing combinatorial or medicinal chemistry libraries for drug discovery. 1. A qualitative and quantitative characterization of known drug databases. J Comb Chem 1999; 1(1): 55-68. [http://dx.doi.org/10.1021/cc9800071] [PMID: 10746014]

[3]   Pardridge WM. Strategies for the delivery of drugs through the blood-brain barrier Annual Reports in Medicinal Chemistry-20, Academic Press. 1985; pp. 305-13.

[4]   Frank HJ, Pardridge WM, Morris WL, Rosenfeld RG, Choi TB. Binding and internalization of insulin and insulin-like growth factors by isolated brain microvessels. Diabetes 1986; 35(6): 654-61. [http://dx.doi.org/10.2337/diab.35.6.654] [PMID: 3011572]

[5]   Urayama A, Grubb JH, Sly WS, Banks WA. Developmentally regulated mannose 6-phosphate receptor-mediated transport of a lysosomal enzyme across the blood-brain barrier. Proc Natl Acad Sci USA 2004; 101(34): 12658-63. [http://dx.doi.org/10.1073/pnas.0405042101] [PMID: 15314220]

[6]   Banks WA, Lebel CR. Strategies for the delivery of leptin to the CNS. J Drug Target 2002; 10(4): 297-308. [http://dx.doi.org/10.1080/10611860290031895] [PMID: 12164378]

[7]   Neeper M, Schmidt AM, Brett J, *et al*. Cloning and expression of a cell surface receptor for advanced glycosylation end products of proteins. J Biol Chem 1992; 267(21): 14998-5004.

[PMID: 1378843]

[8]   Bu G, Geuze HJ, Strous GJ, Schwartz AL. 39 kDa receptor-associated protein is an ER resident protein and molecular chaperone for LDL receptor-related protein. EMBO J 1995; 14(10): 2269-80. [PMID: 7774585]

[9]   Zhang Y, Pardridge WM. Rapid transferrin efflux from brain to blood across the blood-brain barrier. J Neuroimmunol 2001; 76(5): 1597-600. [http://dx.doi.org/10.1016/S0165-5728(01)00242-9] [PMID: 11240028]

[10]   Skarlatos S, Yoshikawa T, Pardridge WM. Transport of [125I]transferrin through the rat blood-brain barrier. Brain Res 1995; 683(2): 164-71. [http://dx.doi.org/10.1016/0006-8993(95)00363-U] [PMID: 7552351]

[11]   Huwyler J, Wu D, Pardridge WM. Brain drug delivery of small molecules using immunoliposomes. Proc Natl Acad Sci USA 1996; 93(24): 14164-9. [http://dx.doi.org/10.1073/pnas.93.24.14164] [PMID: 8943078]

[12]   Zhang Y, Calon F, Zhu C, Boado RJ, Pardridge WM. Intravenous nonviral gene therapy causes normalization of striatal tyrosine hydroxylase and reversal of motor impairment in experimental parkinsonism. Hum Gene Ther 2003; 14(1): 1-12. [http://dx.doi.org/10.1089/10430340360464660] [PMID: 12573054]

[13]   Ko YT, Bhattacharya R, Bickel U. Liposome encapsulated polyethylenimine/ODN polyplexes for brain targeting. J Control Release 2009; 133(3): 230-7. [http://dx.doi.org/10.1016/j.jconrel.2008.10.013] [PMID: 19013203]

[14]   Shi N, Zhang Y, Zhu C, Boado RJ, Pardridge WM. Brain-specific expression of an exogenous gene after i.v. administration. Proc Natl Acad Sci USA 2001; 98(22): 12754-9. [http://dx.doi.org/10.1073/pnas.221450098] [PMID: 11592987]

[15]   Zheng C, Ma C, Bai E, Yang K, Xu R. Transferrin and cell-penetrating peptide dual-functioned liposome for targeted drug delivery to glioma. Int J Clin Exp Med 2015; 8(2): 1658-68. [PMID: 25932094]

[16]   Ying X, Wen H, Yao HJ, *et al*. Pharmacokinetics and tissue distribution of dual-targeting daunorubicin liposomes in mice. Pharmacology 2011; 87(1-2): 105-14. [http://dx.doi.org/10.1159/000323222] [PMID: 21282968]

[17]   Qin L, Wang CZ, Fan HJ, *et al*. A dual-targeting liposome conjugated with transferrin and arginine-glycine-aspartic acid peptide for glioma-targeting therapy. Oncol Lett 2014; 8(5): 2000-6. [PMID: 25289086]

[18]   Ding H, Sagar V, Agudelo M, *et al*. Enhanced blood-brain barrier transmigration using a novel transferrin embedded fluorescent magneto-liposome nanoformulation. Nanotechnology 2014; 25(5): 055101. [http://dx.doi.org/10.1088/0957-4484/25/5/055101] [PMID: 24406534]

[19]   Markoutsa E, Papadia K, Giannou AD, *et al*. Mono and dually decorated nanoliposomes for brain targeting, *in vitro* and *in vivo* studies. Pharm Res 2014; 31(5): 1275-89. [http://dx.doi.org/10.1007/s11095-013-1249-3] [PMID: 24338512]

[20]   Zhao H, Li GL, Wang RZ, *et al*. A comparative study of transfection efficiency between liposomes, immunoliposomes and brain-specific immunoliposomes. J Int Med Res 2010; 38(3): 957-66. [http://dx.doi.org/10.1177/147323001003800322] [PMID: 20819432]

[21]   Zhang Y, Schlachetzki F, Pardridge WM. Global non-viral gene transfer to the primate brain following intravenous administration. Mol Ther 2003; 7(1): 11-8. [http://dx.doi.org/10.1016/S1525-0016(02)00018-7] [PMID: 12573613]

[22]   Méresse S, Delbart C, Fruchart JC, Cecchelli R. Low-density lipoprotein receptor on endothelium of brain capillaries. J Neurochem 1989; 53(2): 340-5. [http://dx.doi.org/10.1111/j.1471-4159.1989.tb07340.x] [PMID: 2746225]

[23]   Lucarelli M, Borrelli V, Fiori A, *et al.* The expression of native and oxidized LDL receptors in brain microvessels is specifically enhanced by astrocytes-derived soluble factor(s). FEBS Lett 2002; 522(1-3): 19-23.
[http://dx.doi.org/10.1016/S0014-5793(02)02857-0] [PMID: 12095612]

[24]   Hülsermann U, Hoffmann MM, Massing U, Fricker G. Uptake of apolipoprotein E fragment coupled liposomes by cultured brain microvessel endothelial cells and intact brain capillaries. J Drug Target 2009; 17(8): 610-8.
[http://dx.doi.org/10.1080/10611860903105986] [PMID: 19694613]

[25]   Pinzón-Daza M, Garzón R, Couraud P, *et al.* The association of statins plus LDL receptor-targeted liposome-encapsulated doxorubicin increases *in vitro* drug delivery across blood-brain barrier cells. Br J Pharmacol 2012; 167(7): 1431-47.
[http://dx.doi.org/10.1111/j.1476-5381.2012.02103.x] [PMID: 22788770]

[26]   Zhang Y, Proenca R, Maffei M, Barone M, Leopold L, Friedman JM. Positional cloning of the mouse obese gene and its human homologue. Nature 1994; 372(6505): 425-32.
[http://dx.doi.org/10.1038/372425a0] [PMID: 7984236]

[27]   Tamaru M, Akita H, Fujiwara T, Kajimoto K, Harashima H. Leptin-derived peptide, a targeting ligand for mouse brain-derived endothelial cells *via* macropinocytosis. Biochem Biophys Res Commun 2010; 394(3): 587-92.
[http://dx.doi.org/10.1016/j.bbrc.2010.03.024] [PMID: 20214882]

[28]   Deane RJ. Is RAGE still a therapeutic target for Alzheimers disease? Future Med Chem 2012; 4(7): 915-25.
[http://dx.doi.org/10.4155/fmc.12.51] [PMID: 22571615]

[29]   Deane R, Du Yan S, Submamaryan RK, *et al.* RAGE mediates amyloid-beta peptide transport across the blood-brain barrier and accumulation in brain. Nat Med 2003; 9(7): 907-13.
[http://dx.doi.org/10.1038/nm890] [PMID: 12808450]

[30]   Deane R, Bell RD, Sagare A, Zlokovic BV. Clearance of amyloid-beta peptide across the blood-brain barrier: implication for therapies in Alzheimers disease. CNS Neurol Disord Drug Targets 2009; 8(1): 16-30.
[http://dx.doi.org/10.2174/187152709787601867] [PMID: 19275634]

[31]   Markoutsa E, Papadia K, Clemente C, Flores O, Antimisiaris SG. Anti-Aβ-MAb and dually decorated nanoliposomes: effect of Aβ142 peptides on interaction with hCMEC/D3 cells. Eur J Pharm Biopharm 2012; 81: 49-56.
[http://dx.doi.org/10.1016/j.ejpb.2012.02.006] [PMID: 22386910]

[32]   Kang YS, Pardridge WM. Brain delivery of biotin bound to a conjugate of neutral avidin and cationized human albumin. Pharm Res 1994; 11(9): 1257-64.
[http://dx.doi.org/10.1023/A:1018982125649] [PMID: 7816753]

[33]   Thöle M, Nobmann S, Huwyler J, Bartmann A, Fricker G. Uptake of cationzied albumin coupled liposomes by cultured porcine brain microvessel endothelial cells and intact brain capillaries. J Drug Target 2002; 10(4): 337-44.
[http://dx.doi.org/10.1080/10611860290031840] [PMID: 12164382]

[34]   Helm F, Fricker G. Liposomal conjugates for drug delivery to the central nervous system. Pharmaceutics 2015; 7(2): 27-42.
[http://dx.doi.org/10.3390/pharmaceutics7020027] [PMID: 25835091]

[35]   Hynynen K, McDannold N, Vykhodtseva N, Jolesz FA. Noninvasive MR imaging-guided focal opening of the blood-brain barrier in rabbits. Radiology 2001; 220(3): 640-6.
[http://dx.doi.org/10.1148/radiol.2202001804] [PMID: 11526261]

[36]   Yang FY, Wang HE, Lin GL, Lin HH, Wong TT. Evaluation of the increase in permeability of the blood-brain barrier during tumor progression after pulsed focused ultrasound. Int J Nanomedicine

2012; 7: 723-30.
[http://dx.doi.org/10.2147/IJN.S28503] [PMID: 22359451]

[37]   Aryal M, Park J, Vykhodtseva N, Zhang YZ, McDannold N. Enhancement in blood-tumor barrier permeability and delivery of liposomal doxorubicin using focused ultrasound and microbubbles: evaluation during tumor progression in a rat glioma model. Phys Med Biol 2015; 60(6): 2511-27.
[http://dx.doi.org/10.1088/0031-9155/60/6/2511] [PMID: 25746014]

[38]   Wang X, Liu P, Yang W, *et al.* Microbubbles coupled to methotrexate-loaded liposomes for ultrasound-mediated delivery of methotrexate across the blood-brain barrier. Int J Nanomedicine 2014; 9: 4899-909.
[PMID: 25364248]

[39]   Abbot NJ, Ronnback L, Hansson E. Astrocyte-endothelial interactions at the blood-brain barrier. Nat Rev Neurosci 2006; 7(1): 41-53.
[http://dx.doi.org/10.1016/j.ejpb.2012.02.006] [PMID: 22386910]

CHAPTER 5

Neuronopathic LSDs: Quest for Treatments Drives Research in Nanomedicine and Nanotechnology

Cinzia M. Bellettato[1], David J. Begley[1,2], Christina Lampe[1,3] and Maurizio Scarpa[1,3,4,*]

[1] *Brains for Brains Foundation, Department of Women and Children Health, Via Giustiniani 3, 35128 Padova, Italy*

[2] *Kings College London, Institute of Pharmaceutical Science, Franklin-Wilkins Building, 150 Stamford Street, London, SE1 9NH, UK*

[3] *Center for Rare Diseases Helios Horst Schmidt Kliniken, Department of Child and Adolescent Medicine, Ludwig-Erhard-Straße 100 65199 Wiesbaden, Germany*

[4] *University of Padova, Department of Women and Children Health, Via Giustiniani 3, 35128 Padova, Italy*

Abstract: Lysosomal storage diseases (LSDs) are due to mutations in genes coding for high molecular weight lysosomal enzymes, which result in a deficiency or complete loss of enzyme activity and the consequent storage of undegraded substrate within lysosomes. Therapeutic approaches capable of modifying the natural history of the disease are available today and many have already entered into clinical practice. Among these, enzyme replacement therapy (ERT) represents an approved key treatment for a number of LSDs. Unfortunately, none of the used therapeutic replacement enzymes have, so far, proved to be effectively able to reach the central nervous system (CNS) in significant amounts and arrest neurodegeneration. Thus, currently, only the peripheral disease can be treated with ERT while storage product continues to accumulate in the CNS, resulting in severe neurodegeneration and premature death in childhood for all neurologically affected patients. In recent years, scientific advances in nanotechnology have led to development of revolutionary approaches potentially capable to provide a solution to the still unmet problem of increasing drug delivery across the Blood Brain Barrier. In particular, the growing interest in the medical applications of nanotechnology has contributed to the advent of a new field of applied science named nanomedicine that offers promising strategies to overcome several of the current impediments and disadvantages of ERT. The combination of existing nanotechnology with already available enzymes can, in fact, significantly improve the enzyme delivery opening a promising new era in the treatment of LSDs. This chapter aims to review the most recent advancement in nanomedicine and nanotechnology presenting novel therapeutic approaches designed to address neuronopathic LSDs.

* **Corresponding author Maurizio Scarpa:** Center for Rare Diseases Helios Horst Schmidt Kliniken, Department of Child and Adolescent Medicine, Ludwig-Erhard-Straße 100 65199 Wiesbaden, Germany; Tel/Fax: 0049 611 43 2325; E-mail: Maurizio.Scarpa@helios-kliniken.de

Keywords: Blood-brain barrier, Enzyme replacement, Lysosomal storage diseases, Neurodegeneration, Nanomedicine, Nanotechnology.

INTRODUCTION

Lysosomal Storage Diseases (LSDs) are a heterogeneous group of more than 50 inherited metabolic disorders characterized by the absence or deficiency of a functional lysosomal enzyme or lysosomal component implicated in the degradation and recycling of macromolecules, or due to errors in enzyme trafficking/targeting or defective function of non-enzymatic lysosomal proteins [1]. Any of these defects prevents the complete breakdown and recycling of target macromolecules that consequently accumulate inside the lysosome. Such accumulation of undegraded compounds alters lysosomal function resulting in a progressive and systemic disease process commonly affecting multiple organs and tissues including the central nervous system (CNS). These conditions, each year, affect the lives of numerous children worldwide with an overall prevalence of about 1 every 5000 live birth affected by LSDs, but these data are probably underestimated because LSDs heterogeneous phenotypes make the diagnosis complicated [2]. In fact, for reasons still not well understood, age of onset and clinical manifestations may vary widely among patients affected by a given LSD, and significant phenotypic heterogeneity between family members carrying identical mutations has been reported. Commonly LSDs clinical phenotypes range from classical severe forms to very attenuated ones with limited disability. Main classic symptomatology includes organomegaly (mainly hepatosplenomegaly), connective-tissue and ocular pathology, musculoskeletal abnormalities, coarsening of hair and facial features and, in the neuronopathic forms, CNS pathology [3]. Typically, in the absence of a family history of the disease, pregnancy develops in an uneventful natural manner, and the infant appears normal at delivery and develops normally during the first year of life. Nevertheless, progressive lysosomal accumulation of stored undegraded product starts very early in life and usually begins to affect normal neurological development by the first - second year of life. The fundamental neuro-developmental steps of childhood are not attained as expected and mental retardation becomes recognized by both parents and medical staff. It has been estimated that more than 70% of LSD affected children suffer from different grades of CNS involvement with various grades of neurodegeneration and CNS cell death depending on their disease phenotype [4]. Generally affected children manifest a progressive deterioration of movement, skills, speech and cognition. Communicating hydrocephalus and progressive profound mental retardation are the major CNS features and are frequently responsible for the demise of these children in early childhood with devastating consequences on their immediate environment and relatively high costs for society [1].

The science of the last decade has provided more tools, discoveries and scientific insights to develop novel LSDs treatments but the translation of ideas into drugs that are available is a greater challenge as the drug development system was not designed for these complex and heterogeneous diseases, most of which have never before been studied. For long time LSDs have not been considered a public health priority by the pharmaceutical industry, since, because of their low-prevalence, the market was usually seen as unprofitable. Before 1987, no treatments were in fact available and therapeutic management essentially consisted of simple symptomatic care of disease manifestations, with no possibility for cure. The introduction across the globe of special Orphan Drug legislations, providing incentives for pharmaceutical companies to develop and market needed medicinal products to treat rare diseases, marked important milestones for LSDs patients. Since 1983, year in which the first Orphan Drug Act was signed, these Regulations have been, and continue to be, an important force in driving treatment innovation for rare diseases mostly stimulating the research toward the clinical development of drugs for rare diseases, including LSDs [5]. In addition, recent advances in molecular biology and biochemistry have allowed a very thorough knowledge of the basic genetic mechanisms responsible of LSDs pathology, further contributing a rise in bio-company attention to LSDs [6]. Rare disease research has particularly exploded in recent decades with the development of several new therapeutic strategies capable of modifying a disease's natural history and improving a patient's quality of life [7]. A number of different approaches for treating the LSDs already exist and are commercially available. They include: (i) Substrate reduction therapies (SRT) consisting of the administration of a drug that inhibits an early stage in the degradation pathway and reduces the production of the accumulating substrate and (ii) Small molecules, named chaperones, which play an essential role in the regulation of protein conformation states. Chaperones bind to the active site of the defective and misfolded enzyme and induce its proper conformational folding, stabilizing it and preventing its degradation and restoring enzymic activity, thus ensuring the proper intracellular trafficking and delivery of the functional enzyme to the lysosomal compartment and last but not least, (iii) enzyme replacement therapy (ERT), consisting in the replacement of the defective enzyme by the regular intravenous infusions of the functional enzyme [7, 8]. ERT is considered a key treatment which has already proved to be safe and effective for peripheral manifestations in patients with Gaucher disease (GD), Fabry disease, mucopolysaccharidosis (MPS) types I, II, and VI, and Pompe disease (PD) where it reduces the lysosomal substrate load positively modifying the natural course of the disease as confirmed by the many extensive clinical trials (see www.ClinicalTrials.gov). However, there are still several obstacles that have to be overcome for the achievement of successful ERT since benefits are only evident when irreversible organ damage has not occurred underscoring the importance of early intervention with ERT. Unfortunately, the currently available replacement enzymes are not capable of being effectively delivered to all the affected organs and in particular, they do not cross the blood-brain barrier (BBB) and are thus unavailable to the brain.

Consequently, currently, only the peripheral disease can be treated with ERT and storage product continues to accumulate in CNS resulting in severe neurodegeneration and premature death in childhood for all neurologically affected patients.

Hence, there is an urgent need to direct efforts to overcome the BBB developing new therapeutic strategies, which are capable of effectively targeting the drug to the brain compartment and at the same time are able of overcoming several of the many other issues associated with the administration of protein drugs and therapeutic peptides. Compared with the administration of conventional pharmaceutical compounds, the use of therapeutically manufactured peptides poses some extra concerns related to their high molecular weight and their rather short half-life in plasma attributable to opsonisation processes [9]. Protein delivery can be significantly improved by nanotechnology offering promising strategies to overcome several of the current impediments and drawbacks related to the use of proteins as therapeutic agents, including exposure to proteases, high potential for immunogenicity, and also a decreased penetration within tissues and cells in the body [10]. This chapter aims to review the most recent advancement in nanomedicine and nanotechnology and will present novel therapeutic approaches designed to address neuronopathic LSDs.

NANOTECHNOLOGY AND NANOMEDICINE

The term nanotechnology commonly refers to the designing, characterisation, fabrication and use of molecularly precise structures, devices and systems by controlling shape and dimension at the nanometre scale (typically 0.1 mm or smaller) [11]. In the recent past, nanotechnology was a field of research largely dominated by physicists and material scientists. Today it encompasses a broader area in which physicists, engineers, chemists, biochemists, biologists and physicians work in concert fusing their expertise. In particular, the increasing attention in the field of nanotechnology medical applications has contributed to the advent of a new discipline called nanomedicine. As stated by the European Science Foundation in its report, nanomedicine consists in the application of nanotechnology to medicine using molecular tools and molecular knowledge of the human body aiming at preserving and improving human health [12, 13]. In the last years, there has been an enormous interest in the application of nanotechnology for more efficient therapeutic drug delivery systems. Significant progress in the area of novel drug delivery systems and their related medical applications in cancer and neurological disorders have already been made. Above all, the use of designed nanosized drug delivery systems with targeting properties, and in particular of nanoparticles (NPs) for drug delivery to the brain, offers great hope for the treatment of several neurodegenerative disorders and in particular LSDs. Since these disorders are caused by well-documented monogenic mutations responsible of a single lysosomal enzyme dysfunction, it is plausible that a single perfect therapy can cure these neurodegenerative diseases. LSDs therefore constitute the ideal environment in which further investigate the relationship between lysosomal dysfunction-related neurodegeneration and the development of new therapies able to cross the BBB and reach the brain compartment thus overcoming some of the main limits and impediments of the ERT.

BLOOD BRAIN BARRIER AND DRUG DELIVERY

In recent years, scientific progress in nanotechnology has led to the advent of revolutionary approaches potentially capable of solving the unmet problem of increasing the transport of drugs across the BBB. The BBB, being the tightest endothelium in the body, constitutes the major obstacle to drug delivery. This unique membranous barrier is composed of a complex system of endothelial cells, astroglia, pericytes, and perivascular mast cells. Thanks to its characteristic anatomical structure, the BBB tightly regulates the transport of drugs and other solutes between the blood and the brain accurately controlling the internal brain compartment. Brain capillaries are, in fact, anatomically different from peripheral systemic capillaries, as they are composed of a continuous endothelium, which is not fenestrated, and which exhibits a relatively low endocytic activity. In addition, the presence of tight junctions between the endothelial cells lining the brain microvessels effectively seals the paracellular aqueous diffusional pathways and prevents polar solutes and proteins from passively diffusing in and out of the CNS, forcing any molecular traffic to be predominantly across the cell (transcellular) [14]. In other words, the properties of the luminal (blood facing) and abluminal (brain facing) membranes of the cerebral endothelial cells determine the properties of the BBB.

Transport Properties of the Blood-Brain Barrier

Passive Diffusion

Lipid soluble molecules can enter the cell membrane and diffuse passively across the endothelium into brain. A correlation between increased lipid solubility and the level of penetration into brain exists. Blood gases and a number of drugs, for example anaesthetics and heroin enter the brain in this fashion [8, 14].

Solute Carriers

Polar molecules may be transported across the endothelial cell membrane by solute carriers (SLC transporters), embedded in the membrane, of which more than 360 transporters have been cloned. These transporters may simply facilitate, *via* the energy of binding producing conformational change, bi-directional passage of solute across the cell membrane (depending on the concentration gradient), or be secondary transporters relying on, for example, an existing ion gradient to co-transport in the same direction, or to exchange a solute for a particular ion across the cell membrane. Nutrients like glucose, nucleosides and amino acids are taken up by cells in this manner as are a number of drugs which serendipitously "fit" and have affinity for an existing transporter, for example L-DOPA which is carried into brain by a large neutral amino acid transporter [8, 14].

Efflux Transporters

ATP-Binding cassette transporters (ABC transporters) transport their substrates out of cells and the lipid core of the cell membrane and have a wide affinity for solutes with varied properties and structures. The favoured substrates are usually large, lipid soluble molecules with a number of nitrogen and oxygen atoms in their structure. These ABC transporters power their activity by hydrolysing ATP and so are primarily energy-driven and can thus transport against a concentration gradient. They thus reduce the brain entry of a large number of lipid soluble solutes, which might otherwise passively diffuse into the brain and at the same time they limit the entry of many potentially useful drugs into the brain. P-glycoprotein (Pgp:ABCB1) and Breast Cancer-Related Protein (BCRP:ABCG2) are the principal ABC efflux transporters in the BBB. A number of cytotoxic drugs are substrates, which confounds the treatment of brain tumours and brain metastases [15].

Receptor-Mediated Transcytosis

Large molecules like peptides and proteins may bind to a receptor on the endothelial cell membrane and induce an endocytic event. The protein is moved across the endothelial cell by vesicular transport and is exocytosed on the brain side of the barrier. Cerebral endothelial cells are more selective in the variety of receptors that they express than peripheral endothelia, for example, they do not endocytose albumin, hence the observed reduction in the number of endocytic profiles seen in the BBB. Proteins such as transferrin, which carries iron into the brain, and lipoprotein receptors (LRP receptors), are able to induce receptor-mediated transcytosis (RMT) of these large molecules [7, 16].

Diapedesis of Mononuclear Leukocytes

Circulating leukocytes bind to cell adhesion molecules on the cerebral endothelial cells and then invasively enter the endothelial cell in a process that can best be called macrocytosis. They consequently then leave the endothelial cell on the brain side. Once in the brain they transform and become microglia, the main immune competent cells of the CNS. Hence, there is always a slow but continuous turnover of the microglial cells during life. This turnover of microglia may be significant in some approaches to treatment for LSDs, which involve bone marrow transplantation, as the transplanted hematogenous cells are capable of manufacturing active enzyme. Bone marrow transplantation does not appear to halt neurodegeneration in many LSDs and this may be because the entry of leucocytes into the brain is slow and the generation of active enzyme in the CNS is thus inadequate to prevent storage. Only in MPSI it does seem to be effective and then it must be performed at a very early stage of the disease. In inflammatory conditions of the brain, leucocytes appear to also move additionally through the tight junctions and infiltrate at a faster rate [17].

In the last decades, research efforts have been particularly focused at overcoming the BBB and developing new therapeutic strategies capable of successfully targeting the drug to the brain compartment. Many strategies have been explored including both invasive and non invasive approaches. The invasive ones include, for instance, osmotic opening of the BBB, the intraventricular or intrathecal injection of the drug and the use of intracerebral polymeric transplants. Among the non invasive ones, efforts have been directed toward the exploitation of chemical modifications that make the drug suitable for the physiological carrier-mediated transport; the development of engineered "Trojan horse" technology and the linking of BBB-impermeant pharmaceutical to molecules capable to overcome the barrier by means of receptor-mediated transport systems [18, 19]. However, brain drug delivery using nanosized drug delivery systems constitutes a new intriguing tool that holds great promises to be the more effective delivery systems, improving product efficacy and safety, as well as patient convenience and compliance.

NANOSIZED DRUG DELIVERY SYSTEMS

One of the main challenges of today's pharmacological research consists of the exploitation of attractive strategies for an effective delivery of drugs to the desired site of action, in particular to the CNS, and most important to specific target areas of the CNS. In this sense, encouraging promises come from the use of nanosized carriers as drug delivery vehicles. The term nanosized drug delivery systems, also referred as nanocarriers or nanoparticles (NPs), commonly includes a wide variety of drug delivery platforms, mainly colloid carriers, like liposomes, polymeric NPs, solid lipid NPs, polymeric micelles, dendrimers etc. that protecting loaded drugs from being metabolized and assure a controlled release of their cargo [20]. Research efforts are particularly directed at increasing the ability of NPs to effectively target the therapeutic site minimizing the doses of drugs released at undesired sites [21]. Although this constitutes one of the biggest challenges that still partially limits the successful commercial use of NPs, nanosized drug delivery system approaches hold great promise to transform medicine and raise significant hopes not just for LSD patients, but for all those suffering from neurological problems. The possibility of boosting therapeutic drugs level in the target tissues, particularly the CNS where delivery takes place without damaging the barrier, while reducing its relative concentration in the other parts of the body offers, in fact, enormous benefits including a substantial decrease of drug toxicity and side effects and an increase of a drugs efficacy and bioavailability [20]. Thanks to their subcellular size that ranges from 1 to 100 nm these NPs can easily overcome the vascular endothelium and enter into tissues. However, their most attractive feature is represented by their capacity to load and carry other active compounds maintaining their stability and efficacy. NPs in fact, thanks to their quite large functional area, are able to encapsulate, bind, adsorb and carry diverse types of compounds (ranging from high to low molecular weight, hydrophilic, lipophilic and genetic material) [20]. The fact that they can be conjugated to several ligands (including antibodies, proteins, or aptamers) for targeting specific tissues and further enhancing their binding affinity, and the availability of several routes of administration exist (including oral, inhalational, and parenteral) makes NPs particularly attractive for medical applications [22]. Several nanosized drug delivery systems have been applied to different body systems, but only those pertinent to the crossing of the BBB for the treatment of LSDs neurological pathology will be described in this chapter.

LIPIDIC FORMULATIONS FOR DELIVERY OF LYSOSOMAL ENZYMES

Lipid-based micelles (encapsulating vesicles formed by a single lipid layer that usually range from 5 to 20 nm in diameter ~) [23] and in particular liposomes (vesicles composed by one or more amphiphilic bilayer resembling plasma membranes, with sizes from 50 nm to several micrometers) [24] constitute the first generation of nanosized drug delivery systems [10]. Conventionally they consist of artificially-prepared spherical vesicular lipid bilayers composed of biocompatible and biodegradable lipids, delimiting an internal aqueous compartment [19]. Considering the large amounts of drug or therapeutic that can be encapsulated into or absorbed onto these vectors, liposomes potentially represent a more efficient system for enzyme delivery. Thanks to a liposomes natural features a hydrophilic drug can be entrapped inside their aqueous core, or anchored on the liposome surface while hydrophobic molecules can be embedded within the membrane [25]. Since liver cells are the natural target for liposomes, initial attempts used liposome-immobilized enzymes for treating diseases localized in the liver cells. Data from mice experiments using liposomes containing β-Glucuronidase demonstrated a longer longevity in the blood (the enzyme stayed active for more than a week) that allows their fast accumulation in the pathological areas, in particular into the lysosomes of liver cells [26]. The possibility of using liposome entrapped enzymes as opened new horizons for LSDs treatment, hence biodistribution investigations using liposomes made of phosphaticlylcholine, phosphatidic acid and cholesterol containing alpha mannosidase [27] and neuraminidases [28] were performed leading in late 1970's to the first attempt to treat a patient with type 1 glycogenesis (Pompe's disease, PD) [29]. PD is LSD caused by amyloglucosidase deficiency and characterized by accumulation of its substrate, glycogen, resulting in muscular hypotonia, weakness, and finally death. Researchers performed daily intravenous infusion of liposomes containing amyloglucosidase for just a week since, unfortunately, the patients then died. Biopsy results showed that the enlarged liver significantly decreased in size during therapy and presented half of the expected glycogen level. Nevertheless, the level of glycogen storage in the other tissues appeared to be unaltered by the ERT indicating that targeting of liposomes containing enzyme to specific tissues, apart from liver, was only minimally successful [30]. Impressive results using liposome-entrapped human glucocerebrosidase (β-glucosidase.), were anyway achieved by treatment of a patient affected by Gaucher's disease, a LSDs caused by glucocerebrosidase deficiency producing undesired glucocerebroside accumulation, particularly in the lysosomes of the liver [31].

These preliminary data encouraged further exploitation of such intriguing strategy and gave rise to the first models for the development of therapeutic endeavours in selected LSDs. In particular it was showed that liposome entrapped β-galactosidase injected into rats *via* the tail vein can penetrate the BBB and is more efficient in reaching the lysosomes of the CNS tissue than the free enzyme [32]. Further confirmations regarding the increased efficacy in correcting the lysosomal storage using exogenous enzyme trapped in liposomes derived from experiments on murine model of globoid cell leukodystrophy, also known as Krabbe disease. Krabbe disease is a LSD caused by genetic defects in the lysosomal enzyme galactocerebrosidase characterized by severe alteration in the motor function and mental status, accompanied by hypertonicity and, often, blindness and deafness. Authors, using enzyme replacement with liposomes containing β-galactosidase obtained from Charonia lumpas, proved that the use of exogenous enzyme encapsulated in liposomes could be helpful for reducing the amount of stored compound [33]. Similarly, studies *ex vivo* with cultured neurons and *in vivo* into the brain of mice, showed that intravenously administered dioleoyl-phosphatidylserine (DOPS) liposomes containing saposin C could cross the BBB ameliorating the neurological condition caused by saposin C deficiency. This enzyme deficiency, in fact induces a rare neurological variant form of Gaucher disease, characterized by the accumulation of glucosylceramide and other lipids in the lysosomes of monocyte/macrophage system clinically resulting in enlargement of both the liver and the spleen, anemia, bone crisis, respiratory problems and neurological involvement [34]. These studies opened new horizons not just for the treatment of the variant form of Gaucher disease but also for new therapeutic strategies to protect, preserve and regenerate neurons. Indeed saposin C, derives from a precursor protein, so-called prosaposin, which is capable of exerting several beneficial effects on neurons, including neurite outgrowth stimulation, neuron preservation, and nerve regeneration enhancement [34].

Recently, attempts to improve lysosome-targeted drug delivery system based on liposomes have been performed, taking into account that the incorporation on the liposome surface of targeting moieties with high binding affinity to the specific receptors in target cells can further increase their circulation time, furnishing selective targeting, and boosting their transport [10]. In particular, to obtain a specific intracellular lysosomal targeting, researchers used liposomes modified with a lysosome-specific ligand, the lysosomotropic octadecyl-rhodamine B (RhB), and loaded with therapeutic glucocerebrosidase. Results from this *in vitro* investigation clearly indicated that surface modification of liposomes with a lysosomotropic ligand significantly increases lysosomal delivery of the enzyme into Gaucher's cells, and in particular to the liver which is among the primary target of this and many other LSDs therapy [35]. Yet, most of the antibody-modified liposomes accumulate in the liver, hampering their important accumulation in target tissues, thus the same group of researchers is now planning to use PEGylated liposomes conjugated with different lysosomotropic targeting ligands to study the delivery in targets other than the liver, possibly the brain [36]. The use of liposomes for delivering therapeutic enzymes through the 'BBB seems very attractive and data showed that horseradish peroxidase packaged into liposomes is able to cross the barrier, while the native enzyme is not, as demonstrated by histochemical analysis of rat brain tissue samples [37]. Similar results, obtained in rats injected with liposomes containing fl-galactosidase purified from Aspergillus oryzae, showed that liposomes can penetrate the BBB and access the brain, indicating liposomes as effective vehicles for the delivery of exogenous enzyme to the CNS of LSDs patients [38]. Such results encouraged the exploitation of sugar residues of glycolipids to increase liposomes delivery into specific organs, the brain in particular, and later the same authors showed that among all the components examined, sulfatide was the best for delivering the liposome entrapped beta-galactosidase enzyme into the brain of rats [39]. Nevertheless, such modifications of liposomes makes them quite unstable and inclined to prompt clearance, mostly because of plasma immunoglobulins and proteins depositing on their surface, promoting uptake by reticuloendothelial macrophages [40]. New pathways coupling polymers with liposomal formulations are therefore being explored to improve these aspects [41].

POLYMERIC FORMULATIONS FOR DELIVERY OF LYSOSOMAL ENZYMES

During the past few decades polymeric NPs have fascinated many scientist and have been applied to an increasing number of areas, especially as drug delivery vehicles because of their, superior efficacy in delivering therapeutic drugs directly into the targeted site of action. Their major favourable properties include wide bioavailability and biocompatibility, easy design and preparation, better encapsulation, controlled release and less toxic properties. In particular, main features of polymeric NPs for drug delivery across the BBB should include biocompatibility, nontoxicity, nonthrombogenicity and nonimmunogenicity. Polymeric NPs are prepared from biocompatible and biodegradable polymers, with size typically ranging from 10 to 1000 nm, into which the drug is dissolved, entrapped, encapsulated or attached to the matrix [13]. Depending upon the method of preparation nanoparticles, nanospheres or nanocapsules can be obtained [42]. Chitosans, dendrimers, nanogels, poly(D,L-lactide-co-glycolide (PLGA), poly(D,L-lactide) (PLA) and polybutylcyanoacrylate (PBCA) constitute some typical examples of polymeric NPs. Polymeric NPs represent promising brain drug delivery system not only for their potential in encapsulating drugs, but also for their capacity of preserving them from excretion and degradation, and driving active agents beyond the BBB without injuring it [43]. Different mechanisms have been suggested for nanoparticle mediated drug uptake by the brain, including: i) improved retention in and adsorption onto capillaries of the BBB inducing a high concentration gradient across the BBB; ii) opening of tight junctions attributable to the presence of nanoparticles; iii) transcytosis of nanoparticles through the endothelium [44]. Unfortunately, after intravenous injection, nanoparticles are rapidly taken up by the reticuloendothelial system and distributed to its organs, mainly to the liver and spleen and, in a smaller amount, to the bone marrow [19, 45]. To further boost their functionality, the application of specific ligands to augment the specificity of drugs delivered to the CNS has recently been taken into account [43]. The nanoparticle materials that can be used for brain delivery have to be biocompatible and rapidly biodegradable in order to avoid a prolonged retention of the polymer in the brain and among all poly(lactic-co-glycolic acid) (PLGA) and poly(butyl cyanoacrylate) (PBCA) are the polymers that best meet these requirements.

Poly(lactic-coglycolic) Acid (PLGA) Nanocarriers

Biodegradable and biocompatible polymeric nanoparticles prepared from poly (D,L-lactide-co-glycolide) (PLGA), a biocompatible member of the aliphatic polyester family of biodegradable polymers, have been widely explored for prolonged and specific delivery of different agents, comprising recombinant proteins, plasmid DNA, and low molecular weight substances [46].

Thanks to their characteristic properties including biodegradability and biocompatibility, protection of drug from degradation, possibility of extended release, and the chance to alter the properties of the surface for a better and specific targeting n, PLGA NPs have received the Food and Drug Administration (FDA) and European Medicine Agency (EMA) approval in drug delivery systems for parenteral administration [46, 47]. PLGA is therefore one of the most used constituents in the preparation of polymeric NPs for therapeutic scopes [48]. In particular, the predisposition of these NPs to end up in the lysosome makes them an ideal tool for delivering drugs to lysosomes [30].

Initial attempts were therefore directed at demonstrating that the conjugation of enzyme to NPs can represent a proper alternative to classical ERT, especially increasing the efficiency of cellular internalization, and possibly reducing the immunoreaction against the recombinant enzyme. In this regard, the use of PLGA NPs has been recently taken into consideration also for the delivery of the enzyme acid alpha-glucosidase, whose deficiency causes the Glycogen storage disease type II, also known as Pompe Disease (PD). Intravenous ERT for PD exists but, although its efficacy, the therapeutic drug presents several limitations including the fact that it is expensive, requires long-term infusions and causes immunoreaction in many patients, particularly those with the classic infantile-onset form who demonstrate a diminished response to ERT and often suffer from a rapidly progressive clinical deterioration. The study therefore was aimed at investigating possibly advantages related to deliver the drug by mean of PLGA NPs such as the possibility of improving enzyme stability/integrity and hence prolonging its lifespan *in vivo* [49]. Authors speculated that polymeric NPs appears as a potentially useful technique able to improve PD treatment efficacy, mainly by virtue of the fact that when rhGAA is loaded to the PLGA-NPs, its low exposition reduces the arising of immunologic reactions. NPs optimization of the NPs formulation [49].

Research efforts are now directed at exploiting new approaches based on PLGA NPs to improve brain drug delivery, in particular by mean of their surface modification and conjugation with targeting ligands to allow spatial and temporal control of drug delivery, further enhancing the therapeutic effect and while reducing harmful adverse events [50].

Recently, PLGA NPs conjugated with an opioid glycopeptide (g7) have shown *in-vitro* and *in-vivo* ability to overcome the BBB delivering inside the CNS different types of molecules, typically unable to cross the BBB [46, 47]. These specifically engineered g7-NPs, in fact, acquired the capacity to overcome the BBB at a level of up to 10% of the injected dose [51]. *In vitro* outcomes proved that g7-NPs are capable of crossing the BBB and, thanks to the intra- and inter-cellular vesicular transport, they can target specific cells in the brain. NPs seem therefore promising tools for the treatment of neurological and neurodegenerative diseases including LSDs [47, 51, 52]. In particular confocal microscopy studies of brain sections from mouse indicated that g7-NPs is able to reach the CNS also *via* intraperitoneal, intranasal and oral administrations [53]. These results open new horizons for therapeutic applications of g7-NPs for neurological diseases [47] as suggested by the preliminary results of *in vivo* experiments on MPS I and II mouse models performed by Tosi *et al.* (http://hdl.handle.net/11380/963705). Accordingly testing the brain drug delivery capacity of enzyme-loaded g7-NPs compartment assessing the related corrective effect on the cellular pathological glycosaminoglycan deposits in MPS I (Hurler Disease) and II (Hunter Disease) is underway (http://hdl.handle.net/11380/973494).

Poly Butyl cyanoacrylate (PBCA) Nanocarriers

It is already well known that several drugs, including peptides, can be targeted to brain using poly butyl cyanoacrylate (PBCA) polymers combined with a non-ionic polysorbate-80 surfactant coating [54 - 58]. Recently research efforts have been directed at exploring the possibility of linking the human recombinant enzyme arylsulfatase B (ASB) to PBCA nanoparticles (NPs). ERT with human ASB has just recently became available for the treatment of MPS VI, a severe LSD also known as Maroteaux-Lamy syndrome, however, it presents several of the previously discussed limitations, including the fact that, being intravenously administered, it is rapidly removed from the circulation [59]. Therefore, up to date, only the direct injection of enzyme into the cerebrospinal fluid has proved to reduce lysosomal accumulation of storage material in the dura mater of MPS VI cats [60]. It is evident that the development of an effective carrier system for enhancing the brain uptake of the lysosomal enzyme would considerably advance its clinical usefulness [57]. To this aim, an original study designed to investigate the effectiveness of arylsulfatase B adsorption onto PBCA NPs as well as *in vitro* release in serum was undertaken. Results showed that with alteration of parameters like temperature, incubation time, pH, and enzyme amount, the adsorption process was stable with a maximum capacity of 67 g/mg NP at a pH of

6.3. *In vitro* release experiments confirmed that the adsorption is steadily maintained for about 60 minutes in human blood serum demonstrating that ASB-loaded PBCA NPs constitute encouraging approaches for ERT of MPS VI [57, 58]. Furthermore, the same group also evaluated the possibility of using PBCA NPs for delivering the enzyme arylsulfatase A (ASA), whose deficiency causes metachromatic leukodystrophy (MLD), a fatal LSD clinically characterized by demyelination in the peripheral and CNS. Similarly to the case of ASB, exogenous ASA is not able to penetrate the BBB because of its high molecular mass and hydrophilicity, and therefore, an effective ERT for MLD is not yet available [61].

As reported in a recent publication, researchers successfully formulated stable ASA-loaded PBCA NPs that could be easily freeze-dried and stored for more than 8 weeks. NPs maximum loading capacity was 59 microg per 1 mg of PBCA and 100% of their activity was recovered in the presence of 3% sucrose [58]. These outcomes suggest that the PBCA-based formulation of ASA is a potential candidate for the ERT of MLD and indicate that, since PBCA NPs can bound large enzymes such as ASB and ASA, it is very likely that other LSDs enzymes can also be bound in a similar way opening up new scenarios for the treatment of other diseases like Fabry, MLD, and others [57].

Polyethylene Glycol (PEG) Nanocarriers

Nanoparticle investigations have been directed to extend the plasma half-life of polymeric NPs and consequently prolong their brain retention using the polyethylene glycol (PEG) approach, which is based on the long-circulating characteristics of NPs coated or linked with PEGs. This approach has been taken into consideration since the 1980s when different techniques were suggested for the alteration of the hydrophobic particle surface, mostly consisting in the physical adsorption of PEG-containing surfactants [45]. It is known that PEG has high hydrophilicity, chain flexibility, electrical neutrality and lack of functional groups [45]. All these features allow it to interact with water molecules producing a protective hydration film that, shielding it, avoid unnecessary interactions with biological structures and activation of the complement and immune system [45]. Proof of concept of the advantages in using PEG NPs to prolong the blood circulation of the recombinant enzymes, was furnished by outcomes derived from a mouse model of LSDs characterized by storage of dextran into lysosomes. The administration of PEG-modified dextranase, in fact, reduced the accumulation of dextran in the liver more efficiently than its non-PEG counterpart, confirming that enzyme delivery was enhanced by PEG coating and its stability was prolonged [62]. Similar effect have also been obtained using PEG immunoliposomes targeted with the 8D3 anti-transferrin receptor antibody for transfecting the luciferase and alpha-galactosidase genes into the brain of rats [63].

CYCLODEXTRINS

In the last decades cyclodextrins (CDs) have revolutionized pharmacology mainly thank to their facility to form inclusion complexes with numerous compounds and the easily forming of supramolecular structures that make then particularly useful for application in drug delivery investigations [64]. CDs are a group of cyclic oligosaccharides characterized by the presence of a hydrophilic outer surface and a hydrophobic interior cavity [65]. The cyclic nature of CDs produces a doughnut-shaped molecule with a hydrophobic "hole" in the centre and a hydrophilic exterior. This enables a lipophilic molecule to inert itself into the centre of the CD thus becoming water soluble. Side groups on the cyclodextrin facing outwards can also be modified to further change the properties of the complex. Natural and modified CDs (α-CD-a ring of 6 glucose molecules, β-CD-a 7 glucose ring and γCD-an 8 glucose ring) have been studied [66], and some obtained US FDA approval for safety [64]. The α, β and γ cyclodextrins have progressively larger cavities in the centre of the molecule which enable them to accommodate larger lipophilic entities [62]. Yet, hydroxypropyl-β-cyclodextrin (HP-β-CD has been considered as an excipient and has not received FDA approval as a pharmaceutical drug product for use as a therapeutic agent [67].

Recent studies have investigated the potential benefits of using CDs as drug carrier vehicles for the treatment of Niemann-Pick type C (NPC), a lethal neurodegenerative disease usually due to defective NPC1 or NPC2 proteins leading to widespread progressive intracellular storage of cholesterol and other lipids in various tissues, mainly in the visceral organs and CNS. Patients normally manifest ataxia, deglutition problems, seizures, and progressive loss of motor and intellectual function in early childhood and premature death [67]. Available therapeutic drugs are limited and no disease modifying therapy so far exists. In the last decades, a small number of drugs have been tested in Npc1(-/-) mice showing limited benefit. A preliminary study investigated the effect of using CD as a drug delivery vehicle for allopregnanolone, which cleared cholesterol from the CNS and extended the life span of NPC1(-/-) mice when administered peripherally. It was expected that CD activity was neutral except for enhancing the delivery of allopregnalone to cells, but preliminary research uncovered evidence for cyclodextrin alone as the active agent in reversing the storage of gangliosides and cholesterol [68]. In particular, from 2007 to date extensive efforts have been devoted to test the efficacy of 2-hydroxypropyl-β-cyclodextrin (HP-β-CD) that is known to have strong cholesterol-binding capacity, in NPC animal models. Outcomes from Npc1(-/-) and Npc2(-/-) mice showed that HP-β-

CD is effective in delaying clinical disease onset, reducing intraneuronal storage and secondary markers of neurodegeneration, and significantly increasing lifespan of both Npc1(-/-) and Npc2(-/-) mice [68, 69]. Moreover, HP-β-CD can ameliorate the cholesterol transport defect by quickly favouring the release of lysosomal cholesterol for normal sterol processing within the cytosol [69, 70]. Further investigations were performed on feline models to test the efficacy of HP-β-CD *in vivo* when dosed either parenterally or directly into the brain, confirming its ability to mobilize cellular cholesterol, decrease GM2 and GM3 ganglioside storage, slow the rate of cerebellar Purkinje cell loss, reduce the clinical effects of the disease, and prolong survival of the affected animals [71, 72]. Since HP-β-CD does not cross the BBB, its efficacy in the treatment of the neurological aspects of NPC in both NPC mice and cats is surprising and alternative mechanisms remain possible [73, 74]. HP-β-CD may be able to remove cholesterol from the CNS directly across the BBB by creating "sink" conditions into blood, or the CD may load-up with a different lipophilic molecule in the circulation which is actually delivered to the CNS and contributes to reverse the storage. Nevertheless, HP-β-CD received orphan drug designation for the treatment of NPC and was authorized by FDA and EMA in 2010 and 2013, respectively and clinical trials in children with NPC have started in several countries [75]. A Phase I clinical trial to prove safety and efficacy is now ongoing (https://clinicaltrials.gov/ct2/show-/NCT01747135). Recently it was also shown that the combined effect of using cyclodextrin, allopregnanolone and miglustat can ameliorate NPC1 disease course in a mutant mouse model and in particular can reduce cerebellar neuro-degeneration [76], further confirming the importance of the therapeutic use of CDs and CD-NPs in neurodegenerative LSDs. In this perspective, their biological effects, their ability to interact with plasma membranes and their capacity to extract different lipids play a significant role at the level of the BBB [75].

On the basis of the successful results achieved with NPC1 and NPC2 therapy, the efficacy of HP-β-CDs have been assessed in animal models of other neurological LSDs characterized by cholesterol and glycosphingolipids accumulation. Surprisingly HP-β-CDs administration in mouse models of GM1 gangliosidosis and mucopolysaccharidosis type III A, both severe forms of LSDs characterized by accumulation of cholesterol and glycosphingolipids, had no measureable benefit [68]. Likewise, HP-β-CDs did not reduce motor impairment and Purkinje cell loss in a knock-out mouse model of mucolipidosis II, another progressively debilitating LSDs due to defect in N-acetylglucosamine-1-phosphotransferase [77]. These results underline the necessity to further explore the molecular mechanisms by which CDs exert their effect ameliorating cholesterol accumulation and the complex interaction interplayed at the BBB level among CDs and brain endothelial cells.

PEPTIDE TARGETED NANOPARTICLES

Conjugation among biodegradable copolymers as PLGA NPs and short peptides constitutes a system able to deliver NPs to the CNS as showed in a study aimed at assessing the ability of five short opioid-like peptides to cross the BBB. Investigation were undertaken using the *in vivo* rat brain perfusion technique and through rat femoral vein injection for a systemic administration [78]. Peptides fluorescence allows the tracking of their passage across the BBB after intravenous injection by fluorescent microscopy [79]. Outcomes proved that peptide conjugated PLGA-NPs could cross the BBB while simple PLGA NPs could not [78].

Recently efforts have been directed to further explore the effect of coupling therapeutic proteins to targeted nanocarriers for enhancing their biodistribution. In particular attention has been paid to the use of intercellular adhesion molecule-I (ICAM·l), an endothelial surface protein which is commonly up-regulated and functionally involved in inflammation, a key symptom of several lysosomal disorders, including Niemann-Pick type B (NPB) [80]. NPB is a LSD disorder due to the deficiency of the enzyme acid sphingomyelinase (ASM) which induces abnormal storage of sphingomyelin and cholesterol in several organs, in particular the lungs. NPB does not present any neurological involvement and therefore the development of efficient ERT should constitute an apparently easy goal. Unfortunately preclinical studies in ASM knockout mice revealed that, while ERT reduces the sphingomyelin amount and leads to histological and biochemical improvements in heart, liver, and spleen, amelioration of lung pathology remains an important therapeutic challenge [81, 82].

The evidence that strategies for enhancing drug delivery to specific target organs may strengthen ERT efficacy for NPB to further boosts the planning of new studies aimed at evaluating new delivery vehicles for ERT NPB and other LSDs. In this regard recombinant ASM was coupled to anti-ICAM model polystyrene and PLGA NPs and anti-ICAM nanocarriers showing that ICAM-targeted nanocarriers are an effective and useful tool capable of bypassing glycosylation and clathrin-mediated pathways and significantly enhancing enzyme' (i) binding, (ii) internalization, (iii) lysosomal delivery, and (iv) *in situ* catalytic activity (in endothelial cells and NPD patient fibroblasts) [80]. The vascular targeting, organ accumulation and effects of these nanocarriers were also tested *in vivo* in a mouse model of NPB. Results showed that lysosomal enzyme delivery aided by targeted polymer nanocarriers constitutes a valuable expedient to improve recombinant ASM delivery to vascular endothelium, enabling, in particular, an efficient pulmonary uptake probably related to the augmented levels of inflammatory elements in NPD pathology. Research effort are therefore now directed at modulating the biodistribution properties of these nanomaterials developing novel strategies for targeting multiple receptors or different epitopes on the same receptor [83]. Recent studies using apolipoprotein A and E to target 200nm human serum albumin (HSA) NPs to the CNS have shown that the particles are transported across the BBB endothelial cells by receptor-mediated transcytosis (RMT) by a mechanism involving lipoprotein receptors, probably LRP1. Within 30 minutes, the nanoparticles are present in the cell bodies of neurones and if the particles are not linked with the apolipoprotein transcytosis and accumulation in neurones does not occur [84 - 86]. Similar examination have been performed with antibody-directed targeting of polymer nanocarriers to transferrin receptor (TfR) and intercellular adhesion molecule 1 (ICAM-1) [83]. Similarly ICAM-1 targeted nanocarriers were used to deliver the enzyme acid alpha glucosidase (GAA) restoring the specific enzyme deficiency typical of PD and Fabry diseases. Outcomes confirmed the enhanced enzyme delivery of ICAM-1 targeting holding promises for an extended application in the LSDs field [87, 88].

It became then clear that this strategy might render ERT more effective and efficient enhancing the peripheral and brain targeting and allowing the use of reduced doses of enzyme to get therapeutic benefits. Further investigations are ongoing, including the exploitation of the possibility of targeting nanocarriers to multiple epitopes or receptors as in the case of the NPs dual targeting combined-with antibody anti-ICAM and anti transferrin receptor (TfR) or NPs triple targeting combined with antibody anti intercellular, platelet-endothelial, and/or vascular CAMs (ICAM-1, PECAM-1, VCAM-1) that proved to further enhance the delivery of the ASA enzyme [83, 89]. It has been shown that in cellular barriers, such as those pertinent to the BBB and the gastrointestinal epithelium, ICAM-1-targeted nanocarriers are transported across cells by a transcytosis mechanism preserving the maintenance of the barrier permeability and enhancing lysosomal enzyme delivery to peripheral organs and the brain, with improved internalization within cells, transport to lysosomes, and enhanced biochemical effects. Thus ICAM-1-targeted NPs holds considerable potential regarding the development of effective therapy for Niemann-Pick and other LSDs including the neuronopathic ones (Muro, personal communication).

LIPOSOMES AND NPS AS NON VIRAL GENE TRANSFER APPROACHES FOR LSDS

The discovery of the phenomenon of cross-correction has also important implication for the development of gene therapy opening the possibility of genetically modifying a relatively small number of cells capable of correcting several cell types at a distance, if enzyme is secreted into the blood from the genetically modified cells. According to this concept, in fact, the therapeutic delivery of a corrected copy of a defective gene into a patient's cell assures, not just a long-term supply of an otherwise deficient enzyme in the transduced cells, but and at the same time, thanks to the existence of a secretion/uptake process of the expressed protein, it provides cross-correction to neighbouring cells improving also untransduced cells. This has enormously stimulated research toward the development of new gene therapy approaches capable of transferring recombinant DNA with therapeutic function directly into the cells of specific organs, in particular the brain, arresting the related pathology. To this aim viral and non viral vectors delivery system have been investigated. Some concerns have emerged that viral vectors present some risks of insertional mutagenesis and of a strong immune response. On the other side, the application of non-viral vectors to humans has, however, been hampered by the reduced efficacy in cell delivery and the transient expression of the transgenes [90]. Liposomes and NPs offer a valuable solution for increasing efficacy of non viral gene transfer and hold promise for future therapeutic applications of these nucleic acid delivery systems. Attention has been recently paid to cationic nano-emulsions, which, thanks to their biocompatible structure and to the fact that they can easy form complexes with DNA protecting it from enzymatic degradation, constitute useful tool for the treatment of LSDs as potential systems for nucleic acid delivery [91, 92]. Liposome-mediated gene transfer for LSDs has been tested *in vitro* using, as target cells, fibroblasts derived from patients affected by Fucosidosis and Fabry disease [93]. The delivery system was composed by lipofectin and a peptide including both an integrin-targeting domain and a poly-lysine domain. The vector was also bound to plasmid DNA, containing transfer genes encoding the lysosomal enzymes alpha-L-fucosidase (LID-alpha-Fuc) or alpha-galactosidase A (LID-alpha-Gal), whose deficiency respectively causes Fucosidosis or Fabry disease [93]. Transfection in cultures of normal fibroblasts gave rise to the secretion of great quantities of functional enzyme that, being taken up by other cells allowed the recovery of 10-40% of the total activity [93]. Similarly the use of lipofectamine to transduce the beta-galactosidase gene into fibroblasts from patients affected by GM1 Gangliosidosis showed a 33 to 100- fold increase in enzyme activity [94].

Liposome-mediated gene transfer has been performed also *in vivo* in a mouse model of MPS VII by intravenous injection of liposomes conjugated to monoclonal antibody directed against the mouse transferrin receptor and containing DNA plasmids encoding the lysosomal enzyme, beta-glucuronidase (GUSB) [95]. Outcomes confirmed that the delivery of plasmid DNA encoding lysosomal enzymes is possible in many organs including the brain [95]. Moreover, results showed that liposome injection could increase the activity of GUSB enzyme in the brain achieving the therapeutic level. Thus further confirmed that brain delivery of non-viral plasmid DNA encoding lysosomal enzyme activity is possible by mean of intravenous injection of receptor-specific Trojan horse liposomes [95]. In the last year, much attention has been given to the PLGA NPs as non-viral gene transfer systems thanks to their sustained ability to deliver several different therapeutic agents.

Also, the possibility of using g7-PLGA NPs, able to cross the BBB, enhance the possibility of developing .an efficient gene therapy approach with non-viral vectors also for LSDs with neurological involvement [46]. Moreover, in the recent years, magnetic NPs have been identified as a novel main family of non-viral gene delivery agents. They consist of innovative components including a core of magnetic elements coated with biocompatible polymers capable of binding/condensing DNA for biomedical applications [96]. They particularly enhanced the transfection of neural stem cells (NSCs) (propagated as neurospheres) which constitute an important approach for regenerative neurology. Clinical trials using human fetal NSCs have been already initiated for Pelizaeus-Merzbacher disease, chronic spinal cord injury, amyotrophic lateral sclerosis, stroke and Batten's disease, showing that this methodology offer a means to augment the therapeutic potential of NSCs by safe genetic manipulation [97].

CONCLUSION

Despite the enormous progress towards exploitation and development of different approaches for effectively treating LSDs, current therapies still present several serious limitations. Among the many different techniques of therapeutic use so far developed, ERT has proved to be effective and highly beneficial since it can greatly modify or attenuate the phenotype, especially when applied before irreversible damage to organs has occurred. It has already undergone numerous clinical trials and has thus been adopted into standard clinical management for several LSDs, being the most commonly commercially available treatment for this class of disorders. It thus represents currently the most important advance in the treatment of LSDs. Nevertheless, it has many limitations mainly due to the fact that enzymes cannot cross the BBB and therefore, ERT cannot relieve CNS dysfunction and brain pathologies [59]. Enzyme biodistribution represents another main problem. In fact because of their relative large dimension, recombinant enzymes do not spontaneously cross membranes and their delivery to lysosomes requires binding to mannose-6-phosphate receptor systems [59]. Moreover, heart valves and bones are relatively resistant to ERT and thus important causes of major morbidity such as skeletal muscle pathology, cardiac valve disease and joint disease are not fully stopped or reversed by ERT, if pathological changes are already established [98]. Finally, ERT can trigger immune responses, resulting in sub-optimal treatment. Moreover, being a very expensive therapy, this limits patients access to treatment in those countries that cannot afford expensive health care [59]. Nowadays, the application of nanotechnology is being evaluated to primarily improve ERT, but also other existing therapeutic methods. Research is mainly focused at using new advances in nanotechnology for increasing selectivity and specificity of drug action, in order to deliver drugs or replacement enzymes directly to the lysosomes. Today, nanomedicine is increasingly showing its outstanding advantages for the treatment of the LSDs as well as for many other disorders. Several drug delivery systems, lipid- or polymer-based NPs, have been developed to increase drugs pharmacological and therapeutic effects. In particular, the development of newly designed peptide targeted NPs, maximizing the biodistribution and bioavailability of the therapeutic lysosomal enzyme can further potentiate the beneficial effect of ERT. Efforts are actually particularly directed at assuring the efficient delivery of each specifically designed nanocarrier, assuring its therapeutic effect in specific organs in the body, in particular the brain, over an extended period of time, increasing the stability of the infused enzymes and providing a solution to avoid immune system detection and prolong their blood circulation avoiding early clearance [99]. Although the field is still in its infancy, it has been observed that NPs can constitute promising tools for the advancement of drug delivery and hopefully, in a near future, nanotechnology will assume an essential role in brain drug delivery of human therapeutics for LSDs and many other diseases. Nevertheless, while research progresses proves that drug targeting and distribution can be effectively increased using nanocarriers, many related controversial topics needs to be clarified, including the potential risks deriving from unwanted adverse effects and the likely associated nanotoxicity. The vagueness of this last concept and the results so far achieved highlights the necessity to further explore the potential effect of nanotechnology on human health and in particular, to deeply investigate their important role as crucial tools for a more efficient brain drug delivery, especially for the patients affected by neuronopathic LSDs.

CONFLICT OF INTEREST

The authors confirm that they have no conflict of interest to declare for this publication.

ACKNOWLEDGMENTS

This chapter arises from the project Inherited NeuroMetabolic Diseases Information Network (InNerMeD-I-Network, agreement no. 2012 12 12) which has received funding from the European Union, in the framework of the Health Programme.

The sole responsibility for the content of this review lies with the authors. The Executive Agency is not responsible for any use that may be made of the information contained therein.

REFERENCES

[1]   Bellettato CM, Scarpa M. Pathophysiology of neuropathic lysosomal storage disorders. J Inherit Metab Dis 2010; 33(4): 347-62.
[http://dx.doi.org/10.1007/s10545-010-9075-9] [PMID: 20429032]

[2]   Fuller M, Meikle PJ. JJ H. Epidemiology of lysosomal storage diseases: an overview. In: Mehta A, Beck M, Sunder-Plassmann G, Eds. Fabry Disease: Perspectives from 5 Years of FOS Oxford: Oxford PharmaGenesis. 2006.

[3]   Ortolano S, Viéitez I, Navarro C, Spuch C. Treatment of lysosomal storage diseases: recent patents and future strategies. Recent Pat Endocr Metab Immune Drug Discov 2014; 8(1): 9-25.
[http://dx.doi.org/10.2174/1872214808666140115111350] [PMID: 24433521]

[4]   Platt FM, Walkley SU. Lysosomal Disorders of the Brain: Recent Advances in Molecular and Cellular Pathogenesis and Treatment. Oxford University Press 2004.
[http://dx.doi.org/10.1093/acprof:oso/9780198508786.001.0001]

[5]   Mechler K, Mountford WK, Hoffmann GF, Ries M. Pressure for drug development in lysosomal storage disorders - a quantitative analysis thirty years beyond the US orphan drug act. Orphanet J Rare Dis. 10. England 2015. p. 46.

[6]   Parenti G, Pignata C, Vajro P, Salerno M. New strategies for the treatment of lysosomal storage diseases (review). Int J Mol Med 2013; 31(1): 11-20. [review].
[PMID: 23165354]

[7]   Scarpa M, Bellettato CM, Lampe C, Begley DJ. Neuronopathic lysosomal storage disorders: Approaches to treat the central nervous system. Best Pract Res Clin Endocrinol Metab 2015; 29(2): 159-71.
[http://dx.doi.org/10.1016/j.beem.2014.12.001] [PMID: 25987170]

[8]   Begley DJ, Pontikis CC, Scarpa M. Lysosomal storage diseases and the blood-brain barrier. Curr Pharm Des 2008; 14(16): 1566-80.
[http://dx.doi.org/10.2174/138161208784705504] [PMID: 18673198]

[9]   Rangelov S, Pispas A. Polymer and Polymer-Hybrid Nanoparticles: From Synthesis to Biomedical Applications. CRC Press 2013.
[http://dx.doi.org/10.1201/b15390]

[10]   Hsu J, Muro S. Nanomedicine and Drug Delivery Strategies for Treatment of Genetic Diseases. In: Plaseska-Karanfilska DD, Ed. Human Genetic Diseases: InTech. 2011.
[http://dx.doi.org/10.5772/24366]

[11]   Union européenne. Direction générale de la r. Nanotechnologies: Principles, Applications, Implications and Hands-on Activities: Office for Official Publications of the European Communities 2012.

[12]   Saha M. Nanomedicine: promising tiny machine for the healthcare in future-a review. Oman Med J 2009; 24(4): 242-7.
[PMID: 22216376]

[13]   Nanomedicine. Strasbourg: European Science Foundation 2005.

[14]   Abbott NJ, Patabendige AA, Dolman DE, Yusof SR, Begley DJ. Structure and function of the blood-brain barrier. Neurobiol Dis 2010; 37(1): 13-25.
[http://dx.doi.org/10.1016/j.nbd.2009.07.030] [PMID: 19664713]

[15]   Begley DJ. ABC transporters and the blood-brain barrier. Curr Pharm Des 2004; 10(12): 1295-312.
[http://dx.doi.org/10.2174/1381612043384844] [PMID: 15134482]

[16]   Preston JE, Joan Abbott N, Begley DJ. Transcytosis of macromolecules at the blood-brain barrier. Adv Pharmacol 2014; 71: 147-63.
[http://dx.doi.org/10.1016/bs.apha.2014.06.001] [PMID: 25307216]

[17]   Muldoon LL, Alvarez JI, Begley DJ, *et al.* Immunologic privilege in the central nervous system and the blood-brain barrier. J Cereb Blood Flow Metab 2013; 33(1): 13-21.
[http://dx.doi.org/10.1038/jcbfm.2012.153] [PMID: 23072749]

[18]   Pardridge WM. Drug transport across the blood-brain barrier. J Cereb Blood Flow Metab 2012; 32(11): 1959-72.
[http://dx.doi.org/10.1038/jcbfm.2012.126] [PMID: 22929442]

[19]   Masserini M. Nanoparticles for brain drug delivery ISRN Biochem. 2013;2013:238428.
[http://dx.doi.org/10.1155/2013/238428]

[20]   Grabrucker A, Chhabra R, Belletti D, Forni F, Vandelli M, Ruozi B, *et al.* Nanoparticles as Blood–Brain Barrier Permeable CNS Targeted Drug Delivery Systems. In: Fricker G, Ott M, Mahringer A, Eds. The Blood Brain Barrier (BBB) Topics in Medicinal Chemistry 10: Springer Berlin Heidelberg. 2014; pp. 71-89.

[21]   Barua S, Mitragotri S. Challenges associated with Penetration of Nanoparticles across Cell and Tissue Barriers: A Review of Current Status and Future Prospects. Nano Today 2014; 9(2): 223-43.
[http://dx.doi.org/10.1016/j.nantod.2014.04.008] [PMID: 25132862]

[22]   Petkar KC, Chavhan SS, Agatonovik-Kustrin S, Sawant KK. Nanostructured materials in drug and gene delivery: a review of the state of the art. Crit Rev Ther Drug Carrier Syst 2011; 28(2): 101-64.
[http://dx.doi.org/10.1615/CritRevTherDrugCarrierSyst.v28.i2.10] [PMID: 21663574]

[23]   Salunkhe SS. Development of Lipid Based Nanoparticulate Drug Delivery Systems and Drug Carrier Complexes for Delivery to Brain. Journal of Applied Pharmaceutical Science 2015; 5(5): 110-29.
[http://dx.doi.org/10.7324/JAPS.2015.50521]

[24]   Tiwari A. Bioengineered Nanomaterials. CRC Press 2013.
[http://dx.doi.org/10.1201/b15403]

[25]   Kraft JC, Freeling JP, Wang Z, Ho RJ. Emerging research and clinical development trends of liposome and lipid nanoparticle drug delivery systems. J Pharm Sci 2014; 103(1): 29-52.
[http://dx.doi.org/10.1002/jps.23773] [PMID: 24338748]

[26]   Steger LD, Desnick RJ. Enzyme therapy. VI: Comparative *in vivo* fates and effects on lysosomal integrity of enzyme entrapped in negatively and positively charged liposomes. Biochim Biophys Acta 1977; 464(3): 530-46.
[http://dx.doi.org/10.1016/0005-2736(77)90028-1] [PMID: 836826]

[27]   Patel HM, Rayman BE. Alpha-mannosidase in zinc-deficient rats. Possibility of liposomal therapy in mannosidosis. Bioch Soc Trans 1974; pp. 1014-7.

[28]   Gregoriadis G, Ryman BE. Fate of protein-containing liposomes injected into rats. An approach to the treatment of storage diseases. Eur J Biochem 1972; 24(3): 485-91.
[http://dx.doi.org/10.1111/j.1432-1033.1972.tb19710.x] [PMID: 4500958]

[29]   Finkelstein M, Weissmann G. The introduction of enzymes into cells by means of liposomes. J Lipid Res 1978; 19(3): 289-303.
[PMID: 349106]

[30]   Panyam J, Zhou WZ, Prabha S, Sahoo SK, Labhasetwar V. Rapid endo-lysosomal escape of poly(DL-lactide-co-glycolide) nanoparticles: implications for drug and gene delivery. FASEB J 2002; 16(10): 1217-26.
[http://dx.doi.org/10.1096/fj.02-0088com] [PMID: 12153989]

[31]   Belchetz PE, Crawley JC, Braidman IP, Gregoriadis G. Treatment of Gauchers disease with liposome-entrapped glucocerebroside: beta-glucosidase. Lancet 1977; 2(8029): 116-7.
[http://dx.doi.org/10.1016/S0140-6736(77)90123-4] [PMID: 69198]

[32]   Onodera H, Takada G, Tada K, Desnick RJ. Microautoradiographic study on the tissue localization of liposome-entrapped or unentrapped 3H-labeled beta-galactosidase injected into rats. Tohoku J Exp Med 1983; 140(1): 1-13.
[http://dx.doi.org/10.1620/tjem.140.1] [PMID: 6408762]

[33]   Umezawa F, Eto Y, Tokoro T, Ito F, Maekawa K. Enzyme replacement with liposomes containing beta-galactosidase from Charonia lumpas in murine globoid cell leukodystrophy (twitcher). Biochem Biophys Res Commun 1985; 127(2): 663-7.
[http://dx.doi.org/10.1016/S0006-291X(85)80212-6] [PMID: 3919736]

[34]   Chu Z, Sun Y, Kuan CY, Grabowski GA, Qi X. Saposin C: neuronal effect and CNS delivery by liposomes. Ann N Y Acad Sci 2005; 1053: 237-46.
[http://dx.doi.org/10.1196/annals.1344.021] [PMID: 16179529]

[35]   Thekkedath R, Koshkaryev A, Torchilin VP. Lysosome-targeted octadecyl-rhodamine B-liposomes enhance lysosomal accumulation of glucocerebrosidase in Gauchers cells *in vitro.* Nanomedicine (Lond) 2013; 8(7): 1055-65.
[http://dx.doi.org/10.2217/nnm.12.138] [PMID: 23199221]

[36]   Torchilin V. Intracellular delivery of protein and peptide therapeutics. Drug Discov Today Technol 2008; 5(2-3): e95-e103.
[http://dx.doi.org/10.1016/j.ddtec.2009.01.002] [PMID: 24981097]

[37]   K Y. Incorporation of enzyme through bloodbrainbarrier into the brain by means of liposomes. Biochem Int 1980; 1: 591-.

[38]   Takada G, Onodera H, Tada K. Delivery of fungal beta-galactosidase to rat brain by means of liposomes. Tohoku J Exp Med 1982; 136(2): 219-29.
[http://dx.doi.org/10.1620/tjem.136.219] [PMID: 7071841]

[39]   Naoi M, Yagi K. Effect of glycolipids contained in liposomes on the incorporation of beta-galactosidase into specific organs. Biochem Int 1984; 9(2): 267-72.
[PMID: 6435635]

[40]   Moghimi SM, Szebeni J. Stealth liposomes and long circulating nanoparticles: critical issues in pharmacokinetics, opsonization and protein-binding properties. Prog Lipid Res 2003; 42(6): 463-78.
[http://dx.doi.org/10.1016/S0163-7827(03)00033-X] [PMID: 14559067]

[41]   Musacchio T, Torchilin VP. Recent developments in lipid-based pharmaceutical nanocarriers. Front Biosci (Landmark Ed) 2011; 16: 1388-412.
[http://dx.doi.org/10.2741/3795] [PMID: 21196238]

[42]   Grumezescu A. Nanobiomaterials in Medical Imaging: Applications of Nanobiomaterials. Elsevier Science 2016.

[43]   Tosi G, Costantino L, Ruozi B, Forni F, Vandelli MA. Polymeric nanoparticles for the drug delivery to the central nervous system. Expert Opin Drug Deliv 2008; 5(2): 155-74.
[http://dx.doi.org/10.1517/17425247.5.2.155] [PMID: 18248316]

[44]   Chen Y, Dalwadi G, Benson HA. Drug delivery across the blood-brain barrier. Curr Drug Deliv 2004; 1(4): 361-76.
[http://dx.doi.org/10.2174/1567201043334542] [PMID: 16305398]

[45]   Neha B, Ganesh B, Preeti K. Drug Delivery to The Brain Using Polymeric Nanoparticles: A Review. International Journal of Pharmaceutical and Life Sciences 2013; 2(3): 2013.
[http://dx.doi.org/10.3329/ijpls.v2i3.15457]

[46]   Tosi G, Bortot B, Ruozi B, *et al.* Potential use of polymeric nanoparticles for drug delivery across the blood-brain barrier. Curr Med Chem 2013; 20(17): 2212-25.
[http://dx.doi.org/10.2174/09298673113201700006] [PMID: 23458620]

[47]   Tosi G, Ruozi B, Belletti D, *et al.* Brain-targeted polymeric nanoparticles: *in vivo* evidence of different routes of administration in rodents. Nanomedicine (Lond) 2013; 8(9): 1373-83.
[http://dx.doi.org/10.2217/nnm.12.172] [PMID: 23565661]

[48]   Pascolo L, Bortot B, Benseny-Cases N, *et al.* Detection of PLGA-based nanoparticles at a single-cell level by synchrotron radiation FTIR spectromicroscopy and correlation with X-ray fluorescence microscopy. Int J Nanomedicine 2014; 9: 2791-801.
[PMID: 24944512]

[49]   Tancini B, Tosi G, Bortot B, *et al.* Use of Polylactide-Co-Glycolide-Nanoparticles for Lysosomal Delivery of a Therapeutic Enzyme in Glycogenosis Type II Fibroblasts. J Nanosci Nanotechnol 2015; 15(4): 2657-66.
[http://dx.doi.org/10.1166/jnn.2015.9251] [PMID: 26353478]

[50]   Swami A, Shi J, Gadde S, Votruba A, Kolishetti N, Farokhzad O. Nanoparticles for Targeted and Temporally Controlled Drug Delivery. In: Svenson S, Prud'homme RK, Eds. Multifunctional Nanoparticles for Drug Delivery Applications. Springer, US: Nanostructure Science and Technology 2012; pp. 9-29.
[http://dx.doi.org/10.1007/978-1-4614-2305-8_2]

[51]   Vilella A, Tosi G, Grabrucker AM, *et al.* Insight on the fate of CNS-targeted nanoparticles. Part I: Rab5-dependent cell-specific uptake and distribution. J Control Release 2014; 174: 195-201.
[http://dx.doi.org/10.1016/j.jconrel.2013.11.023] [PMID: 24316476]

[52]   Tosi G, Vilella A, Chhabra R, *et al.* Insight on the fate of CNS-targeted nanoparticles. Part II: Intercellular neuronal cell-to-cell transport. J Control Release 2014; 177: 96-107.
[http://dx.doi.org/10.1016/j.jconrel.2014.01.004] [PMID: 24417968]

[53]   Vilella A, Ruozi B, Belletti D, *et al.* Endocytosis of Nanomedicines: The Case of Glycopeptide Engineered PLGA Nanoparticles. Pharmaceutics 2015; 7(2): 74-89.
[http://dx.doi.org/10.3390/pharmaceutics7020074] [PMID: 26102358]

[54]   Kreuter J. Nanoparticulate systems for brain delivery of drugs. Adv Drug Deliv Rev 2001; 47(1): 65-81.
[http://dx.doi.org/10.1016/S0169-409X(00)00122-8] [PMID: 11251246]

[55]   Alyautdin RN, Kreuter J, Kharkevich DA. [Drug delivery to the brain with nanoparticles]. Eksp Klin Farmakol 2003; 66(2): 65-8. [Drug delivery to the brain with nanoparticles].
[PMID: 12962052]

[56]   Ramge P, Unger RE, Oltrogge JB, *et al.* Polysorbate-80 coating enhances uptake of polybutylcyanoacrylate (PBCA)-nanoparticles by human and bovine primary brain capillary endothelial cells. Eur J Neurosci 2000; 12(6): 1931-40.
[http://dx.doi.org/10.1046/j.1460-9568.2000.00078.x] [PMID: 10886334]

[57]   Mühlstein A, Gelperina S, Kreuter J. Development of nanoparticle-bound arylsulfatase B for enzyme replacement therapy of mucopolysaccharidosis VI. Pharmazie 2013; 68(7): 549-54.
[PMID: 23923636]

[58]   Mühlstein A, Gelperina S, Shipulo E, Maksimenko O, Kreuter J. Arylsulfatase A bound to poly(butyl cyanoacrylate) nanoparticles for enzyme replacement therapyphysicochemical evaluation. Pharmazie 2014; 69(7): 518-24.
[PMID: 25073397]

[59]   Wraith JE. Limitations of enzyme replacement therapy: current and future. J Inherit Metab Dis 2006; 29(2-3): 442-7.
[http://dx.doi.org/10.1007/s10545-006-0239-6] [PMID: 16763916]

[60]   Auclair D, Finnie J, Walkley SU, *et al.* Intrathecal recombinant human 4-sulfatase reduces accumulation of glycosaminoglycans in dura of mucopolysaccharidosis VI cats. Pediatr Res 2012; 71(1): 39-45.
[http://dx.doi.org/10.1038/pr.2011.13] [PMID: 22289849]

[61]   Beck M. Therapy for lysosomal storage disorders. IUBMB Life 2010; 62(1): 33-40.
[PMID: 20014233]

[62]   Muro S. New biotechnological and nanomedicine strategies for treatment of lysosomal storage disorders. Wiley Interdiscip Rev Nanomed Nanobiotechnol 2010; 2(2): 189-204.
[http://dx.doi.org/10.1002/wnan.73] [PMID: 20112244]

[63]   Pardridge WM. Drug and gene targeting to the brain with molecular Trojan horses. Nat Rev Drug Discov 2002; 1(2): 131-9.
[http://dx.doi.org/10.1038/nrd725] [PMID: 12120094]

[64]   Lakkakula JR, Maçedo Krause RW. A vision for cyclodextrin nanoparticles in drug delivery systems and pharmaceutical applications. Nanomedicine (Lond) 2014; 9(6): 877-94.
[http://dx.doi.org/10.2217/nnm.14.41] [PMID: 24981652]

[65]   Loftsson T, Vogensen SB, Brewster ME, Konrádsdóttir F. Effects of cyclodextrins on drug delivery through biological membranes. J Pharm Sci 2007; 96(10): 2532-46.
[http://dx.doi.org/10.1002/jps.20992] [PMID: 17630644]

[66]   Szejtli J. Introduction and General Overview of Cyclodextrin Chemistry. Chem Rev 1998; 98(5): 1743-54.
[http://dx.doi.org/10.1021/cr970022c] [PMID: 11848947]

[67]   Ottinger EA, Kao ML, Carrillo-Carrasco N, *et al.* Collaborative development of 2-hydroxypropyl-β-cyclodextrin for the treatment of Niemann-Pick type C1 disease. Curr Top Med Chem 2014; 14(3): 330-9.
[http://dx.doi.org/10.2174/1568026613666131127160118] [PMID: 24283970]

[68]   Davidson CD, Ali NF, Micsenyi MC, *et al.* Chronic cyclodextrin treatment of murine Niemann-Pick C disease ameliorates neuronal cholesterol and glycosphingolipid storage and disease progression. PLoS One 2009; 4(9): e6951.
[http://dx.doi.org/10.1371/journal.pone.0006951] [PMID: 19750228]

[69]   Liu B, Ramirez CM, Miller AM, Repa JJ, Turley SD, Dietschy JM. Cyclodextrin overcomes the transport defect in nearly every organ of NPC1 mice leading to excretion of sequestered cholesterol as bile acid. J Lipid Res 2010; 51(5): 933-44.
[http://dx.doi.org/10.1194/jlr.M000257] [PMID: 19965601]

[70]   Taylor AM, Liu B, Mari Y, Liu B, Repa JJ. Cyclodextrin mediates rapid changes in lipid balance in Npc1-/- mice without carrying cholesterol through the bloodstream. J Lipid Res 2012; 53(11): 2331-42.
[http://dx.doi.org/10.1194/jlr.M028241] [PMID: 22892156]

[71]   Ward S, ODonnell P, Fernandez S, Vite CH. 2-hydroxypropyl-beta-cyclodextrin raises hearing threshold in normal cats and in cats with Niemann-Pick type C disease. Pediatr Res 2010; 68(1): 52-6.
[http://dx.doi.org/10.1203/PDR.0b013e3181df4623] [PMID: 20357695]

[72]   Vite CH, Bagel JH, Swain GP, *et al.* Intracisternal cyclodextrin prevents cerebellar dysfunction and Purkinje cell death in feline Niemann-Pick type C1 disease. Sci Transl Med 2015; 7(276): 276ra26.
[http://dx.doi.org/10.1126/scitranslmed.3010101] [PMID: 25717099]

[73]   Pontikis CC, Davidson CD, Walkley SU, Platt FM, Begley DJ. Cyclodextrin alleviates neuronal storage of cholesterol in Niemann-Pick C disease without evidence of detectable blood-brain barrier permeability. J Inherit Metab Dis 2013; 36(3): 491-8.
[http://dx.doi.org/10.1007/s10545-012-9583-x] [PMID: 23412751]

[74]   Vite CH, Bagel JH, Swain GP, Prociuk M, Sikora TU, Stein VM, *et al.* Intracisternal cyclodextrin prevents cerebellar dysfunction and purkinje cell death in feline niemann-pick type C1 disease Science translational medicine. 2015;7(276):276ra26-ra26.
[http://dx.doi.org/10.1126/scitranslmed.3010101]

[75]   Vecsernyés M, Fenyvesi F, Bácskay I, Deli MA, Szente L, Fenyvesi É. Cyclodextrins, blood-brain barrier, and treatment of neurological diseases. Arch Med Res 2014; 45(8): 711-29.
[http://dx.doi.org/10.1016/j.arcmed.2014.11.020] [PMID: 25482528]

[76]   Maass F, Petersen J, Hovakimyan M, *et al.* Reduced cerebellar neurodegeneration after combined therapy with cyclodextrin/allopregnanolone and miglustat in NPC1: a mouse model of Niemann-Pick type C1 disease. J Neurosci Res 2015; 93(3): 433-42.
[http://dx.doi.org/10.1002/jnr.23509] [PMID: 25400034]

[77]   Paton L, Bitoun E, Kenyon J, *et al.* A novel mouse model of a patient mucolipidosis II mutation recapitulates disease pathology. J Biol Chem 2014; 289(39): 26709-21.
[http://dx.doi.org/10.1074/jbc.M114.586156] [PMID: 25107912]

[78]   Costantino L, Gandolfi F, Tosi G, Rivasi F, Vandelli MA, Forni F. Peptide-derivatized biodegradable nanoparticles able to cross the blood-brain barrier. J Control Release 2005; 108(1): 84-96.
[http://dx.doi.org/10.1016/j.jconrel.2005.07.013] [PMID: 16154222]

[79]   Jain KK. Nanobiotechnology-based strategies for crossing the blood-brain barrier. Nanomedicine (Lond) 2012; 7(8): 1225-33.
[http://dx.doi.org/10.2217/nnm.12.86] [PMID: 22931448]

[80]   Muro S, Schuchman EH, Muzykantov VR. Lysosomal enzyme delivery by ICAM-1-targeted nanocarriers bypassing glycosylation- and clathrin-dependent endocytosis. Mol Ther 2006; 13(1): 135-41.
[http://dx.doi.org/10.1016/j.ymthe.2005.07.687] [PMID: 16153895]

[81]   Garnacho C, Dhami R, Simone E, *et al.* Delivery of acid sphingomyelinase in normal and niemann-pick disease mice using intercellular adhesion molecule-1-targeted polymer nanocarriers. J Pharmacol Exp Ther 2008; 325(2): 400-8.
[http://dx.doi.org/10.1124/jpet.107.133298] [PMID: 18287213]

[82]   Miranda SR, He X, Simonaro CM, *et al.* Infusion of recombinant human acid sphingomyelinase into niemann-pick disease mice leads to visceral, but not neurological, correction of the pathophysiology. FASEB J 2000; 14(13): 1988-95.
[http://dx.doi.org/10.1096/fj.00-0014com] [PMID: 11023983]

[83]   Papademetriou IT, Garnacho C, Schuchman EH, Muro S. *In vivo* performance of polymer nanocarriers dually-targeted to epitopes of the same or different receptors. Biomaterials 2013; 34(13): 3459-66.
[http://dx.doi.org/10.1016/j.biomaterials.2013.01.069] [PMID: 23398883]

[84]   Wagner S, Zensi A, Wien SL, *et al.* Uptake mechanism of ApoE-modified nanoparticles on brain capillary endothelial cells as a blood-brain barrier model. PLoS One 2012; 7(3): e32568.
[http://dx.doi.org/10.1371/journal.pone.0032568] [PMID: 22396775]

[85]   Zensi A, Begley D, Pontikis C, *et al.* Albumin nanoparticles targeted with Apo E enter the CNS by transcytosis and are delivered to neurones. J Control Release 2009; 137(1): 78-86.
[http://dx.doi.org/10.1016/j.jconrel.2009.03.002] [PMID: 19285109]

[86]   Zensi A, Begley D, Pontikis C, *et al.* Human serum albumin nanoparticles modified with apolipoprotein A-I cross the blood-brain barrier and enter the rodent brain. J Drug Target 2010; 18(10): 842-8.
[http://dx.doi.org/10.3109/1061186X.2010.513712] [PMID: 20849354]

[87]   Hsu J, Northrup L, Bhowmick T, Muro S. Enhanced delivery of α-glucosidase for Pompe disease by ICAM-1-targeted nanocarriers: comparative performance of a strategy for three distinct lysosomal storage disorders. Nanomedicine (Lond) 2012; 8(5): 731-9.
[PMID: 21906578]

[88]   Hsu J, Serrano D, Bhowmick T, *et al.* Enhanced endothelial delivery and biochemical effects of α-galactosidase by ICAM-1-targeted nanocarriers for Fabry disease. J Control Release 2011; 149(3): 323-31.
[http://dx.doi.org/10.1016/j.jconrel.2010.10.031] [PMID: 21047542]

[89]   Papademetriou I, Tsinas Z, Hsu J, Muro S. Combination-targeting to multiple endothelial cell adhesion molecules modulates binding, endocytosis, and *in vivo* biodistribution of drug nanocarriers and their therapeutic cargoes. J Control Release 2014; 188: 87-98.
[http://dx.doi.org/10.1016/j.jconrel.2014.06.008] [PMID: 24933603]

[90]   Fraga M, Bruxel F, Lagranha VL, Teixeira HF, Matte U. Influence of phospholipid composition on cationic emulsions/DNA complexes: physicochemical properties, cytotoxicity, and transfection on Hep G2 cells. Int J Nanomedicine 2011; 6: 2213-20.
[PMID: 22114484]

[91]   Nam HY, Park JH, Kim K, Kwon IC, Jeong SY. Lipid-based emulsion system as non-viral gene carriers. Arch Pharm Res 2009; 32(5): 639-46.
[http://dx.doi.org/10.1007/s12272-009-1500-y] [PMID: 19471876]

[92]   Matte U, Baldo G, Giugliani R. Non Viral gene transfer approaches for lysosomal storage disorders, non-viral gene therapy In: X-b Yuan, Ed. Non viral gene therapy. InTech 2011.

[93]   Estruch EJ, Hart SL, Kinnon C, Winchester BG. Non-viral, integrin-mediated gene transfer into fibroblasts from patients with lysosomal storage diseases. J Gene Med 2001; 3(5): 488-97.
[http://dx.doi.org/10.1002/jgm.214] [PMID: 11601762]

[94]   Balestrin RC, Baldo G, Vieira MB, *et al.* Transient high-level expression of beta-galactosidase after transfection of fibroblasts from GM1 gangliosidosis patients with plasmid DNA. Braz J Med Biol Res 2008; 41(4): 283-8.
[http://dx.doi.org/10.1590/S0100-879X2008000400005] [PMID: 18392450]

[95]   Zhang Y, Wang Y, Boado RJ, Pardridge WM. Lysosomal enzyme replacement of the brain with intravenous non-viral gene transfer. Pharm Res 2008; 25(2): 400-6.
[http://dx.doi.org/10.1007/s11095-007-9357-6] [PMID: 17602284]

[96]   McBain SC, Yiu HH, Dobson J. Magnetic nanoparticles for gene and drug delivery. Int J Nanomedicine 2008; 3(2): 169-80.
[PMID: 18686777]

[97]   Pickard MR, Adams CF, Barraud P, Chari DM. Using magnetic nanoparticles for gene transfer to neural stem cells: stem cell propagation method influences outcomes. J Funct Biomater 2015; 6(2): 259-76.
[http://dx.doi.org/10.3390/jfb6020259] [PMID: 25918990]

[98]   Muenzer J. Early initiation of enzyme replacement therapy for the mucopolysaccharidoses. Mol Genet Metab 2014; 111(2): 63-72.
[http://dx.doi.org/10.1016/j.ymgme.2013.11.015] [PMID: 24388732]

[99]   Kateb B, Heiss JD. The Textbook of Nanoneuroscience and Nanoneurosurgery. CRC Press 2013.
[http://dx.doi.org/10.1201/b15274]

CHAPTER 6

Targeting Brain Disease in Mucopolysaccharidoses

Marika Salvalaio[1,2], **Laura Rigon**[1,2], **Francesca D'Avanzo**[1,2,3], **Elisa Legnini**[1,2], **Valeria Balmaceda Valdez**[1], **Alessandra Zanetti**[1,2] and **Rosella Tomanin**[1,2,*]

[1] *Laboratory of Diagnosis and Therapy of Lysosomal Disorders, Department of Women's and Children's Health, University of Padova, Padova, Italy*

[2] *Pediatric Research Institute "Città della Speranza", Padova, Italy*

[3] *Brains for Brain Foundation Onlus, Padova, Italy*

Abstract: Mucopolysaccharidoses (MPSs) are a group of inherited disorders due to the deficit of the lysosomal enzymes involved in the degradation of the mucopolysaccharides, which thus accumulate within different organs, taking to a heavy progressive malfunctioning. The disorders involve most of the organ-systems and in the patients affected by MPS I, II, III and VII, also the neurological compartment may be severely affected. Many therapeutic strategies have been proposed along the years, and, following the identification of the genes underlying each disorder, in the last decade some MPSs have taken advantage on the availability of the recombinant enzymes, systemically administered to the patients. Such treatment, however, has hardly shown any effects on the CNS disease, given the inability of the enzymes to efficiently cross the blood-brain barrier. Therefore, the efforts of the last years have been focused on developing new therapeutic strategies targeting this aspect. This chapter summarizes the most relevant proposed, discussing their advantages, limitations and potential applications. Treatment of the brain disease in neuronopathic MPSs, conjugated with an early diagnosis, would represent a milestone in the improvement of patients' and families' life condition.

Keywords: Brain therapy, Blood-brain barrier, BBB, BBB crossing, Brain, Enzyme Replacement Therapy, ERT, Mucopolysaccharidosis, Neurological disease.

INTRODUCTION

Mucopolysaccharidoses (MPSs) are a group of inherited metabolic diseases belonging to the wider group of the Lysosomal Storage Disorders (LSDs), of which they represent about 30% of the patients [1]. Each MPS is due to the deficit

---
* **Corresponding author Rosella Tomanin:** Laboratory of Diagnosis and Therapy of Lysosomal Disorders, Dept. of Women's and Children's Health, University of Padova, Via Giustiniani 3, 35128 Padova, Italy; Tel: +39 0498211264; Fax: +39 0498217478; E-mail: rosella.tomanin@unipd.it

of one of the eleven lysosomal hydrolases, normally degrading mucopo-lysaccharides or glycosaminoglycans (GAGs), this determining a pathological accumulation of such molecules inside cell lysosomes as well as in the extracellular matrix (Table **1**).

Table 1. Mucopolysaccharidoses classification.

Type	Eponym	MIM ID	Gene	Enzyme Name	EC number	Stored GAG
I	Hurler syndrome	#607014	IDUA	Alpha-L-iduronidase	3.2.1.76	DS, HS
	Hurler/Scheie syndrome	#607015				
	Scheie syndrome	#607016				
II	Hunter syndrome	#309900	IDS	Iduronate 2-sulfatase	3.1.6.13	DS, HS
IIIA	Sanfilippo A syndrome	#252900	SGSH	N-sulphoglucosamine sulphohydrolase	3.10.1.1	HS
IIIB	Sanfilippo B syndrome	#252920	NAGLU	Alpha-N-acetylglucosaminidase	3.2.1.50	HS
IIIC	Sanfilippo C syndrome	#252930	HGSNAT	Heparan-alpha-glucosaminide N-acetyltransferase	2.3.1.78	HS
IIID	Sanfilippo D syndrome	#252940	GNS	N-acetylglucosamine-6-sulfatase	3.1.6.14	HS
IVA	Morquio A syndrome	#253000	GALNS	N-acetylgalactosamine-6-sulfatase	3.1.6.4	C6S, KS
IVB	Morquio B syndrome	#253010	GLB1	Beta-galactosidase	3.2.1.23	KS
VI	Maroteaux-Lamy syndrome	#253200	ARSB	Arylsulfatase B	3.1.6.12	DS
VII	Sly syndrome	#253220	GUSB	Beta-glucuronidase	3.2.1.31	C4S, C6S, DS, HS
IX	Hyaluronidase deficiency	#601492	HYAL1	Hyaluronidase-1	3.2.1.35	HYAL

C4S=chondroitin-4-sulfate; C6S=chondroitin-6-sulfate; DS=dermatan sulfate; GAG=glycosaminoglycan; HS=heparan sulfate; HYAL=hyaluronan KS= keratan sulfate.

Being lysosomal hydrolases housekeeping enzymes, accumulation of GAGs affects most of the organ-systems, although each MPS is characterized by specific clinical manifestations, variously affecting liver, spleen, heart, bones, joints, eyes, ears and other organs [2, 3]. Many MPS patients also present an important, progressive neurological deficit. In particular, brain disease affects at various degree of severity MPSs type I, II, III and VII. Nevertheless, a general brain involvement including brain and spinal cord compression may be observed in all MPSs [4].

MPS treatment has been at first attempted by procedures of bone marrow transplantation and, more recently, of hematopoietic stem cell transplantation, mainly applied to Mucopolysaccharidosis type I. Following the identification and cloning of the genes underlying each disease, in the last 10-15 years many efforts have been directed to the production of the recombinant functional forms of some enzymes. Starting 2004, protocols of Enzyme Replacement Therapy (ERT) have been settled for MPS I, II, VI and, more recently, for MPS IVA [5]. A trial has recently reached phase III for MPSIIIA [6] and preliminary encouraging data derive from a pilot study for the treatment of MPS VII [7]. Some efficacy has been demonstrated for the systemic compartment involved in these disorders, although with alternative success [8]. However, the present formulations and therapeutic concentration of the recombinant lysosomal hydrolases have shown to be unable to target the brain involvement, due to their inability to efficiently cross the blood-brain barrier (BBB). Therefore, most of the efforts of the last years have been dedicated to the design of new drug formulations or to the set-up of safe and efficient delivery systems, hopefully able to render the drugs available to the neurological compartment.

The present chapter reviews several of these approaches, underlying their features, perspectives and limitations of their potential applicability to the treatment of these diseases.

LYSOSOMES AND LYSOSOMAL STORAGE DISORDERS

Lysosomes are intracellular organelles deputized to several functions within the cell compartment. They have been long considered mainly the site of cellular degradation of materials deriving from the outside as well as from the inside of the cell. Along the years these functions have been progressively widen and it is now quite clear of how many other functions these organelles are entitled. They exert a control on cellular homeostasis, they are implied in the repair of plasmatic membrane and also in cell signalling and, together with mitochondria, in the energy metabolism of the cell. For their degrading functions, lysosomes contain about 50 hydrolases able to "digest" complex macromolecules as proteins, lipids and sugars, as well as "multiple-molecule" complexes, as small organisms endocytosed by the cell, or cellular debris.

All these functions help to maintain a healthy cellular environment and a correct equilibrium between anabolic and catabolic reactions of biological macro-molecules. Lysosomal hydrolases are coded by housekeeping genes, therefore their deficit functionally involves most tissues and organs. In a few cases, non-enzymatic lysosomal or non-lysosomal proteins involved in lysosomal biogenesis are deficient [9], as lysosomal membrane proteins or proteins involved in the hydrolases trafficking. Overall, more than 50 Lysosomal Storage Disorders have been so far identified, commonly grouped either by the type of substrate accumulated, as sphingolipidoses, olygosaccharidoses, mucolipidoses, glycogenoses, lipidoses, mucopo- lysaccharidoses, or, more recently, on the molecular basis of the disease.

MUCOPOLYSACCHARIDOSES

Features and Treatment

All MPSs are rare monogenic conditions, inherited as autosomal recessive disorders, except for MPS II, which presents an X-linked transmission. As presented in Table 1, each MPS is due to the deficit of one lysosomal enzyme, which takes to a pathologic accumulation of one or more of the 6 mucopolysaccharides. Although such accumulation variously affects most of the organ-systems in all MPSs, each syndrome may be characterized by the prevalent impairment of specific tissues or organs, as bones and joints in MPSs IV and VI, liver, spleen and cardiac valves in MPS I, II and VII, in most cases also associated to a deep neurocognitive decline. In the paediatric forms, representing the vast majority, symptoms typically develop within the second year of life, often rapidly progressing in the absence of a correct diagnosis and adequate treatment.

In some MPSs no residual activity can be detected, as in the vast majority of the MPS II patients, while in others a low level of enzyme activity can be measured, as in MPS VI patients. It has been previously reported that the amount of enzyme activity required for correction may be only 1 to 10% of normal value [10, 11]. Certainly, in MPSs levels of recombinant enzymes administered for therapeutic purposes do not seem to be critical. Indeed, *in vitro* evaluations have shown that levels higher than physiological ones are apparently well tolerated by the cells [12] and levels of enzyme activity much lower than normal may be sufficient to normalize the pathological phenotype [13].

During the trafficking process of newly synthesized lysosomal enzymes, a small percentage of them escapes from the endosomal/lysosomal compartment [14] and is released from the cell into the systemic circulation. This makes the corrective enzymes available for adjacent and distant cells that can take them up primarily by endocytosis mediated by the cation-independent mannose-6-phosphate receptor (M6PR), ubiquitously present in the plasma membrane on almost every cell [15] (Fig. 1). This process, which was highlighted by the discovery in MPS fibroblasts of the phenomenon known as "cross-correction" [16], makes gene therapy, as well as cell and enzyme therapy, viable and promising options in the case of MPSs, and in most LSDs. Therefore, in the last decades not only the enzyme replacement therapy (ERT) has been widely developed and clinically applied, but also several gene and cell therapy strategies, employing different delivery systems, have been tested for these pathologies [13].

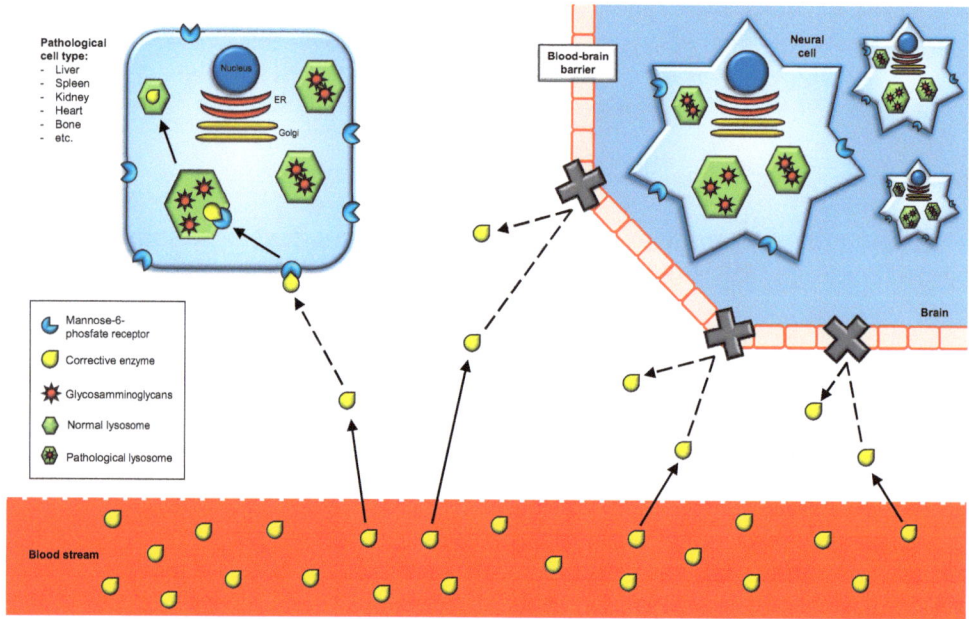

Fig. (1). Pathological cells of different tissues are able to internalize the therapeutic enzyme through the mannose 6-phosphate receptor located on the plasma membrane. The lack of this receptor on the BBB endothelial cells surface makes enzyme-based therapy unfeasible for the treatment of the neurological compartment in lysosomal disorders.

Based on severity of symptoms and progression of the disease, each MPS is generally classified in 2-3 different forms, although since many years a continuum of progressive severity is clinically recognized for these diseases. In particular, MPS I, II and VII present a systemic involvement, conjugated with a neurological progressive deficit in most patients. In MPS III the systemic involvement is rather mild, while brain disease is quite dramatic in all patients. Finally, in MPS IV and VI systemic involvement is quite heavy, especially regarding bones, but neurological involvement is mainly limited to brain and spinal cord compression.

The wide spectrum of clinical severity makes diagnosis often complicated in these patients, especially for the non-neurological forms. Not rarely the diagnosis of MPSs takes several months or years from the onset of the first symptoms, thus reducing the benefits of the available therapies [17]. This delay is caused by a difficult differential diagnosis *versus* other more common diseases and by the fact that MPSs, being rare, are not clinically well known, although an increasing knowledge has been achieved in the last few years.

The Neurological Involvement

Many MPS patients present a progressive neurological disease including mental delay and behavioural problems which have been very rarely associated with MPS IV and VI. Whereas MPS III is mainly described as a neurological disorder, MPS I, II and VII present different forms, ranging from the attenuated to the severe, based on the severity of such neurological impairment. In particular, the attenuated forms are defined based on the absence of a progressive intellectual deterioration, which is often coupled with a slower progression of the systemic symptoms.

Neurological involvement in MPSs also includes general non-cognitive neurological manifestations such as brain and spinal cord compression, present in most patients, if not all. Although not usually compromising the neurological functions, also brain and/or spinal cord compression, due to the progressive accumulation of GAGs in the anterior extradural space [18], needs to be treated, since it may even take to death in the severe cases. At present, the only available procedure with which most patients are treated is the surgical decompression, by implantation of ventricular-peritoneal shunts to relieve CSF pressure.

Pathogenesis of neurological decline in MPSs remains quite unknown. This is due to the extreme rarity of the disorders, together with the unavailability of case studies and autoptic human specimens, which could result extremely useful to the brain disease comprehension. In addition for most cases it is hard to correlate genotype findings with phenotype, and in some examples same mutations have been associated to severe as well as to mild symptoms. In MPS II, which is commonly characterized by extremely low/totally absent residual enzyme activity, same enzymatic deficit may be associated with neuronopathic or non-neuronopathic forms. All this makes the comprehension of the physiopathogenetic mechanisms very difficult. Other molecular factors, presently of unknown origin, are likely involved in the onset of brain disease. These aspects, however, cannot be investigated taking advantage of the animal models, which, being clones, cannot present an important phenotypic variety. Animal models have allowed the understanding of other neurological aspects, for examples they have shown that, in the case of the brain, not only primary deposits, represented by GAG, but also secondary deposits, as gangliosides GM2 and GM3, can be accumulated, likely contributing to the final neurological phenotype [19]. Full understanding of the brain disease would also hopefully allow the identification of involved molecular species potentially representing new therapeutic targets.

BRAIN-TARGETED THERAPEUTIC APPROACHES

Systemic High Dosage Enzyme Replacement Therapy (ERT)

Systemic administration of enzyme is usually performed intravenously; however, whichever systemic way we choose, also considering intramuscular, intraperitoneal or subcutaneous, the present formulations of the recombinant lysosomal hydrolases are unable to target the brain compartment. Lysosomal enzymes need mannose 6-phosphate receptor to be internalized into lysosomes, but this receptor is missing on the BBB endothelial cells surface. Moreover, these enzymes are too big and hydrophilic to be passively transported through the BBB. Also, it has been demonstrated that efficient BBB crossing and efficacy could be reached in mice only if treatment was performed within the second week of life [20], and this is hardly feasible in patients commonly showing their disease later on in life. Finally, at present the same dosage is commonly administered to all patients, the minimum systemic effective/safe dose, which may not be sufficient to reach the neurological compartment. Therapeutic protocols based on higher enzyme dosages have been preclinically assayed in MPSs. A study conducted in 2005 showed that in the MPS VII mouse model high dosages of the corrective enzyme, constantly administered for a prolonged period of time, could correct the brain diseased phenotype by clearing the CNS storage [21]. In 2010, a study showed that systemic infusion of young IDS knock-out mice with a slightly higher dosage of enzyme, with respect to present therapeutic concentrations, could ameliorate and apparently stop brain disease progression in this mouse model [22]. In 2014, another study performed in the MPS I mouse model has demonstrated a reduced GAG accumulation within the CNS and rescued cognitive impairment following high-dose enzyme therapy [23]. These results, however, did not lead to any future application in MPS patients. Still unclear are, in fact, the immune reactions and other side-effects raised by increased dosages of enzyme. Also from the economic point of view, a further increment of the costs sustained for these diseases by the community might become debatable. Therefore efforts need to be concentrated either on slight, targeted modifications of the presently used molecules, rendering them proficient of BBB crossing, or on the set up of proper delivery systems able to carry over molecules administered systemically in a standard dose or, again, on the evaluation of new routes of administration, allowing the direct reaching of the brain district.

Direct Administration to the Brain Compartment

Due to the extreme difficulty of the therapeutic enzyme intravenously administred in bypassing efficiently the BBB, several groups have evaluated different ways to directly target to the brain the recombinant lysosomal hydrolases, or engineered vectors delivering their cDNA sequences. These procedures include the direct intracranial injection into the brain parenchyma, the intracerebroventricular administration, the intracisternal infusion, the lumbar CSF (intrathecal) and the intranasal administration (Fig. **2**). Same ways have been sometimes used to administer either purified free proteins or recombinant viral vectors as DNA delivery systems. We will describe in this paragraph the direct brain administration of free enzymes, instead all deliveries performed by means of recombinant viral vectors will be reported in the gene therapy paragraph.

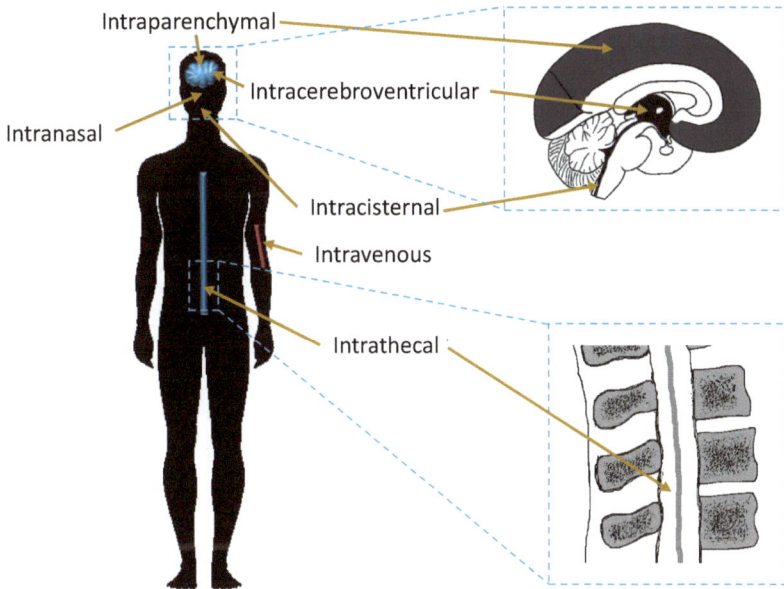

Fig (2). Schematic representation of the different routes of administration for brain-targeted therapies.

Intraparenchymal Injection

Direct administration of therapeutic molecules into the brain is usually an invasive procedure, including first of all intraparenchymal injection which, bypassing directly BBB, is thought to increase efficiency of delivery. Along with invasiveness, this procedure, however, is limited also by the extremely reduced diffusion of the drugs around the area of injection, which may be as little as 2 mm. Therefore, such procedure, which may found some applications for tumour treatment, is difficultly applied to disorders as MPSs in which the entire or most of the brain area needs to be reached by the drug to obtain therapeutic efficacy.

Direct Administration into the Cerebrospinal Fluid (CSF): ICV, IT and IC Injections

Direct administration of therapeutic proteins into the CSF includes injection in the lumbar spine (intrathecal, IT), into the lateral ventricles of the brain (intracerebrocentricular, ICV) and into the subarachnoid space (intracisterna magna, IC) (Fig. **2**).

The principle of direct injection of therapeutic enzymes into the CSF is based on the concept that these recombinant proteins are taken up by neurons and glia from the cerebrospinal fluid through the mannose-6-phosphate receptors which, on the contrary, are missing on the BBB endothelial cells [24, 25].

Since 2004 [26] many studies have been carried out on small and large animals (Table **2**). In general, these studies revealed a wide distribution of the enzymes into the brain (including parenchyma, hippocampus, thalamus, basal ganglia, periventricular white matter, caudate) with a dose-dependent penetration and a higher enzyme levels closest to the CSF [27]. A detectable enzyme activity and a significant reduction of GAG storage have been described in different cell types and in the brain, with a general improvement in neuropathology and in lifespan of MPS animal models [27].

Table 2. **Studies of direct administration into the CSF in MPS animal models.**

Disease	Organism	Administration route	Study
MPS I	Rat	ICV	[28]
MPS I	Cat	IT	[29]
MPS I	Dog	IT	[24, 26, 30 - 32]
MPS II	Mouse	ICV	[33]
MPS II	Mouse	IT	[25, 34]
MPS II	Dog	IT	[25]
MPS II	Monkey	IT	[25, 35]
MPS IIIA	Mouse	IC	[36 - 38]
MPS IIIA	Dog	IC	[39 - 41]
MPS IIIA	Dog	IT	[41]
MPS IIIA	Dog	ICV	[41]
MPS IIIA	Monkey	IT	[42]
MPS IIIB	Mouse	ICV	[43]
MPS VI	Cat	IT	[44]

IC=intracisternal. ICV=intracerebroventricular; IT=intrathecal.

Although most of these studies registered some success in reducing brain pathological phenotype and/or behavioural alterations, the invasiveness of the procedures, which need anyway to be performed at least once a month, has so far discouraged a large clinical application. In particular, the intracisternal infusion is considered quite risky for humans, given the possible mechanical damage that could be caused to areas of the brain involved in cardiac or pulmonary functions [1]. Therefore, almost all clinical trials have focused on intrathecal administration.

In MPS III, brain is the most affected compartment, being peripheral districts relatively poorly involved. Hence, the application of procedures aimed to target the brain disease in this syndrome has always been the major goal to reach and trials considering systemic administration of recombinant enzymes as ERT have never been taken into account. The only clinical trial lately conducted has been directed to the evaluation of safety and tolerability aspects related to the administration of the enzyme directly into the neurological compartment by an intrathecal drug delivery device (U.S. NIH, 2014 [6];). A similar protocol had been carried out few years before in one MPS I patient [45] and in one MPS VI patient [46] with some success. Recently, Dickson and co-workers evaluated the intrathecal injection in 5 MPS I patients for 4 months; even if they were unable to demonstrate a therapeutic efficacy, they showed that IT injection is a feasible approach to treat neurological involvement in MPS I [47]. These data have not been fully supported by Muenzer and co-workers [48], who evaluated the safety of the intrathecal delivery of a modified formulation of idursulfase, the idursulfase-IT, in 12 cognitive impaired MPS II patients, who continued to receive weekly systemic administration of idursulfase as well. The study, conducted for 6 months, registered no serious adverse events related to the procedure, although the drug delivery device had to be removed in advance in half of the patients, due to different complications including infection at the implant site. An extension trial and a phase II/III study are ongoing to evaluate the long-term outcomes.

In conclusion, a number of clinical trials using IT injection are ongoing for MPS patients (Table **3**) and, following evaluation of safety and establishment of less invasive protocols, direct administration of the therapeutic enzymes to the brain may be more widely considered.

Table 3. Summary of the clinical trials proposed for the treatment of the brain involvement in MPSs by enzyme-based therapies.

Disease	Year	Administration route	Therapy	Identifier code	Phase	Status	Country
MPS I	2005	IT	ERT	NCT00215527[a]	I	Terminated (due to slow enrolment)	USA
MPS I	2007	IT	ERT	2006-005216-27[b]	-	Ongoing	Finland

(Table 3) contd.....

Disease	Year	Administration route	Therapy	Identifier code	Phase	Status	Country
MPS I	2008	IT	ERT	NCT00786968[a]	I	Terminated (due to slow enrolment)	USA
MPS I	2008	IT	ERT	NCT00638547[a]	I	Ongoing, but not recruiting participants	USA
MPS I	2009	IT	ERT	NCT00852358[a]	-	Recruiting participants	USA
MPS II	2010	IT + IV	ERT	2010-020048-36[b]	I/II	Ongoing	UK
MPS II	2012	IT + IV	ERT	2011-000212-25[b]	-	Ongoing	UK
MPS II	2013	IT + IV	ERT	2013-002885-38[b]	-	Ongoing	UK
MPS II	2014	IT + IV	ERT	NCT02055118[a]	II/III	Recruiting participants.	USA
MPS IIIA	2010	Intrathecal Drug Delivery Device	ERT	2009-015984-15[b]	I/II	Ongoing	UK
MPS IIIA	2010	IT	ERT	2010-021348-16[b]	-	Ongoing	UK
MPS IIIA	2010	Intrathecal Drug Delivery Device	ERT	NCT01155778[a]	I/II	Completed	USA
MPS IIIA	2013	Intrathecal Drug Delivery Device	ERT	2013-003450-24[b]	IIb	Ongoing	UK
MPS IIIA	2014	Intrathecal Drug Delivery Device	ERT	2014-003960-20[b]	-	Ongoing	Spain
MPS IIIA	2014	Intrathecal Drug Delivery Device	ERT	NCT02060526[a]	II	Ongoing, but not recruiting participants	USA

[a]source: www.clinicaltrial.gov, [b]source EU clinical trial register (www.clinical- trialsregister.eu). IT= intrathecal, IV:intravenous, ERT:Enzyme replacement therapy.

Gene Therapy-Mediated by Viral Vectors

To pursue brain therapy in neuronopathic MPSs, numerous viral vectors have been preclinically tested in the last decades, exploring several ways of administration as systemic, intranasal, intraparenchymal, intratechal, intracerebroventricular (Table **4**). Efficiency of delivery varies considerably based on the way chosen and the type of vector used, which will also condition the target site reached within the neurological compartment.

Interestingly, systemic gene therapy, beside reaching the brain district, may also represent an effective delivery to the skeletal structures [49], severely involved in most MPSs.

Table 4. The table reports some examples of recombinant viral vectors administered to obtain treatment of brain disease in the MPSs indicated.

Disease	Viral Vector	Animal model	Administration route	Study
MPS I	AAV5	Mouse	IV or ICV	[50]
MPS I	AAV	Mouse	IT	[51]
MPS I	AAV8	Mouse	ICV	[52]
MPS IIIA	AAV9	Mouse	IC	[53]
MPS IIIA	LV	Mouse	ICV	[54]
MPS IIIB	LV	Mouse	IP	[55]
MPS IIIB	AAV2/AAV5 + LV	Mouse	IC AAV + IV LV	[56]
MPS VII	AAV1/AAV2	Mouse	IV	[57]
MPS VII	AAV9 or AAVrh10	Dog	IV and/or IC	[58]
MPS VII	AAV	Mouse	IT	[59, 60]
MPS VII	LV	Mouse	IV or ICV	[61]
MPS VII	LV	Mouse	IV	[62]
MPS VII	AV	Mouse	ICV	[63]
MPS VII	AV	Dog	IV	[64 - 66]
MPS VII	HSV-1	Mouse	IP	[67]

AAV=adeno-associated virus; AV=adenovirus; HSV-1= Herpes Simplex virus type 1; ICV=intracerebroventricular; IP=intraparenchymal; IT=intrathecal; IV=intravenous; LV=lentivirus.

Systemic delivery may be obtained through *ex-vivo* procedures, mainly consisting in recovery of the cells from the organism, their modification through the insertion of DNA sequences, often by means of retroviral vectors, and re-implantation into the same patient. Instead, procedures planning the use of delivery vectors which do not need any integrations into cell DNA to express the transgene, as recombinant adenoviruses or adeno-associated viruses, may be conducted by direct administration of the viral suspension by IV injection. Intranasal delivery, rarely evaluated, may provide a rapid and relatively safe access to the brain, although widespread distribution of the vector to the brain is difficult to reach. Instead, a distribution of the viral vector proximal to the rostral area may be more likely obtained. Finally, intraparenchymal, intratechal and intracerebroventricular administrations, being direct brain delivery procedures, may represent the most efficient methods.

To treat brain disease in MPSs mainly retroviral vectors and adeno-associated viral vectors have been generated and preclinically assayed. At the beginning of the viral gene therapy era, also recombinant vectors deriving from adenoviruses and from herpes simplex virus 1 have been exploited; however, these vectors have been progressively less and less considered, due to their possible pathogenicity and inflammatory potential. Very recently, a canine adenovirus vector [64 - 66] and a non-pathogenic herpes simplex virus 1 vector [67] have been investigated for MPS VII brain targeting.

Except few studies employing gamma retroviruses [68], MPSs have been more recently approached by using lentiviral vectors [54, 62, 69] or, more commonly, adeno-associated vectors (AAV) [50, 53, 58] (Table **4**). Also, most of the preclinical studies conducted relate to MPS VII, for which the murine model has been early available [70], followed by MPS IIIA and IIIB, characterized by deep neurological symptoms [53 - 56].

Sometimes two ways of administration have been contemporary assayed, as in Bielicki *et al*, 2010 where the same lentiviral vector expressing the murine β-glucuronidase has been administered either by intravenous injection, or by ventricular infusion obtaining, respectively, a reduction and a normalization of brain pathology in the MPS VII mouse model.

Finally, a combination of both an intracranial AAV vector and a systemic lentiviral vector have been administered to treat a 2-4 days old mouse model for MPS IIIB. The simultaneous treatment has obtained the best results in terms of improved motor function and hearing, also determining an increased lifespan of the animals [56].

Summarizing, while gamma retroviruses are limited by the inability to efficiently transduce non-dividing cells, as most brain cells, therefore they are mainly used in *ex-vivo* procedures, lentiviral-derived vectors can transduce dividing and non-dividing cells, including terminally differentiated cells as neurons [1]. However, both classes of vectors still present several risks including, first of all, phenomena of insertional mutagenesis into cell DNA [71]. At present, most preclinical studies and also some clinical trials are conducted by employing AAV vectors, which have shown to be able to safely and efficiently transduce a variety of cell types, for which the vector has tropism [72]. In particular, as for clinical trials employing viral vectors (Table **5**), recent studies have been directed to treat MPS IIIA and B by using AAV vectors either ICV or IV administered.

Table 5. Summary of the clinical trials proposed for the treatment of MPSs by gene therapy approaches.

Disease	Year	Administration route	Vector	Identifier code	Phase	Status	Country
MPS IIIA	2010	ICV	AAV	2010-019962-10[b]	I/II	Completed	France
MPS IIIA	2011	ICV	AAV	NCT01474343[a]	I/II	Completed	USA

(Table 5) contd.....

Disease	Year	Administration route	Vector	Identifier code	Phase	Status	Country
MPS IIIB	2013	ICV	AAV	FR-0056[c]	I/II	Ongoing	France
MPS IIIB	2014	IV	AAV	US-1289[c]	I/II	Ongoing	USA
MPS IIIA	2014	IV	AAV	US-1354[c]	I/II	Ongoing	USA

[a]source: www.clinicaltrial.gov, [b]source EU clinical trial register (www.clinical- trialsregister.eu). [c]source: Whiley database (http://www.abedia.com/wiley/). ICV=intracerebro- ventricular; IV=intravenous.

Cell Therapy Approaches

Bone Marrow Transplantation (BMT) and, more recently, Haematopoietic Stem Cell Transplantation (HSCT) would be the ideal approach to be used to target the brain deficits. It is, in fact, well-known that bone marrow and blood cells, differentiating in microglial cells, can cross BBB, although slowly, thus reaching the brain compartment. Therefore, besides being enzyme suppliers, healthy cells in themselves can represent the therapeutic agent, if supplementation consists of stem cells able to engraft and replace diseased ones.

Attempts to apply this procedure to MPSs have, however, encountered several drawbacks or failures, some of them of totally ignored origin, which have limited its application mainly to MPS I and rarely to MPS II. The elevated risk/benefit ratio, due to the still high morbidity of the transplant procedure [73], makes it worthwhile only in cases of possible success.

Given the progressive and irreversible deterioration of the brain district, due to the very poor replacement of the damaged brain cells, it is extremely important that the transplant is performed as early as possible, with in the first year of life. This way, if successful the intervention would provide a sufficient stop to cells' degeneration, allowing an almost normal intellectual development.

Hobbs and colleagues first proposed hematopoietic stem cells transplantation (HSCT) for the treatment of MPSs in 1981, as a permanent source of enzyme for Hurler syndrome.

At first bone marrow and afterwards umbilical cord blood transplantations were then attempted in several MPSs with various outcomes [74]. In fact, until supplementation of purified enzymes became available for MPSs, cell transplantation was the only way to provide patients with the correct enzyme. However, HSCT has shown to be largely ineffective in MPS II, where in addition the application evidenced a risks/benefits ratio too high to be pursued, as well as in MPS III [75] and in MPS IV [76]. Indeed, besides the difficult finding of histocompatible donors, transplantation was not successful in treating bone deformities, heavily affecting MPS IV, while brain improvements or a delayed progression of its impairment could only be obtained when applied very early in patients' life, but not for all MPSs [75]. At present the procedure is still successfully applied to MPS I [77], with good results obtained in various districts including the brain, while the inability of HSCT to reach and treat the brain compartment in MPS II and III remains unclear. A study published few years ago, evaluating cord blood stem cell transplantation in a Hunter patient, described no phenotypic and behavioural recovery following treatment. The authors suggested that the general poor success obtained by transplantation in MPS II *vs* MPS I subjects might be due to the later diagnosis of MPS II patients with respect to the MPS I, due to the slower disease progression [78].

After 30 years of HSCT in MPSs the scenario of its possible application, including efficacy and safety considerations, seems to be quite clear. Where the procedure has shown a good clinical success, namely in MPS I, it has been applied to hundreds of cases [77], due to its undeniable advantages with respect to Enzyme Replacement Therapy (ERT), including the potentiality to treat the brain disease. Also, it generally needs to be performed only once in life, thus greatly reducing the need of hospitalization for the patients, although a standardized follow-up of transplanted patients, to monitor post-HSCT complications, is strongly recommended [73].

Modified Proteins, Receptor-targeting

Conjugation with Brain-Targeting Ligands and Chemical Modification

Other strategies to improve the delivery of therapeutic enzymes to the brain include conjugation with brain-targeting ligands, with the final aim to exploit the physiological receptor-mediated transcytosis (RMT) mechanism to reach the therapeutic target. Suitable candidate ligands could be antibodies against transferrin or insulin receptor, ligands of LDL receptor-related proteins 1 and 2, melanotransferrin, etc.

Employing a monoclonal antibody against the transferrin receptor escapes competition with endogenous circulating transferrin and the interaction of antibody with the receptor seems to initiate RMT [79]. β-galactosidase (GLB1, EC 3.2.1.23), the missing enzyme in MPS IVB, has been directly coupled to an 8D3 monoclonal antibody to the transferrin receptor and successfully carried to the mouse brain after intravenous injection [80].

The successful BBB crossing and brain uptake of IDS and SGSH enzymes conjugated with a monoclonal antibody against human insulin receptor (HIRMAb) was demonstrated by Boado and colleagues by radiolabeling the fusion protein and injecting it in rhesus monkeys [81, 82]. Moreover for

HIRMAb-IDS the safety of a prolonged treatment was confirmed in the same animals [82].

Cationizing lysosomal hydrolases or attaching a cationic cell penetrating peptide may also possibly offer an opportunity to cross the BBB, by exploiting absorptive-mediated transcytosis (AMT) mechanisms. A cell penetrating peptide, the N-terminal 11–amino acid protein transduction domain (PTD) from HIV TAT protein, has been employed to deliver also large proteins as the enzyme β-galactosidase (GLB1) across the BBB of healthy mice [83]. The same peptide, attached to β-glucuronidase (GUSB) and intravenously injected in MPS VII mice, has been shown to improve the clearance of storage also in bone osteoblasts and osteoclasts, as well as in cardiac valve tissue [84]. In some cases also chemical modifications have been experimented. A form of β-glucuronidase chemically modified to inactivate terminal sugars so as to eliminate the enzyme uptake and clearance by sugar-dependent receptors, is able to cross the BBB and to reduce neuronal storage in an immunotolerant MPS VII mouse and in adult mouse [85, 86]. However, modifications of recombinant hydrolases may not be riskless, since the increased ability to cross BBB may not be conjugated with a maintenance of its ability to cross neuronal cell membrane through the natural receptors, or modifications may change somehow protein folding or conformation, thus decreasing or losing its enzymatic activity [27].

Combined Approaches

Several approaches combining different type of strategies and/or routes of administrations have also been exploited in the last years.

Wang and colleagues evaluated the efficiency of delivery to the brain of a plasmid coding for IDUA enzyme fused with a receptor-binding peptide from apolipoprotein E (apoE) which use LDLR-mediated transcytosis system for BBB crossing. Transient IDUA hepatic expression led to elevated enzyme levels in capillary-depleted brain tissues and protein delivery into non-endothelium perivascular cells, neurons, and astrocytes after 2 days of treatment. Moreover, 5 months after HSC-mediated gene transfer by using a hybrid promoter for restricted expression in maturing red blood cells, 2% to 3% of normal brain IDUA activities were obtained as well as normalization of brain glycosaminoglycans [87]. The same group evaluated the long-term CNS biodistribution, dose-correlation, and therapeutic benefits of this strategy after systemic delivery *via* HSC-mediated gene therapy, with expression restricted to erythroid/ megakaryocyte lineages evidencing an increased correction of brain pathology and behavioural deficits in MPS I mice with respect to wild-type HSCs fully engrafted in MPS I chimeras [88]. The approach evaluated by Rozaklis combines a high dosage with a chemical modification of the SGSH enzyme to reduce its carbohydrate-dependent clearance; repeated IV administration of modified SGSH to adult MPS IIIA mice significantly reduced lysosomal storage in somatic organs, but with no beneficial effects on brain [89].

Liposomes and Nanoparticles

In the last years, a great effort has been dedicated to develop biomaterials and nanotechnologies to ameliorate the distribution of therapeutic agents. Liposomes and nanoparticles (NPs) may represent valid approaches for targeted enzyme delivery as theoretically significant amounts of therapeutics can be encapsulated into or absorbed onto these carriers. The small size and/or the lipid solubility of these particles enable them to cross the BBB and facilitate the delivery beyond the barrier. The main advantage of this carrier technology is that nanoparticles disguise the characteristics of the therapeutic enzyme as for size, polarity and a possible immuno-response, all features limiting the efficiency of enzyme BBB crossing [90, 91]. Moreover, inclusion increases the drug stability, therefore prolonging the circulation timing and the bioavailability.

For example, the therapeutic enzyme may be encapsulated in a liposome that has been conjugated with polyethylenglycol (PEG) strands, constituting a stable nanocontainer for enzyme delivery, as PEG prolongs the circulation time by reducing the uptake of the liposomes by the rough endoplasmic reticulum. To permit the delivery across the BBB, the top of 1–2% of the PEG strands must be conjugated to a targeting monoclonal antibody, to form a PEGylated immunoliposome [92]. In 2008 Zhang and coworkers described the successful delivery of the cDNA for GUSB, the enzyme missing in MPS VII, to the brain of the animal model, by using a PEGylated immunoliposome. The cDNA was inserted in a plasmid and encapsulated in liposomes carrying antibodies for the transferrin receptor [93]. Results showed a quite efficient delivery of the plasmid in all districts analysed, except heart, but including the brain.

A Brazilian group has recently reported the *in vivo* testing of laronidase-functionalized multiple wall lipid-core nanocapsules describing an enhanced enzyme delivery to different organs in comparison to free enzyme, except for the brain district, where nanocapsules could not access [94].

Nanoparticles can be functionalized to favour brain targeting: apolipoprotein E attached to the surface of nanoparticles improves the transport of drugs across the BBB by interaction with lipoprotein receptors on the brain capillary endothelial cell membranes [95]; PLGA nanoparticles were decorated with simil-opioid peptides [96, 97].

Polymeric nanoparticles have been already employed to deliver small therapeutics into the brain [98] but also large ones could be incorporated into their structure [99, 100]. Our group is presently conducting a series of evaluations on the ability of PLGA nanoparticles to efficiently deliver high molecular weight molecules across the blood-brain barrier [101].

Several features may render particularly advantageous the use of nanoparticles for the delivery of lysosomal enzymes to the brain compartment and mainly their spontaneous cellular destination to the lysosomes and the possibility to modify their composition according to the need of release of the enclosed drug(s). Nanoparticles, likely crossing the BBB through a clathrin-mediated endocytosis mechanism [102], are rapidly internalized by the endosomes of the cell and afterwards by the lysosomes (Fig. **3**). Therefore, differently to what happens when nanoparticles are used to deliver other drugs to the brain compartment, as for MPSs and for most LSDs, nanoparticles, which spontaneously localize inside lysosomes, represent the ideal drug delivery system. This also prevents nanoparticles dispersion in the cytoplasm as well as their destruction. The second important point is that NPs composition, and the molecular weight of the different components can be varied, thus their opening can be somehow modulated.

However, more experiments are necessary to evaluate the impact of nanoparticles on the brain if they lodge there; the potential toxic outcomes of this need to be settled prior to their therapeutic employment in clinics for the treatment of CNS disorders, as for some MPSs [90].

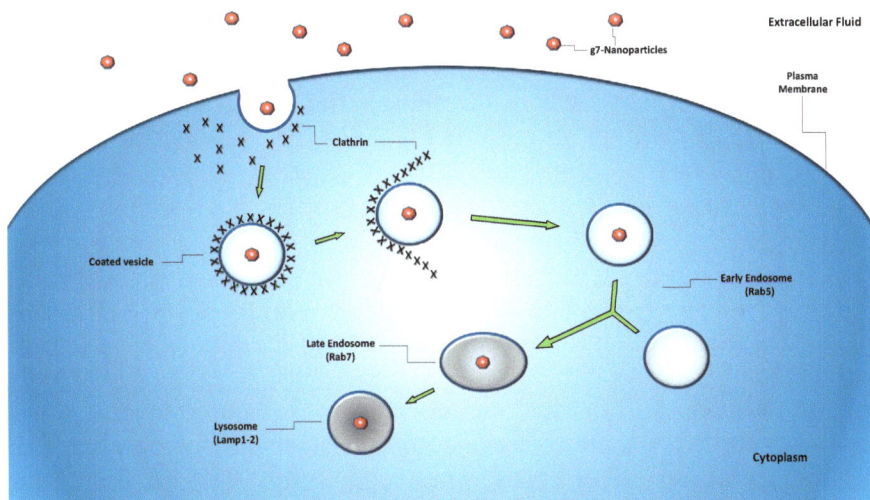

Fig. (3). Mechanism of nanoparticles' uptake. Once internalized through clathrin-mediated endocytosis, nanoparticles follow the endosomal pathway and spontaneously localize in the lysosomes.

THE NEED AND THE (NEAR) FUTURE

Brain therapy remains the real open question of MPSs. In fact, not only the neurological disease needs to find an efficacious therapy, but also the particular compartment requires a special care. Therefore, more than in other districts, safety and efficacy need to particularly combine when treating the brain compartment. The use of intrathecal or intracerebroventricular ways of delivery should be proposed after reducing the invasiveness of the procedures applied to patients already suffering from a highly invalidating disease. In a few words, the risk/benefit ratio should be importantly reduced for these procedures, thus possibly increasing patients' compliance. Ideally, our goal would be the employment of non-invasive delivery systems allowing drug administration to the CNS with no alterations or damage of the blood-brain barrier. With such systems we should achieve a satisfactory efficiency of drug transport, together with an appropriate "bio-disposal" of the material used for the delivery and in a relatively short time, thus impeding its storage.

In this respect, the generation of modified drugs able in itself to cross the BBB, would represent a further step forward, bypassing the problem of the delivery system accumulation. However, such modified molecules would also require a specific brain targeting system to be able to reach the neurological compartment in critical amount, impeding that other districts, although taking advantage of the treatment, may use most of the therapeutic molecule, thus subtracting it to the main target.

Moreover, such modified molecule should be somehow protected by a possible plasmatic degradation, which could verify rather quickly, once reached the bloodstream. Delivery systems as the nanoparticles could help drug delivery beyond the BBB, allowing brain targeting and also protection of the content from the plasmatic degradation. In addition, district-targeting would permit the use of relatively low dosages of recombinant enzymes, thus also reducing the risk of immune reaction against the drug. The reduction of such risk is now proposed in MPS I dogs and non-human primates by procedures of immunological tolerance to be performed in the neonatal phase [103], which are, however, not applicable to MPS patients, who show signs and symptoms of the disorders later on during life.

Finally, more will come from the comprehension of the pathogenetic mechanisms of brain disease in MPSs and from the complete structural comprehension of the blood-brain barrier in these diseases. Knowledge might help the identification of unexpected ways of BBB crossing as well as of new therapeutic targets to address in a near future.

CONFLICT OF INTEREST

The authors confirm that they have no conflict of interest to declare for this publication.

ACKNOWLEDGEMENT

This work was partly supported by Fondazione Cassa di Risparmio di Padova e Rovigo - Fondazione Istituto di Ricerca Pediatrica "Città della Speranza" (Bando Ricerca Pediatrica 2012/2014, grant #13/09) and by Brains for Brain Foundation Onlus.

ABBREVIATIONS

AAV = Adeno-Associated Virus

BBB = Blood-Brain Barrier

CSF = Cerebrospinal Fluid

CNS = Central Nervous System

ERT = Enzyme Replacement Therapy

GAGs = Glycosamminoglycans

HSCT = Haematopoietic Stem Cell Transplantation

IC = Intracisternal

ICV = Intracerebroventricular

IP = Intraparenchymal

IT = Intrathecal

IV = Intravenous

PEG = Polyethylenglycol

RMT = Receptor-Mediated Transcytosis

REFERENCES

[1]   Wolf DA, Banerjee S, Hackett PB, Whitley CB, McIvor RS, Low WC. Gene therapy for neurologic manifestations of mucopolysaccharidoses. Expert Opin Drug Deliv 2015; 12(2): 283-96.
       [http://dx.doi.org/10.1517/17425247.2015.966682] [PMID: 25510418]

[2]   Muenzer J. Overview of the mucopolysaccharidoses. Rheumatology (Oxford) 2011; 50 (Suppl. 5): v4-v12.
       [http://dx.doi.org/10.1093/rheumatology/ker394] [PMID: 22210669]

[3]   Neufeld EF, Muenzer J. The mucopolysaccharidoses. In: Scriver CR, Beaudet AL, Sly WS, Valle D, Childs B, Kinzler KW, Eds. The Metabolic and Molecular Bases of Inherited Disease. 8th ed. New York, NY, USA: McGraw-Hill 2001; pp. 3421-52.

[4]   Leone A, Rigante D, Amato DZ, *et al.* Spinal involvement in mucopolysaccharidoses: a review. Childs Nerv Syst 2015; 31(2): 203-12.
       [http://dx.doi.org/10.1007/s00381-014-2578-1] [PMID: 25358811]

[5]   Hendriksz CJ, Burton B, Fleming TR, *et al.* Efficacy and safety of enzyme replacement therapy with BMN 110 (elosulfase alfa) for Morquio A syndrome (mucopolysaccharidosis IVA): a phase 3 randomised placebo-controlled study. J Inherit Metab Dis 2014; 37(6): 979-90.
[http://dx.doi.org/10.1007/s10545-014-9715-6] [PMID: 24810369]

[6]   Andrade F, Aldámiz-Echevarría L, Llarena M, Couce ML. Sanfilippo syndrome: Overall review. Pediatr Int 2015; 57(3): 331-8.
[http://dx.doi.org/10.1111/ped.12636] [PMID: 25851924]

[7]   Fox JE, Volpe L, Bullaro J, Kakkis ED, Sly WS. First human treatment with investigational rhGUS enzyme replacement therapy in an advanced stage MPS VII patient. Mol Genet Metab 2015; 114(2): 203-8.
[http://dx.doi.org/10.1016/j.ymgme.2014.10.017] [PMID: 25468648]

[8]   Noh H, Lee JI. Current and potential therapeutic strategies for mucopolysaccharidoses. J Clin Pharm Ther 2014; 39(3): 215-24.
[http://dx.doi.org/10.1111/jcpt.12136] [PMID: 24612142]

[9]   Filocamo M, Morrone A. Lysosomal storage disorders: molecular basis and laboratory testing. Hum Genomics 2011; 5(3): 156-69.
[http://dx.doi.org/10.1186/1479-7364-5-3-156] [PMID: 21504867]

[10]   Cheng SH, Smith AE. Gene therapy progress and prospects: gene therapy of lysosomal storage disorders. Gene Ther 2003; 10(16): 1275-81.
[http://dx.doi.org/10.1038/sj.gt.3302092] [PMID: 12883523]

[11]   Muenzer J. The mucopolysaccharidoses: a heterogeneous group of disorders with variable pediatric presentations. J Pediatr 2004; 144(5) (Suppl.): S27-34.
[http://dx.doi.org/10.1016/j.jpeds.2004.01.052] [PMID: 15126981]

[12]   Di Francesco C, Cracco C, Tomanin R, *et al. In vitro* correction of iduronate-2-sulfatase deficiency by adenovirus-mediated gene transfer. Gene Ther 1997; 4(5): 442-8.
[http://dx.doi.org/10.1038/sj.gt.3300411] [PMID: 9274721]

[13]   Tomanin R, Zanetti A, Zaccariotto E, DAvanzo F, Bellettato CM, Scarpa M. Gene therapy approaches for lysosomal storage disorders, a good model for the treatment of mendelian diseases. Acta Paediatr 2012; 101(7): 692-701.
[http://dx.doi.org/10.1111/j.1651-2227.2012.02674.x] [PMID: 22428546]

[14]   Vladutiu GD, Rattazzi MC. Excretion-reuptake route of beta-hexosaminidase in normal and I-cell disease cultured fibroblasts. J Clin Invest 1979; 63(4): 595-601.
[http://dx.doi.org/10.1172/JCI109341] [PMID: 438323]

[15]   Sands MS, Davidson BL. Gene therapy for lysosomal storage diseases. Mol Ther 2006; 13(5): 839-49.
[http://dx.doi.org/10.1016/j.ymthe.2006.01.006] [PMID: 16545619]

[16]   Fratantoni JC, Hall CW, Neufeld EF. Hurler and Hunter syndromes: mutual correction of the defect in cultured fibroblasts. Science 1968; 162(3853): 570-2.
[http://dx.doi.org/10.1126/science.162.3853.570] [PMID: 4236721]

[17]   Marsden D, Levy H. Newborn screening of lysosomal storage disorders. Clin Chem 2010; 56(7): 1071-9.
[http://dx.doi.org/10.1373/clinchem.2009.141622] [PMID: 20489136]

[18]   Hendriksz CJ. Mucopolysaccharidoses (MPS). J Inherit Metab Dis 2013; 36(2): 177-8.
[http://dx.doi.org/10.1007/s10545-013-9596-0] [PMID: 23412752]

[19]   Wilkinson FL, Holley RJ, Langford-Smith KJ, *et al.* Neuropathology in mouse models of mucopolysaccharidosis type I, IIIA and IIIB. PLoS One 2012; 7(4): e35787.
[http://dx.doi.org/10.1371/journal.pone.0035787] [PMID: 22558223]

[20]   Vogler C, Levy B, Galvin N, Lessard M, Soper B, Barker J. Early onset of lysosomal storage disease in a murine model of mucopolysaccharidosis type VII: undegraded substrate accumulates in many tissues in the fetus and very young MPS VII mouse. Pediatr Dev Pathol 2005; 8(4): 453-62.
[http://dx.doi.org/10.1007/s10024-005-0025-8] [PMID: 16222480]

[21]   Vogler C, Levy B, Grubb JH, *et al.* Overcoming the blood-brain barrier with high-dose enzyme replacement therapy in murine mucopolysaccharidosis VII. Proc Natl Acad Sci USA 2005; 102(41): 14777-82.
[http://dx.doi.org/10.1073/pnas.0506892102] [PMID: 16162667]

[22]   Polito VA, Abbondante S, Polishchuk RS, Nusco E, Salvia R, Cosma MP. Correction of CNS defects in the MPSII mouse model *via* systemic enzyme replacement therapy. Hum Mol Genet 2010; 19(24): 4871-85.
[http://dx.doi.org/10.1093/hmg/ddq420] [PMID: 20876612]

[23]   Ou L, Herzog T, Koniar BL, Gunther R, Whitley CB. High-dose enzyme replacement therapy in murine Hurler syndrome. Mol Genet Metab 2014; 111(2): 116-22.
[http://dx.doi.org/10.1016/j.ymgme.2013.09.008] [PMID: 24100243]

[24]   Dickson P, McEntee M, Vogler C, *et al.* Intrathecal enzyme replacement therapy: successful treatment of brain disease *via* the cerebrospinal fluid. Mol Genet Metab 2007; 91(1): 61-8.
[http://dx.doi.org/10.1016/j.ymgme.2006.12.012] [PMID: 17321776]

[25]   Calias P, Papisov M, Pan J, *et al.* CNS penetration of intrathecal-lumbar idursulfase in the monkey, dog and mouse: implications for neurological outcomes of lysosomal storage disorder. PLoS One 2012; 7(1): e30341.
[http://dx.doi.org/10.1371/journal.pone.0030341] [PMID: 22279584]

[26]   Kakkis E, McEntee M, Vogler C, *et al.* Intrathecal enzyme replacement therapy reduces lysosomal storage in the brain and meninges of the canine model of MPS I. Mol Genet Metab 2004; 83(1-2): 163-74.
[http://dx.doi.org/10.1016/j.ymgme.2004.07.003] [PMID: 15464431]

[27]   Calias P, Banks WA, Begley D, Scarpa M, Dickson P. Intrathecal delivery of protein therapeutics to the brain: a critical reassessment. Pharmacol Ther 2014; 144(2): 114-22.
[http://dx.doi.org/10.1016/j.pharmthera.2014.05.009] [PMID: 24854599]

[28]   Belichenko PV, Dickson PI, Passage M, Jungles S, Mobley WC, Kakkis ED. Penetration, diffusion, and uptake of recombinant human alpha-L-iduronidase after intraventricular injection into the rat brain. Mol Genet Metab 2005; 86(1-2): 141-9.
[http://dx.doi.org/10.1016/j.ymgme.2005.04.013] [PMID: 16006167]

[29]   Vite CH, Wang P, Patel RT, *et al.* Biodistribution and pharmacodynamics of recombinant human alpha-L-iduronidase (rhIDU) in mucopolysaccharidosis type I-affected cats following multiple intrathecal administrations. Mol Genet Metab 2011; 103(3): 268-74.
[http://dx.doi.org/10.1016/j.ymgme.2011.03.011] [PMID: 21482164]

[30]   Dickson PI, Hanson S, McEntee MF, *et al.* Early *versus* late treatment of spinal cord compression with long-term intrathecal enzyme replacement therapy in canine mucopolysaccharidosis type I. Mol Genet Metab 2010; 101(2-3): 115-22.
[http://dx.doi.org/10.1016/j.ymgme.2010.06.020] [PMID: 20655780]

[31]   Chen A, Vogler C, McEntee M, *et al.* Glycosaminoglycan storage in neuroanatomical regions of mucopolysaccharidosis I dogs following intrathecal recombinant human iduronidase. APMIS 2011; 119(8): 513-21.
[http://dx.doi.org/10.1111/j.1600-0463.2011.02760.x] [PMID: 21749451]

[32]   Dickson PI, Ellinwood NM, Brown JR, *et al.* Specific antibody titer alters the effectiveness of intrathecal enzyme replacement therapy in canine mucopolysaccharidosis I. Mol Genet Metab 2012; 106(1): 68-72.
[http://dx.doi.org/10.1016/j.ymgme.2012.02.003] [PMID: 22402327]

[33]   Higuchi T, Shimizu H, Fukuda T, *et al.* Enzyme replacement therapy (ERT) procedure for mucopolysaccharidosis type II (MPS II) by intraventricular administration (IVA) in murine MPS II. Mol Genet Metab 2012; 107(1-2): 122-8.
        [http://dx.doi.org/10.1016/j.ymgme.2012.05.005] [PMID: 22704483]

[34]   Sohn YB, Lee J, Cho SY, *et al.* Improvement of CNS defects *via* continuous intrathecal enzyme replacement by osmotic pump in mucopolysaccharidosis type II mice. Am J Med Genet A 2013; 161A(5): 1036-43.
        [http://dx.doi.org/10.1002/ajmg.a.35869] [PMID: 23529876]

[35]   Felice BR, Wright TL, Boyd RB, *et al.* Safety evaluation of chronic intrathecal administration of idursulfase-IT in cynomolgus monkeys. Toxicol Pathol 2011; 39(5): 879-92.
        [http://dx.doi.org/10.1177/0192623311409595] [PMID: 21628718]

[36]   Hemsley KM, King B, Hopwood JJ. Injection of recombinant human sulfamidase into the CSF *via* the cerebellomedullary cistern in MPS IIIA mice. Mol Genet Metab 2007; 90(3): 313-28.
        [http://dx.doi.org/10.1016/j.ymgme.2006.10.005] [PMID: 17166757]

[37]   Mader KM, Beard H, King BM, Hopwood JJ. Effect of high dose, repeated intra-cerebrospinal fluid injection of sulphamidase on neuropathology in mucopolysaccharidosis type IIIA mice. Genes Brain Behav 2008; 7(7): 740-53.
        [http://dx.doi.org/10.1111/j.1601-183X.2008.00413.x] [PMID: 18518922]

[38]   Hemsley KM, Hopwood JJ. Delivery of recombinant proteins *via* the cerebrospinal fluid as a therapy option for neurodegenerative lysosomal storage diseases. Int J Clin Pharmacol Ther 2009; 47 (Suppl. 1): S118-23.
        [PMID: 20040322]

[39]   Hemsley KM, Norman EJ, Crawley AC, *et al.* Effect of cisternal sulfamidase delivery in MPS IIIA Huntaway dogsa proof of principle study. Mol Genet Metab 2009; 98(4): 383-92.
        [http://dx.doi.org/10.1016/j.ymgme.2009.07.013] [PMID: 19699666]

[40]   Crawley AC, Marshall N, Beard H, *et al.* Enzyme replacement reduces neuropathology in MPS IIIA dogs. Neurobiol Dis 2011; 43(2): 422-34.
        [http://dx.doi.org/10.1016/j.nbd.2011.04.014] [PMID: 21550404]

[41]   Marshall NR, Hassiotis S, King B, *et al.* Delivery of therapeutic protein for prevention of neurodegenerative changes: comparison of different CSF-delivery methods. Exp Neurol 2015; 263: 79-90.
        [http://dx.doi.org/10.1016/j.expneurol.2014.09.008] [PMID: 25246230]

[42]   Pfeifer RW, Felice BR, Boyd RB, *et al.* Safety evaluation of chronic intrathecal administration of heparan N-sulfatase in juvenile cynomolgus monkeys. Drug Deliv Transl Res 2012; 2(3): 187-200.
        [http://dx.doi.org/10.1007/s13346-011-0043-1] [PMID: 25786866]

[43]   Kan SH, Aoyagi-Scharber M, Le SQ, *et al.* Delivery of an enzyme-IGFII fusion protein to the mouse brain is therapeutic for mucopolysaccharidosis type IIIB. Proc Natl Acad Sci USA 2014; 111(41): 14870-5.
        [http://dx.doi.org/10.1073/pnas.1416660111] [PMID: 25267636]

[44]   Auclair D, Finnie J, Walkley SU, *et al.* Intrathecal recombinant human 4-sulfatase reduces accumulation of glycosaminoglycans in dura of mucopolysaccharidosis VI cats. Pediatr Res 2012; 71(1): 39-45.
        [http://dx.doi.org/10.1038/pr.2011.13] [PMID: 22289849]

[45]   Munoz-Rojas MV, Vieira T, Costa R, *et al.* Intrathecal enzyme replacement therapy in a patient with mucopolysaccharidosis type I and symptomatic spinal cord compression. Am J Med Genet A 2008; 146A(19): 2538-44.
        [http://dx.doi.org/10.1002/ajmg.a.32294] [PMID: 18792977]

[46]   Muñoz-Rojas MV, Horovitz DD, Jardim LB, *et al.* Intrathecal administration of recombinant human N-acetylgalactosamine 4-sulfatase to a MPS VI patient with pachymeningitis cervicalis. Mol Genet Metab 2010; 99(4): 346-50.
[http://dx.doi.org/10.1016/j.ymgme.2009.11.008] [PMID: 20036175]

[47]   Dickson PI, Kaitila I, Harmatz P, *et al.* Safety of laronidase delivered into the spinal canal for treatment of cervical stenosis in mucopolysaccharidosis I. Mol Genet Metab 2015; 116(1-2): 69-74.
[http://dx.doi.org/10.1016/j.ymgme.2015.07.005] [PMID: 26260077]

[48]   Muenzer J, Hendriksz CJ, Fan Z, *et al.* A phase I/II study of intrathecal idursulfase-IT in children with severe mucopolysaccharidosis II. Genet Med 2015.
[PMID: 25834948]

[49]   Noh H, Lee JI. Current and potential therapeutic strategies for mucopolysaccharidoses. J Clin Pharm Ther 2014; 39(3): 215-24.
[http://dx.doi.org/10.1111/jcpt.12136] [PMID: 24612142]

[50]   Janson CG, Romanova LG, Leone P, *et al.* Comparison of Endovascular and Intraventricular Gene Therapy With Adeno-Associated Virus-α-L-Iduronidase for Hurler Disease. Neurosurgery 2014; 74(1): 99-111.
[http://dx.doi.org/10.1227/NEU.0000000000000157] [PMID: 24077583]

[51]   Watson G, Bastacky J, Belichenko P, *et al.* Intrathecal administration of AAV vectors for the treatment of lysosomal storage in the brains of MPS I mice. Gene Ther 2006; 13(11): 917-25.
[PMID: 16482204]

[52]   Wolf DA, Lenander AW, Nan Z, *et al.* Direct gene transfer to the CNS prevents emergence of neurologic disease in a murine model of mucopolysaccharidosis type I. Neurobiol Dis 2011; 43(1): 123-33.
[http://dx.doi.org/10.1016/j.nbd.2011.02.015] [PMID: 21397026]

[53]   Haurigot V, Marcó S, Ribera A, *et al.* Whole body correction of mucopolysaccharidosis IIIA by intracerebrospinal fluid gene therapy. J Clin Invest 2013; 66778.
[PMID: 23863627]

[54]   McIntyre C, Derrick-Roberts AL, Byers S, Anson DS. Correction of murine mucopolysaccharidosis type IIIA central nervous system pathology by intracerebroventricular lentiviral-mediated gene delivery. J Gene Med 2014; 16(11-12): 374-87.
[http://dx.doi.org/10.1002/jgm.2816] [PMID: 25418946]

[55]   Di Domenico C, Villani GR, Di Napoli D, *et al.* Intracranial gene delivery of LV-NAGLU vector corrects neuropathology in murine MPS IIIB. Am J Med Genet A 2009; 149A(6): 1209-18.
[http://dx.doi.org/10.1002/ajmg.a.32861] [PMID: 19449420]

[56]   Heldermon CD, Qin EY, Ohlemiller KK, *et al.* Disease correction by combined neonatal intracranial AAV and systemic lentiviral gene therapy in Sanfilippo Syndrome type B mice. Gene Ther 2013; 20(9): 913-21.
[http://dx.doi.org/10.1038/gt.2013.14] [PMID: 23535899]

[57]   Passini MA, Watson DJ, Vite CH, Landsburg DJ, Feigenbaum AL, Wolfe JH. Intraventricular brain injection of adeno-associated virus type 1 (AAV1) in neonatal mice results in complementary patterns of neuronal transduction to AAV2 and total long-term correction of storage lesions in the brains of beta-glucuronidase-deficient mice. J Virol 2003; 77(12): 7034-40.
[http://dx.doi.org/10.1128/JVI.77.12.7034-7040.2003] [PMID: 12768022]

[58]   Gurda BL, De Guilhem De Lataillade A, Bell P, *et al.* Evaluation of AAV-mediated gene therapy for central nervous system disease in canine mucopolysaccharidosis VII. Mol Ther 2015.
[PMID: 26447927]

[59] Elliger SS, Elliger CA, Aguilar CP, Raju NR, Watson GL. Elimination of lysosomal storage in brains of MPS VII mice treated by intrathecal administration of an adeno-associated virus vector. Gene Ther 1999; 6(6): 1175-8.
[http://dx.doi.org/10.1038/sj.gt.3300931] [PMID: 10455422]

[60] Elliger SS, Elliger CA, Lang C, Watson GL. Enhanced secretion and uptake of beta-glucuronidase improves adeno-associated viral-mediated gene therapy of mucopolysaccharidosis type VII mice. Mol Ther 2002; 5(5 Pt 1): 617-26.
[http://dx.doi.org/10.1006/mthe.2002.0594] [PMID: 11991753]

[61] Bielicki J, McIntyre C, Anson DS. Comparison of ventricular and intravenous lentiviral-mediated gene therapy for murine MPS VII. Mol Genet Metab 2010; 101(4): 370-82.
[http://dx.doi.org/10.1016/j.ymgme.2010.08.013] [PMID: 20864369]

[62] Derrick-Roberts AL, Pyragius CE, Kaidonis XM, Jackson MR, Anson DS, Byers S. Lentiviral-mediated gene therapy results in sustained expression of β-glucuronidase for up to 12 months in the gus(mps/mps) and up to 18 months in the gus(tm(L175F)Sly) mouse models of mucopolysaccharidosis type VII. Hum Gene Ther 2014; 25(9): 798-810.
[http://dx.doi.org/10.1089/hum.2013.141] [PMID: 25003807]

[63] Ghodsi A, Stein C, Derksen T, Martins I, Anderson RD, Davidson BL. Systemic hyperosmolality improves beta-glucuronidase distribution and pathology in murine MPS VII brain following intraventricular gene transfer. Exp Neurol 1999; 160(1): 109-16.
[http://dx.doi.org/10.1006/exnr.1999.7205] [PMID: 10630195]

[64] Ariza L, Giménez-Llort L, Cubizolle A, et al. Central nervous system delivery of helper-dependent canine adenovirus corrects neuropathology and behavior in mucopolysaccharidosis type VII mice. Hum Gene Ther 2014; 25(3): 199-211.
[http://dx.doi.org/10.1089/hum.2013.152] [PMID: 24299455]

[65] Cubizolle A, Serratrice N, Skander N, et al. Corrective GUSB transfer to the canine mucopolysaccharidosis VII brain. Mol Ther 2014; 22(4): 762-73.
[http://dx.doi.org/10.1038/mt.2013.283] [PMID: 24343103]

[66] Serratrice N, Cubizolle A, Ibanes S, et al. Corrective GUSB transfer to the canine mucopolysaccharidosis VII cornea using a helper-dependent canine adenovirus vector. J Control Release 2014; 181: 22-31.
[http://dx.doi.org/10.1016/j.jconrel.2014.02.022] [PMID: 24607662]

[67] Liu W, Griffin G, Clarke T, et al. Bilateral single-site intracerebral injection of a nonpathogenic herpes simplex virus-1 vector decreases anxiogenic behavior in MPS VII mice. Mol Ther Methods Clin Dev 2015; 2: 14059.
[http://dx.doi.org/10.1038/mtm.2014.59] [PMID: 26052529]

[68] Baldo G, Wozniak DF, Ohlemiller KK, Zhang Y, Giugliani R, Ponder KP. Retroviral-vector-mediated gene therapy to mucopolysaccharidosis I mice improves sensorimotor impairments and other behavioral deficits. J Inherit Metab Dis 2013; 36(3): 499-512.
[http://dx.doi.org/10.1007/s10545-012-9530-x] [PMID: 22983812]

[69] Baldo G, Giugliani R, Matte U. Gene delivery strategies for the treatment of mucopolysaccharidoses. Expert Opin Drug Deliv 2014; 11(3): 449-59.
[http://dx.doi.org/10.1517/17425247.2014.880689] [PMID: 24450877]

[70] Birkenmeier EH, Davisson MT, Beamer WG, et al. Murine mucopolysaccharidosis type VII. Characterization of a mouse with beta-glucuronidase deficiency. J Clin Invest 1989; 83(4): 1258-66.
[http://dx.doi.org/10.1172/JCI114010] [PMID: 2495302]

[71] Rothe M, Modlich U, Schambach A. Biosafety challenges for use of lentiviral vectors in gene therapy. Curr Gene Ther 2013; 13(6): 453-68.
[http://dx.doi.org/10.2174/15665232113136660006] [PMID: 24195603]

[72] Mingozzi F, High KA. Therapeutic *in vivo* gene transfer for genetic disease using AAV: progress and challenges. Nat Rev Genet 2011; 12(5): 341-55.
[http://dx.doi.org/10.1038/nrg2988] [PMID: 21499295]

[73] Bhatia S. Long-term health impacts of hematopoietic stem cell transplantation inform recommendations for follow-up. Expert Rev Hematol 2011; 4(4): 437-52.
[http://dx.doi.org/10.1586/ehm.11.39]

[74] Peters C, Steward CG. Hematopoietic cell transplantation for inherited metabolic diseases: an overview of outcomes and practice guidelines. Bone Marrow Transplant 2003; 31(4): 229-39.
[http://dx.doi.org/10.1038/sj.bmt.1703839] [PMID: 12621457]

[75] Klein KA, Krivit W, Whitley CB, *et al.* Poor cognitive outcome of eleven children with Sanfilippo syndrome after bone marrow transplantation and successful engraftment. Bone Marrow Transplant 1995; 15 (Suppl. 1): S176-81.

[76] Krivit W, Sung JH, Lockman L, Shapiro E. Bone marrow transplantation for the treatment of lysosomal and peroxisomal diseases: focus on central nervous system reconstitution. In: Rich RR, Fleisher T, Schwartz BD, Shearer W, Strober W, Eds. Principles of Clinical Immunology St Louis. MO, USA: Mosby 1995; pp. 1852-64.

[77] Aldenhoven M, Wynn RF, Orchard PJ, *et al.* Long-term outcome of Hurler syndrome patients after hematopoietic cell transplantation: an international multicenter study. Blood 2015; 125(13): 2164-72.
[http://dx.doi.org/10.1182/blood-2014-11-608075] [PMID: 25624320]

[78] Araya K, Sakai N, Mohri I, *et al.* Localized donor cells in brain of a Hunter disease patient after cord blood stem cell transplantation. Mol Genet Metab 2009; 98(3): 255-63.
[http://dx.doi.org/10.1016/j.ymgme.2009.05.006] [PMID: 19556155]

[79] Bickel U, Kang YS, Yoshikawa T, Pardridge WM. *In vivo* demonstration of subcellular localization of anti-transferrin receptor monoclonal antibody-colloidal gold conjugate in brain capillary endothelium. J Histochem Cytochem 1994; 42(11): 1493-7.
[http://dx.doi.org/10.1177/42.11.7930531] [PMID: 7930531]

[80] Zhang Y, Pardridge WM. Delivery of beta-galactosidase to mouse brain *via* the blood-brain barrier transferrin receptor. J Pharmacol Exp Ther 2005; 313(3): 1075-81.
[http://dx.doi.org/10.1124/jpet.104.082974] [PMID: 15718287]

[81] Boado RJ, Hui EK, Lu JZ, Sumbria RK, Pardridge WM. Blood-brain barrier molecular trojan horse enables imaging of brain uptake of radioiodinated recombinant protein in the rhesus monkey. Bioconjug Chem 2013; 24(10): 1741-9.
[http://dx.doi.org/10.1021/bc400319d] [PMID: 24059813]

[82] Boado RJ, Ka-Wai Hui E, Zhiqiang Lu J, Pardridge WM. Insulin receptor antibody-iduronate 2-sulfatase fusion protein: pharmacokinetics, anti-drug antibody, and safety pharmacology in Rhesus monkeys. Biotechnol Bioeng 2014; 111(11): 2317-25.
[http://dx.doi.org/10.1002/bit.25289] [PMID: 24889100]

[83] Schwarze SR, Ho A, Vocero-Akbani A, Dowdy SF. *In vivo* protein transduction: delivery of a biologically active protein into the mouse. Science 1999; 285(5433): 1569-72.
[http://dx.doi.org/10.1126/science.285.5433.1569] [PMID: 10477521]

[84] Orii KO, Grubb JH, Vogler C, *et al.* Defining the pathway for Tat-mediated delivery of beta-glucuronidase in cultured cells and MPS VII mice. Mol Ther 2005; 12(2): 345-52.
[http://dx.doi.org/10.1016/j.ymthe.2005.02.031] [PMID: 16043103]

[85] Grubb JH, Vogler C, Tan Y, Shah GN, MacRae AF, Sly WS. Infused Fc-tagged beta-glucuronidase crosses the placenta and produces clearance of storage in utero in mucopolysaccharidosis VII mice. Proc Natl Acad Sci USA 2008; 105(24): 8375-80.
[http://dx.doi.org/10.1073/pnas.0803715105] [PMID: 18544647]

[86] Huynh HT, Grubb JH, Vogler C, Sly WS. Biochemical evidence for superior correction of neuronal storage by chemically modified enzyme in murine mucopolysaccharidosis VII. Proc Natl Acad Sci USA 2012; 109(42): 17022-7.
[http://dx.doi.org/10.1073/pnas.1214779109] [PMID: 23027951]

[87] Wang D, El-Amouri SS, Dai M, *et al.* Engineering a lysosomal enzyme with a derivative of receptor-binding domain of apoE enables delivery across the blood-brain barrier. Proc Natl Acad Sci USA 2013; 110(8): 2999-3004.
[http://dx.doi.org/10.1073/pnas.1222742110] [PMID: 23382178]

[88] El-Amouri SS, Dai M, Han JF, Brady RO, Pan D. Normalization and improvement of CNS deficits in mice with Hurler syndrome after long-term peripheral delivery of BBB-targeted iduronidase. Mol Ther 2014; 22(12): 2028-37.
[http://dx.doi.org/10.1038/mt.2014.152] [PMID: 25088464]

[89] Rozaklis T, Beard H, Hassiotis S, *et al.* Impact of high-dose, chemically modified sulfamidase on pathology in a murine model of MPS IIIA. Exp Neurol 2011; 230(1): 123-30.
[http://dx.doi.org/10.1016/j.expneurol.2011.04.004] [PMID: 21515264]

[90] Jain KK. Nanobiotechnology-based drug delivery to the central nervous system. Neurodegener Dis 2007; 4(4): 287-91.
[http://dx.doi.org/10.1159/000101884] [PMID: 17627131]

[91] Muro S. New biotechnological and nanomedicine strategies for treatment of lysosomal storage disorders. Wiley Interdiscip Rev Nanomed Nanobiotechnol 2010; 2(2): 189-204.
[http://dx.doi.org/10.1002/wnan.73] [PMID: 20112244]

[92] Schlachetzki F, Zhang Y, Boado RJ, Pardridge WM. Gene therapy of the brain: the trans-vascular approach. Neurology 2004; 62(8): 1275-81.
[http://dx.doi.org/10.1212/01.WNL.0000120551.38463.D9] [PMID: 15111662]

[93] Zhang Y, Wang Y, Boado RJ, Pardridge WM. Lysosomal enzyme replacement of the brain with intravenous non-viral gene transfer. Pharm Res 2008; 25(2): 400-6.
[http://dx.doi.org/10.1007/s11095-007-9357-6] [PMID: 17602284]

[94] Mayer FQ, Adorne MD, Bender EA, *et al.* Laronidase-functionalized multiple-wall lipid-core nanocapsules: promising formulation for a more effective treatment of mucopolysaccharidosis type I. Pharm Res 2015; 32(3): 941-54.
[http://dx.doi.org/10.1007/s11095-014-1508-y] [PMID: 25208876]

[95] Michaelis K, Hoffmann MM, Dreis S, *et al.* Covalent linkage of apolipoprotein e to albumin nanoparticles strongly enhances drug transport into the brain. J Pharmacol Exp Ther 2006; 317(3): 1246-53.
[http://dx.doi.org/10.1124/jpet.105.097139] [PMID: 16554356]

[96] Costantino L, Gandolfi F, Tosi G, Rivasi F, Vandelli MA, Forni F. Peptide-derivatized biodegradable nanoparticles able to cross the blood-brain barrier. J Control Release 2005; 108(1): 84-96.
[http://dx.doi.org/10.1016/j.jconrel.2005.07.013] [PMID: 16154222]

[97] Vergoni AV, Tosi G, Tacchi R, Vandelli MA, Bertolini A, Costantino L. Nanoparticles as drug delivery agents specific for CNS: *in vivo* biodistribution. Nanomedicine (Lond) 2009; 5(4): 369-77.
[PMID: 19341816]

[98] Vilella A, Ruozi B, Belletti D, *et al.* Endocytosis of Nanomedicines: The Case of Glycopeptide Engineered PLGA Nanoparticles. Pharmaceutics 2015; 7(2): 74-89.
[http://dx.doi.org/10.3390/pharmaceutics7020074] [PMID: 26102358]

[99] Kreuter J. Nanoparticulate systems for brain delivery of drugs. Adv Drug Deliv Rev 2001; 47(1): 65-81.
[http://dx.doi.org/10.1016/S0169-409X(00)00122-8] [PMID: 11251246]

[100] Kreuter J. Drug delivery to the central nervous system by polymeric nanoparticles: what do we know? Adv Drug Deliv Rev 2014; 71: 2-14.
[http://dx.doi.org/10.1016/j.addr.2013.08.008] [PMID: 23981489]

[101] Salvalaio M, Rigon L, Belletti D, *et al.* Targeted polymeric nanoparticles for brain delivery of high molecular weight molecules in lysosomal storage disorders. PLoS One 2016; 11(5): e0156452.
[http://dx.doi.org/10.1371/journal.pone.0156452] [PMID: 27228099]

[102] Vilella A, Tosi G, Grabrucker AM, *et al.* Insight on the fate of CNS-targeted nanoparticles. Part I: Rab5-dependent cell-specific uptake and distribution. J Control Release 2014; 174: 195-201.
[http://dx.doi.org/10.1016/j.jconrel.2013.11.023] [PMID: 24316476]

[103] Hinderer C, Bell P, Louboutin JP, *et al.* Neonatal Systemic AAV Induces Tolerance to CNS Gene Therapy in MPS I Dogs and Nonhuman Primates. Mol Ther 2015; 23(8): 1298-307.
[http://dx.doi.org/10.1038/mt.2015.99] [PMID: 26022732]

CHAPTER 7

Functional Validation of Drug Nanoconjugates *in vivo*

Ibane Abasolo, Yolanda Fernández and **Simó Schwartz Jr.**[*]

Functional Validation & Preclinical Research (FVPR). Drug Delivery & Targeting. CIBBIM-Nanomedicine. Vall d'Hebron Institut de Recerca (VHIR). Networking Research Center on Bioengineering, Biomaterials and Nanomedicine (CIBER-BBN). Barcelona. Spain

Abstract: Preclinical development of nanotechnology formulated-drugs shares many features with the development of other pharmaceutical products. However, there are some relevant differences. Nanoparticulated therapeutic systems have challenges related to their production, physicochemical characterization, stability and sterilization, but offer special advantages regarding drug solubilization, bioavailability and biodistribution. A good design of the nanoconjugate, should take into account these *pros* and *cons* in the specific setting of the target disease. Moreover, researchers should also bear these in mind when planning *in vitro* and *in vivo* proof-of-concept assays. In this chapter we will focus in assays required to test the efficacy of a therapeutic nanoconjugate and how appropriate animal models and imaging technologies help to speed up preclinical development. In addition, we will also describe how basic *in vivo* pharmacokinetic and biodistribution assays aid researchers to optimize the design of a highly active and non-toxic nanoconjugate.

Keywords: Animal models, *In vivo* preclinical validation, Nanomedicine, Nanoconjugate, Optical bioluminescence and fluorescence imaging, Proof-of-concept, Toxicology, Whole-body biodistribution.

INTRODUCTION

In the last two decades reformulation of drugs by means of nanotechnology has changed the way drug development is foreseen. Nanomedicine holds now the promise to revolutionize medical treatments with more potent, less toxic, and smart therapeutics that could home into disease areas like a magic bullet [1]. Indeed, the number of nanoformulated drugs entering clinical trials is growing exponentially and there are reports indicating that 80% of the pharma market will be related to nanotechnology by 2020.

[*] **Corresponding author Simó Schwartz Jr.:** Drug Delivery & Targeting. CIBBIM-Nanomedicine. Vall d'Hebron Institut de Recerca (VHIR), Barcelona, Spain; Tel: +34-934894053; Fax:+34-932746708; E-mail: simo.schwartz@vhir.og

Giovanni Tosi (Ed)

Nowadays, reformulation of drugs by nanothecnology has open a new door in the treatment of many diseases, especially those neglected by the conventional therapeutic approaches. Many neurological diseases fall into this category because large and hydrophilic drugs are not able to cross the blood brain barrier (BBB) and reach efficiently their target cells in the central nervous system (CNS). Indeed the market of nanotechnology in the area of CNS products is valued for 2016 in almost $30 billions [2].

As with any other drug, nanomedicines also need to prove their efficacy and safety *in vivo*. Once the proof-of-concept is defined and activity of a nanoconjugate is demonstrated, preclinical safety studies help to select the best candidate for development and furthermore, clinical studies. In this chapter we will describe the key steps of this pipeline, from design and manufacturing of the nanoconjugate to its characterization and biological testing.

Finding the Right Nanoconjugate for the Right Disease

Both the indication and the therapeutic agent define the type and characteristics of the nanoconjugate to be used. Intensive work is required to translate such design into a reproducible scale-up and manufacturing process to achieve a consistent product, sterile and endotoxin-free. Reproducibility of the production is extremely important, because safety and efficacy of a given nanoformulated drug can be influenced by small variations along the synthetic, purification or storage procedures. Most *in vitro* proof-of-concept assays are performed during the optimization of the production procedure, which should be well defined when moving to *in vivo* assays. Indeed, many nanoconjugate systems fail *in vivo* after performing excellently in cell culture assays due to uncontrolled deviations on the production and characterization that ultimately change their biodistribution and safety profile [3].

Regarding the physicochemical characterization of nanoconjugates, size/size distribution, shape, charge, composition, purity, and stability are key parameters that should be consistent and reproducible among batches. Whenever possible, characterization must be performed under biologically relevant conditions (*i.e.* human plasma instead of PBS or water), since size, charge and composition of the nanoconjugate may widely vary depending on the dispersion media. Moreover, in the case of drug delivery systems it becomes very relevant to characterize the maximum loading capacity of the system and its kinetic release even before performing any *in vitro* experiment. Many drug conjugates, especially those where the drug is covalently linked to the carrier, are limited by the amount of drug they can incorporate and by the solubility of the nanoconjugate. Such systems might perform well *in vitro*, where achieving a high local drug concentration is feasible, but might fail when tested in animal models because not enough drug concentration is achieved at the target site.

Overall, drugs with a small plasma half-life are usually encapsulated in nanosystems that protect them from fast metabolization and excretion. These nanosystems can be decorated with polyethylene glycol (PEG) or other molecules than reduce opsonization and phagocytosis by the reticuloendothelial system. When nanoconjugates are not efficiently scavenged by macrophages, the resulting increase in blood circulation time and hence bioavailability is expected to extend the duration of the controlled drug delivery or to improve the prospects for nanoparticles to reach target sites by extravasation [4]. This becomes very relevant for targeting solid tumors and inflammation areas, where extended circulation time is combined with a vascular enhanced permeability and retention (EPR) [5, 6]. Accordingly, a leaky vasculature together with a defective lymphatic drainage passively increases the retention of nanoconjugates within the tumor and inflammatory tissues. Nanoconjugates with or without "stealth" modifications that relay in the EPR effect, are considered passive, non-targeted systems. Many of the nanomedicines already marketed are of this type of passive, first generation nanomedicines, including different liposomal formulations, polymers, micelles and nanoparticles, among others [7].

Nowadays however, nanomedicines are evolving towards actively targeted systems. This is particularly relevant for those drugs intended to reach the central nervous system which have no ability to cross the BBB by themselves. In these cases, therapeutic nanoconjugates must use active systems in the form of specific components or mechanisms that will help the drug cross the BBB. Lysosomal storage diseases (LSD) with neurological affectation are the perfect example of how nanotechnology can help to improve current therapies. These diseases are caused by defective lysosomal enzymes or transporters that can be replaced by the exogenous addition of the active proteins. The therapy is known as enzyme replacement therapy (ERT) and it has been successfully applied to six different lysosomal storage diseases. However, classical ERT strategies are suboptimal in attenuating the strong neurological deterioration associated to more than half of LSDs, because enzymes are not able to cross the BBB. Since transient disruptions of BBB by hyperosmolar solutions, solvents, adjuvants, ultrasound, and surgical interventions (intracerebroventricular or intracerebral delivery) have proven to be too invasive and non-efficient, other approaches have been explored. Enzymes have been targeted to certain cell receptors expressed on brain endothelial cells involved in transcytosis. Examples include the use of human insulin receptor [8] and the HIV-1 trans-activator protein transduction domain (TAT) [9]. Further, the aminoacidic sequence known as cRGDfK binding $\alpha v\beta 3$ integrins has also been capable of facilitating the cross of the BBB [10, 11]. Recently, we also found that

RGD-moieties can stabilize liposomal membranes [12], which might have an impact when testing RGD-decorated liposomes for treatment of lysosomal storages disorders. Further, the use of carbohydrates has been also explored based on the higher metabolic activity of the tumoral or inflammation areas [13, 14].

In some CNS diseases, as high-grade gliomas and ischemic strokes, the BBB is broken. Nonetheless, nanomedicines could again concentrate in the site of action due to the EPR effect or by active targeting, which can be used to increase the activity of the nanosystem in the damaged area, either by improving extravassation of the nanoconjugates or their cell internalization. In any case, selection of targeting moieties should depend on the specific tumor, cell type or tissue compartment where the carried drug should be active. Even though the functional validation of the targeting moiety, whether it binds the receptor or it internalizes in the cells, can be done using cell cultures, eventually will require the use of adequate animal models.

Pharmacokinetics and Biodistribution of Drug Nanoconjugates

As explained above, the pharmacokinetic (PK) profile of a drug delivery system is often different from the free naked drug. One great advantage of drug delivery systems is that due to their macromolecular size, it is possible to add labeling agents (fluorochromes or radioactive isotopes) without altering the final size, surface charge and composition of the nanoconjugate. These reporters aid the localization (biodistribution) and quantification of the nanoconjugate in plasma and tissues by conventional quantitative analytical approaches, but also through the use of *in vivo* imaging technologies. As fluorescent dyes, those shifted to the near-infrared (NIR) region are recommended for *in vivo* imaging, to avoid increased tissue absorption and autofluorescente observed with fluorophores with excitation at lower wavelengths. Specifically, Cy.5.5, Alexa750, Alexa680 and DiR fluorophores have been widely used to track nanoconjugate biodistribution [15]. For isotopic tracing with SPECT or micro-PET, ^{99}mTc, ^{111}In, ^{68}Ga or ^{64}Cu are recommended due to their extended half-life. Since fluorescence and isotopic imaging techniques do not give much anatomical information, these techniques are usually combined with other imaging modalities giving more anatomical information [16 - 18].

Knowing that slight changes in a nanoconjugate (size, addition of PEG, functionalization with targeting moieties, etc.) can completely change the biodistribution of the nanoconjugate, imaging guided biodistribution assays might be an adequate methodology to easily screen different prototypes and rapidly optimize their design and the synthesis of a given nanoconjugate. We have recently applied this strategy to screen CD-44 targeted protein nanoparticles produced in bacteria [19] and found that there are substantial differences in the biological performance of a specific protein sequence depending on the purification procedure and the protein source. Moreover, nanoparticle stability and aggregation problems can be easily detected by non-invasive whole body imaging. For instance, in case of nanoparticles designed for intravenous administration, aggregation of nanoparticles within the bloodstream causes a significant accumulation of the nanosystem within the lungs and other RES organs such as liver and spleen (Fig. **1**). Similarly, expected (Fig. **1**, *bottom panel*) or unexpected excretion of the imaging agent could also indicate that the nanoconjugate has been metabolized rapidly *in vivo* as we have demonstrated for Gd labeled gold glyconanopaticles [14]. To avoid this type of situations, image guided nanoparticle biodistribution assays should be correlated with the measurement of the active principle (Fig. **2**). This is especially relevant when using fluorescent imaging because quantitative data from whole organs is dramatically affected by the scattering and the absorption properties of the organ [20].

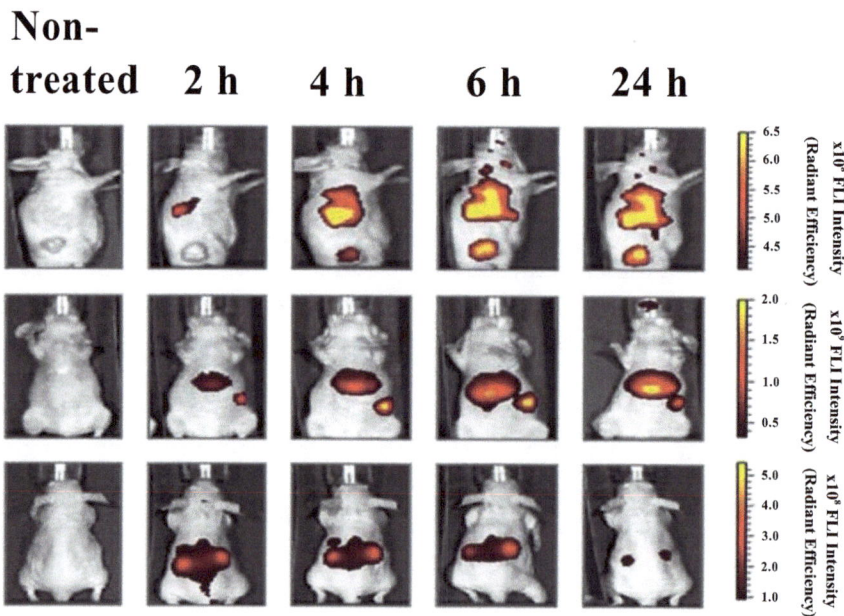

Fig. (1). Biodistribution of fluorescently labeled nanoconjugates. The first two sets of images correspond to the same nanoconjugate. In the first set of images with the animal in a lateral position, tumor-accumulation as well as pulmonary and hepatic accumulations are observed with a maximum intensity at 24 h post-administration. In the second set of images, with the mouse in supine position the liver and spleen accumulations (corresponding to the reticulo-endothelial system; RES sequestration) are visualized. The last row of images corresponds to a different (smaller) nanoconjugate, preferentially excreted through the kidneys at very short times post-administration.

a)   b)

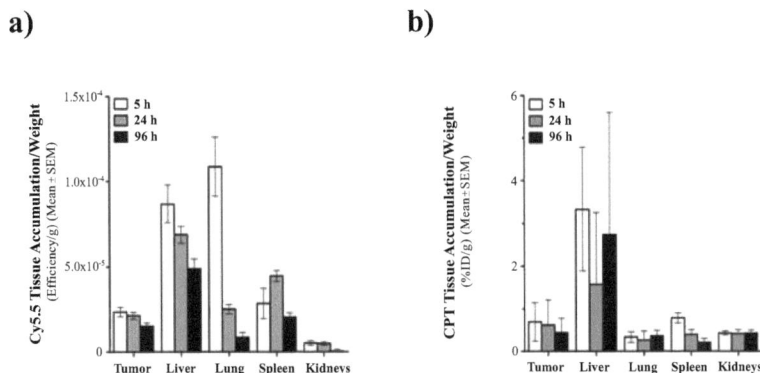

Fig. (2). Tissue biodistribution and quantification assays. Correlation of the fluorescence signal (a) obtained from the *ex vivo* tissue imaging of Cy5.5 labelled and campthotecin loaded silica nanoparticles and the quantification of the drug content (b) by HPLC methods. From (21), with the permission of Elsevier.

Biodistribution assays in animal models should mimic the preferential administration route that the nanomedicine will have in the clinics, because the route and site of administration would determine the final destination and bioavailability of the test compound. There are many examples in the literature where nanoconjugates are intravenously administered in the carotid vein to favor their relative concentration at the brain, but we should bear in mind that this should not be seen as a good administration route for a repeated dose schedule. Another example is the intratumoral administration of nanoconjugates carrying cytotoxic or anti-tumoral agents. Although there are endoscopic and imaging techniques that aid clinicians with the intratumoral injection, it is worth recalling that most of the tumors are not accessible for intratumoral injections and that further, repeated intratumoral administrations are not warranted. In these cases, nanoconjugates should be re-designed to allow better stability and biodistribution, and as explained above, biodistribution assays certainly help on this.

Ideally, macroscopic *in vivo* localization of nanoconjugates should be combined with the direct microscopic visualization of the nanoconjugate in its target tissues, because good tissue accumulation does not always translate in better efficacy. The first liposomal cisplatin formulation tested in clinical trials, SPI-077, accumulated substantially in tumor tissue but exhibited no antitumor efficacy because cisplatin cannot diffuse through the liposomal intact membrane, remaining inside the PEGylated liposome [22]. In preclinical assays of a system that finally went into clinical trials [21], Davis *et al.* [16] labeled siRNA with ^{64}Cu isotope and followed the biodistribution of nanoconjugates with and without a transferrin (Tf) ligand on the surface. They showed that the attachment of Tf targeting ligand has negligible impact on the biodistribution of the nanoconjugates as observed by micro-PET imaging. In this case, the EPR effect worked similarly for both, targeted and non-targeted nanoconjugates, even though higher efficacy and gene-inhibition was achieved with the targeted system. Similar results were obtained by Kirpotin *et al* [24] when using immunoliposomes targeted to HER2 and labeled with [67]Ga. Antibody-directed targeting did not increase tumor localization of immunoliposomes as both, targeted and non-targeted liposomes, achieved similar high levels in tumors. However, nanoconjugates encapsulating colloidal gold showed accumulation of anti-HER2 immunoliposomes inside cancer cells, whereas matched non-targeted liposomes were located predominantly at the extracellular stroma or within macrophages.

Finally, biodistribution and PK assays also give information regarding short and long term toxicities of nanoconjugates, either because they tend to accumulate in non-desired organs, or because their specific drug-release and metabolization kinetics [25]. The first FDA-approved nanoconjugate, doxorubicin containing liposomes, significantly reduced drug's cardiotoxicity, and is currently being used in most chemotherapeutic regimes using doxorubicine, even though Doxil suffers from other side effects, including hand-foot syndrome and mucositis, which indeed could be consequences of the drug slower release [26, 27].

Short and Long Term Toxicology of Novel Nanomedicines

It is important to highlight that, before starting any animal experimentation, the toxicity and the biocompatibility of the nanoconjugate must be addressed *in vitro* taking into account its intended precise use [28] and its potential interference with the chosen technique. Most toxicity assays have been already standardized by the National Characterization Laboratory (NCL) of the National Cancer Institute and the European Nanomedicine Characterisation Laboratory (EU-NCL), and although they only work with nanoconjugates with applications in cancer, the whole nanomedicine community is taking profit of them to facilitate the clinical development of new nanomaterials. Fig. (**3**) summarizes the list of assays to be performed *in vitro* and *in vivo* to evaluate the toxicity of a nanoconjugate to be administered intravenously. Most researchers in the field of nanotechnology perform conventional cytotoxicity and internalization tests, but few laboratories test hemocompatibility and biodegradation (cytotoxicity of degradation products) of their materials. It is essential to choose nanoconjugates that have biologically compatible components as part of their formulation, but it is also a must, to be aware of the potential toxicity of residual products from manufaturing (*i.e.* cetyltrimethylammonium bromide -CTAB- used in the preparation of carbon nanotubes or gold nanorods) or purification procedures (*i.e.* excessive concentration of surfactants such Triton-X). In our hands, *in vitro* hemolysis tests are very useful in order to detect such problems before *in vivo* administration.

In vitro assays
- Efficacy assays
- Cytotoxic drug release kinetics and plasma stability
- Cytotoxicity (MTT, LDH, ROS, etc.)
- Internalization assays (receptor binding and intracellular fate)
- Hemocompatibility (hemolysis, complement activation, etc.)
- Biodegradation (cytotoxicity of degradation products)

Chemistry
- Physico-chemical haracterization (size/size distribution, shape, charge, composition, purity, and stability)
- Batch-to-batch consistency
- Characterization of excipients
- Endotoxin-free production

CLINICAL TRIALS

In vivo assays
- Efficacy assays
- Biodistribution (organ accumulation)
- Pharmacokinetics
- MTD and organ related toxicities
- Long term toxicities
- Immunogenicity
- Metabolic fate

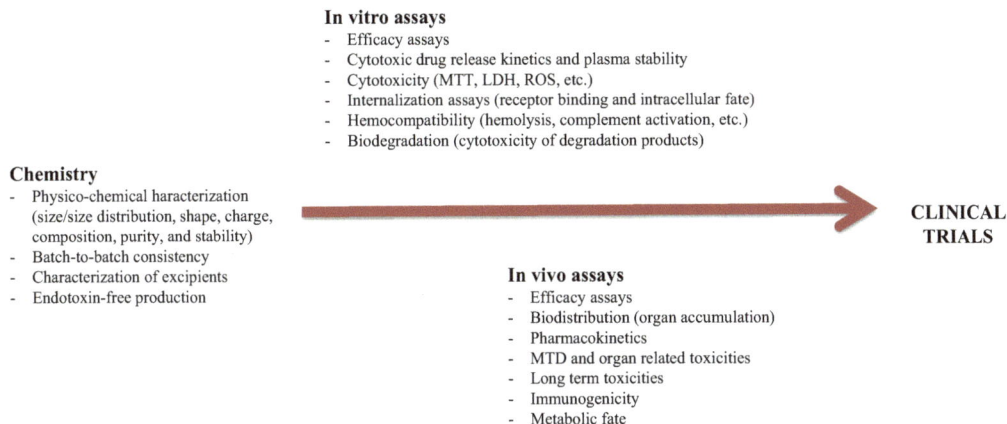

Fig. (3). *In vitro* and *in vivo* pre-clinical toxicological assays.

An increasing concern refers to the inmunotoxicity of the nanoformulated drugs [29], especially in those cases were the nanoconjugate is expected to accumulate in the body for an extended period of time. Such materials tend to distribute to the reticulo-endothelial system (RES) potentially altering the normal function of the phagocytes, and may enhance tumor growth as demonstrated for titanium oxide nanoparticles [30]. On the other hand, some nanoconjugates display reduced immunotoxicity as compared to the free drug. This is the case of Abraxane (nanoformulated Paclitaxel) that eliminates the anaphylaxis derived from the use of Chremophor-EL as a solubilization agent for paclitaxel [31]. Similarly, polyethylene glycol (PEG) coated colloidal gold nanoparticles carrying tumor necrosis alfa (TNFα), significantly reduce systemic imunotoxicity of TNFα [32]. It is important to evaluate the immunocompatiblity of the system as a whole, because the immunotoxicity of the carrier can enhance the immunotoxicity of the active principal ingredient (API). For instance phosphorothioate antisense oligonucleotides and PEGylated liposomes share the ability to activate the complement system, and thus combining them in one formulation is expected to lead to augmented toxicity related to complement activation (*e.g.*, CARPA syndrome) [33, 34].

Ultimately, the maximum tolerated dose (MTD) of the nanoconjugate will have to be determined. As the MTD highly depends on the animal strain and the specific animal model used, we encourage testing the free drug and the vehicle alongside the nanoconjugate. In addition, testing the empty nanocarrier should also be considered. This is straightforwardly done with liposomes, small lipid nanoparticles or systems where the drug is encapsulated without altering the surface (charge and composition) and size of the nanoconjugate. In other systems where the addition of the drug changes the main characteristics of the nanoconjugate (drugs covalently linked to the carrier, self-assembling systems with high molecular weight drugs, etc.), the inclusion of the non-loaded nanocarrier in the MTD assay is not very informative.

Short-term toxicities of the nanoconjugates not attributable to the drug or API are usually due to stability problems. Many nanoconjugates tend to aggregate and intravenous administration in animal models might cause pulmonary collapse and death of the animal. Other nanoconjugates are stable in physiological media, but are prone to aggregate in the bloodstream. Indeed, we have seen that many nanoparticles stable in PBS or saline, tend to accumulate in the lungs after intravenous administration (Fig. **1**). This lung accumulation is very relevant, because although might not be harmful to a healthy patient/animal in the short term, it can cause clotting and vascular problems that could ultimately lead to the death.

Long-term toxicities are more time consuming and difficult to observe, thus such assays are not usually included in scientific publications first describing the use of a novel nanomaterial. Quantum dots (QD) are a good example of potentially toxic nanoparticles with significant applications in nanomedicine. QD are nearly spherical particles with diameters in the range of 2–10 nm composed of toxic heavy metal atoms (*i.e.* Cd, Hg, Pb, As, Pb). Divalent cadmium, present in most QD is nephrotoxic and although this element is incorporated into a nanocrystalline core surrounded by biologicaly inert zinc sulfide and encapsulated within a stable polymer, its clinical use raises much concern. Actually, the only QD approved for a clinical trial are Cornell's C-Dot, composed by a core of gold (more compatible that Cd), silica, a coating of PEG to improve its biocompatibility and renal clearance and an RGD moiety to detect melanoma cells [35]. Biofunctionalization is thought to increase the excretion of the material, but the protein corona formed on the QD surface following intravenous administration can significantly increase its size above the renal filtration threshold, making QD prone to accumulate at the RES. The eventual fate of these nanoparticles is of vital importance, but so far has yet to be elucidated [36].

Therapeutic Efficacy of Novel Nanomedicines

All efficacy and proof-of-concept assays have to be designed carefully, but when dealing with nanopharmaceuticals, this design could be essential to demonstrate that the candidate merits moving into preclinical regulatory assays and subsequent clinical trials.

The key issues in the design of the efficacy assays of a drug loaded drug delivery system are the selection of the animal model, the use of appropriate controls and the choice of the treatment regimens. Animal models routinely used in preclinical trials are far from being representative for the clinical situation. In the field of oncology, xenografted tumors grow more rapidly and tend to have a higher EPR compared to human tumors where the vasculature is less leaky. Subcutaneous tumors, where transformed cells growth in a setting physiologically and anatomically different from its native (ortothopic) origin, are the most commonly used models. Selection of the cell line and the volume of the tumors at the time of the treatment initiation is crucial, since the EPR effect and potential accumulation of the nanoconjugate in the tumor will greatly depend on that [37]. Besides, in most of the cases, these models are merely developed in immunodeficient mice, which rules out the possibility of testing the effect of the immune system in the therapeutic outcome. Once the animal model is chosen, the selection of the control groups is not trivial, since the nanotechnology formulated drug and the free drug may have different excipients or dispersants, and as explained above, those compounds might have an effect in the disease evolution. In addition, whenever possible, the non-loaded nanocarrier should also be tested to show if the therapeutic effect is merely due to the API or if contrarily the nanocarrier is also affecting. Although the principal objective of efficacy assays is to demonstrate that the nanomedines by themselves are effective, it is also important to test the nanomedicine in combination regimens that are clinically relevant. For a certain disease, the standard of care in patients will probably include a combination of agents and it is in these setting where the nanomedicine under development has to show its superiority. Finally, the selection of the doses at which the free drug and the nanothecnology formulate drug are going to be compared is relevant. Many approved nanopharmaceuticals have a higher MTD than their free-drug counterparts, and in these cases, an increase in the efficacy is observed because higher drug doses are administered with less secondary effects. Although this experimental design is correct, it does not give much information on the real effect of the combination drug-carrier. On the other hand, efficacy studies where free drug and nanotechnology formulated drugs are compared at equivalent doses directly reflects the benefit in the efficacy, but do not give much information on the therapeutic window gained (reduction of side effects) when using the nanotechnology-based system. Ideally, both types of assays should be conducted.

Taking all above considerations, efficacy assays become even more expensive and time consuming than in traditional drug development. *In vivo* imaging can again help and accelerate nanotherapeutic drug development process by allowing a longitudinal monitoring of the disease progression as occurs. Due to its easy manipulation and high-throughput capacities bioluminescent imaging (BLI) is a good imaging modality to follow tumor growth and dissemination and treatment efficacy (Fig. **4**). Many cancer models require the inoculation of cells (either from human origin –xenografts- or from the same specie –syngenic models-) than can be previously modified to overexpress the Firefly luciferase reporter. Then, the growth and metastatic dissemination of the disease can be easily followed without the need of euthanizing the animal, as shown in Fig. (**1**) or by a recent publication by our group [38]. Moreover, in the case of fluorescently labeled nanoparticles, it is even possible to co-register the bioluminescent signal (from the cancer cells) with the fluorescent signal (from the nanoparticle) [39].

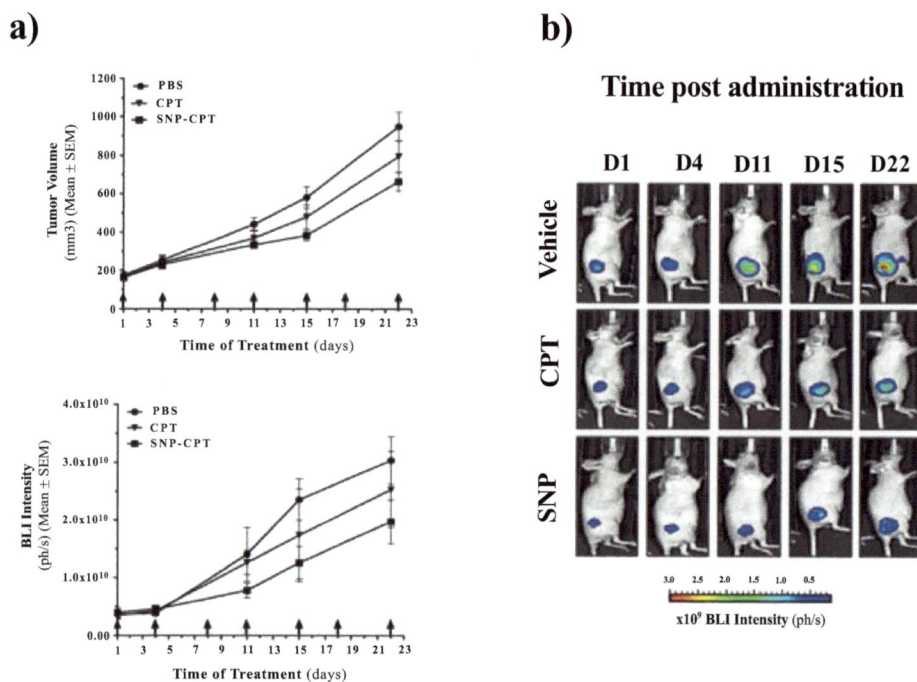

Fig. (4). Tumor growth delay induced by camptothecin (CPT) loaded silica nanoparticles (SNP-CPT) as measured by external measurements of tumor volume (a, *upper panel*), and by tumor bioluminescent signals (a, *lower panel*). Arrows indicate the time of treatment. (b) *In vivo* monitoring of a representative mouse from each treatment group is shown along time. Images are set at the same pseudocolor scale to show relative bioluminescent changes overtime. From (21), with the permission of Elsevier.

CONCLUDING REMARKS

Despite intensive investigation, not many different nanosystems have managed to gain FDA and/or EMA approval. Most of the nanomedicines under development have failed either in late-stage preclinical or in early-stage clinical trials. Error or pitfalls could come from different parts of the nanomedicine developmental pipeline, from the manufacturing, to the scale-up, the poor understanding of the biodistribution and toxicological profiles and the inappropriate design of the preclinical and clinical efficacy and safety assays. Overall, the last two decades in nanomedicine investigation has showed us that a careful characterization and controlled manufacturing process are essential for obtaining a lead nanotherapeutic conjugates with which proceed. Once reaching the state of animal testing, the use of appropriate animal models and imaging technologies could help and accelerate the screening of different prototypes to finally obtain a candidate that could show its benefits in the clinical setting.

CONFLICT OF INTEREST

The authors confirm that they have no conflict of interest to declare for this publication.

ACKNOWLEDGEMENT

We wrote this work while being supported by several grants directly related with the *in vivo* testing of nanomedicines. Specifically we would like to thank the funding from CIBER-BBN's Platform Program, "Fondo de Investigaciones Sanitarias - Instituto de Salud Carlos III" (FIS-ISCIII) (PI14/02079 to SS), Spanish Ministry of Science and Innovation (MICINN, IPT-090000-2010-0001 to IA), all co-funded by FEDER; and "Fundació Marató TV3" (PENTRI project 337/C/2013 to IA).

All *in vivo* experiments developed at Functional Validation & Preclinical Research (FVPR) were performed by the ICTS "NANBIOSIS", more specifically by the CIBER-BBN's *In vivo* Experimental Platform (http://www.nanbiosis.es-/unit/u20-in-vivo-experimentalplatform/).

REFERENCES

[1]   Desai N. Challenges in development of nanoparticle-based therapeutics. AAPS J 2012; 14(2): 282-95.
       [http://dx.doi.org/10.1208/s12248-012-9339-4]

[2]   Bcc Research. Nanotechnology in Medical Applications: The Global Market(HLC069B) [Internet] 2012.
       http://www.bccresearch.com/pressroom/hlc/nanomedical-global-sales-reach-$130.9-billion-2016.

[3]   Crist RM, Grossman JH, Patri AK, *et al.* Common pitfalls in nanotechnology: lessons learned from NCIs Nanotechnology Characterization Laboratory. Integr Biol (Camb) 2013; 5(1): 66-73.
       [http://dx.doi.org/10.1039/C2IB20117H] [PMID: 22772974]

[4]   Moghimi SM, Hunter AC, Murray JC. Long-circulating and target-specific nanoparticles: theory to practice. Pharmacol Rev 2001; 53(2): 283-318.
       [PMID: 11356986]

[5]   Matsumura Y, Maeda H. A new concept for macromolecular therapeutics in cancer chemotherapy: mechanism of tumoritropic accumulation of proteins and the antitumor agent smancs. Cancer Res 1986; 46(12 Pt 1): 6387-92.
       [PMID: 2946403]

[6]   Maeda H, Wu J, Sawa T, Matsumura Y, Hori K. Tumor vascular permeability and the EPR effect in macromolecular therapeutics: a review. J Control Release 2000; 65(1-2): 271-84. [http://dx.doi.org/10.1016/S0168-3659(99)00248-5] [PMID: 10699287]

[7]   Lammers T, Kiessling F, Hennink WE, Storm G. Drug targeting to tumors: principles, pitfalls and (pre-) clinical progress. J Control Release 2012; 161(2): 175-87. [http://dx.doi.org/10.1016/j.jconrel.2011.09.063] [PMID: 21945285]

[8]   Boado RJ, Zhang Y, Zhang Y, Xia C-F, Wang Y, Pardridge WM. Genetic engineering of a lysosomal enzyme fusion protein for targeted delivery across the human blood-brain barrier. Biotechnol Bioeng 2008; 99(2): 475-84. [Internet]. [http://dx.doi.org/10.1002/bit.21602] [PMID: 17680664]

[9]   Lee KO, Luu N, Kaneski CR, Schiffmann R, Brady RO, Murray GJ. Improved intracellular delivery of glucocerebrosidase mediated by the HIV-1 TAT protein transduction domain. Biochem Biophys Res Commun 2005; 337(2): 701-7. [http://dx.doi.org/10.1016/j.bbrc.2005.05.207] [PMID: 16223608]

[10]   Qin J, Chen D, Hu H, Qiao M, Zhao X, Chen B. Body distributioin of RGD-mediated liposome in brain-targeting drug delivery. Yakugaku Zasshi 2007; 127(9): 1497-501. [http://dx.doi.org/10.1248/yakushi.127.1497] [PMID: 17827930]

[11]   Miura Y, Takenaka T, Toh K, *et al.* Cyclic RGD-linked polymeric micelles for targeted delivery of platinum anticancer drugs to glioblastoma through the blood-brain tumor barrier. ACS Nano 2013; 7(10): 8583-92. [http://dx.doi.org/10.1021/nn402662d] [PMID: 24028526]

[12]   Cabrera I, Elizondo E, Esteban O, *et al.* Multifunctional nanovesicle-bioactive conjugates prepared by a one-step scalable method using CO2-expanded solvents. Nano Lett 2013; 13(8): 3766-74. [http://dx.doi.org/10.1021/nl4017072] [PMID: 23829208]

[13]   Irure A, Marradi M, Arnaiz B, Genicio N, Padro D, Penades S. Sugar/gadolinium-loaded gold nanoparticles for labelling and imaging cells by magnetic resonance imaging Biomater Sci 2013; 1(6): 658-8.

[14]   Candiota AP, Acosta M, Simões RV, *et al.* A new *ex vivo* method to evaluate the performance of candidate MRI contrast agents: a proof-of-concept study. J Nanobiotechnology 2014; 12: 12. [http://dx.doi.org/10.1186/1477-3155-12-12] [PMID: 24708566]

[15]   Rao J, Dragulescu-Andrasi A, Yao H. Fluorescence imaging *in vivo*: recent advances. Curr Opin Biotechnol 2007; 18(1): 17-25. [http://dx.doi.org/10.1016/j.copbio.2007.01.003] [PMID: 17234399]

[16]   Bartlett DW, Su H, Hildebrandt IJ, Weber WA, Davis ME. Impact of tumor-specific targeting on the biodistribution and efficacy of siRNA nanoparticles measured by multimodality *in vivo* imaging. Proc Natl Acad Sci USA 2007; 104(39): 15549-54. [http://dx.doi.org/10.1073/pnas.0707461104] [PMID: 17875985]

[17]   Yang X, Hong H, Grailer JJ, *et al.* CRGD-functionalized, DOX-conjugated, and 64Cu-labeled superparamagnetic iron oxide nanoparticles for targeted anticancer drug delivery and PET/MR imaging Biomaterials [Internet] 2011; 32(17): 4151-60. [http://dx.doi.org/10.1016/- j.biomaterials.2011.02.006]

[18]   Zhou M, Zhang R, Huang M, *et al.* A chelator-free multifunctional [64Cu]CuS nanoparticle platform for simultaneous micro-PET/CT imaging and photothermal ablation therapy. J Am Chem Soc 2010; 132(43): 15351-8. [http://dx.doi.org/10.1021/ja106855m] [PMID: 20942456]

[19]   Pesarrodona M, Ferrer-Miralles N, Unzueta U, Gener P, Tatkiewicz W, Abasolo I, *et al.* Intracellular targeting of CD44+ cells with self-assembling, protein only nanoparticles. Int J Pharm Elsevier 2014; 473(1-2): 286-95.

[20]   Liu Y, Tseng Y-C, Huang L. Biodistribution studies of nanoparticles using fluorescence imaging: a qualitative or quantitative method? Pharm Res 2012; 29(12): 3273-7.
[http://dx.doi.org/10.1007/s11095-012-0818-1] [PMID: 22806405]

[21]   Botella P, Abasolo I, Fernández Y, *et al.* Surface-modified silica nanoparticles for tumor-targeted delivery of camptothecin and its biological evaluation. J Control Release 2011; 156(2): 246-57.
[http://dx.doi.org/10.1016/j.jconrel.2011.06.039] [PMID: 21756949]

[22]   Kim ES, Lu C, Khuri FR, *et al.* A phase II study of STEALTH cisplatin (SPI-77) in patients with advanced non-small cell lung cancer. Lung Cancer 2001; 34(3): 427-32.
[http://dx.doi.org/10.1016/S0169-5002(01)00278-1] [PMID: 11714540]

[23]   Davis ME, Zuckerman JE, Choi CHJ, Seligson D, Tolcher A, C a Alabi. Evidence of RNAi in humans from systemically administered siRNA via targeted nanoparticles. Nature 2010; 464(7291): 1067-70.
[http://dx.doi.org/10.1038/nature08956]

[24]   Kirpotin DB, Drummond DC, Shao Y, *et al.* Antibody targeting of long-circulating lipidic nanoparticles does not increase tumor localization but does increase internalization in animal models. Cancer Res 2006; 66(13): 6732-40.
[http://dx.doi.org/10.1158/0008-5472.CAN-05-4199] [PMID: 16818648]

[25]   Li S-D, Huang L. Pharmacokinetics and biodistribution of nanoparticles. Mol Pharm 2008; 5(4): 496-504.
[http://dx.doi.org/10.1021/mp800049w] [PMID: 18611037]

[26]   Gordon KB, Tajuddin A, Guitart J, Kuzel TM, Eramo LR, VonRoenn J. Hand-foot syndrome associated with liposome-encapsulated doxorubicin therapy. Cancer 1995; 75(8): 2169-73.
[http://dx.doi.org/10.1002/1097-0142(19950415)75:8<2169::AID-CNCR2820750822>3.0.CO;2-H] [PMID: 7697608]

[27]   Charrois GJ, Allen TM. Multiple injections of pegylated liposomal Doxorubicin: pharmacokinetics and therapeutic activity. J Pharmacol Exp Ther 2003; 306(3): 1058-67.
[http://dx.doi.org/10.1124/jpet.103.053413] [PMID: 12808004]

[28]   Duncan R, Izzo L. Dendrimer biocompatibility and toxicity. Adv Drug Deliv Rev 2005; 57(15): 2215-37.
[http://dx.doi.org/10.1016/j.addr.2005.09.019] [PMID: 16297497]

[29]   Dobrovolskaia MA. Pre-clinical immunotoxicity studies of nanotechnology-formulated drugs: Challenges, considerations and strategy. J Control Release 2015; 15(S0168-3659): 30098-5.

[30]   Moon EY, Yi GH, Kang JS, Lim JS, Kim HM, Pyo S. An increase in mouse tumor growth by an *in vivo* immunomodulating effect of titanium dioxide nanoparticles. J Immunotoxicol 2011; 8(1): 56-67.
[http://dx.doi.org/10.3109/1547691X.2010.543995] [PMID: 21288165]

[31]   Gradishar WJ, Tjulandin S, Davidson N, *et al.* Phase III trial of nanoparticle albumin-bound paclitaxel compared with polyethylated castor oil-based paclitaxel in women with breast cancer. J Clin Oncol 2005; 23(31): 7794-803.
[PMID: 16172456]

[32]   Libutti SK, Paciotti GF, Byrnes AA, *et al.* Phase I and pharmacokinetic studies of CYT-6091, a novel PEGylated colloidal gold-rhTNF nanomedicine. Clin Cancer Res 2010; 16(24): 6139-49.
[http://dx.doi.org/10.1158/1078-0432.CCR-10-0978] [PMID: 20876255]

[33]   Henry SP, Beattie G, Yeh G, *et al.* Complement activation is responsible for acute toxicities in rhesus monkeys treated with a phosphorothioate oligodeoxynucleotide. Int Immunopharmacol 2002; 2(12): 1657-66.
[http://dx.doi.org/10.1016/S1567-5769(02)00142-X] [PMID: 12469940]

[34]   Szebeni J, Bedocs P, Rozsnyay Z, Weiszhár Z, Urbanics R, Rosivall L, *et al.* Liposome-induced complement activation and related cardiopulmonary distress in pigs: Factors promoting reactogenicity of Doxil and AmBisome. Nanomedicine Nanotechnology. Biol Med 2012; 8(2): 176-84.

[35]   Phillips E, Penate-Medina O, Zanzonico PB, *et al.* Clinical translation of an ultrasmall inorganic optical-PET imaging nanoparticle probe. Sci Transl Med 2014; 6(260): 260ra149.
[http://dx.doi.org/10.1126/scitranslmed.3009524] [PMID: 25355699]

[36]   Choi HS, Liu W, Misra P, *et al.* Renal clearance of quantum dots. Nat Biotechnol 2007; 25(10): 1165-70.
[http://dx.doi.org/10.1038/nbt1340] [PMID: 17891134]

[37]   Duncan R, Sat-Klopsch YN, Burger AM, Bibby MC, Fiebig HH, Sausville EA. Validation of tumour models for use in anticancer nanomedicine evaluation: the EPR effect and cathepsin B-mediated drug release rate. Cancer Chemother Pharmacol 2013; 72(2): 417-27.
[http://dx.doi.org/10.1007/s00280-013-2209-7] [PMID: 23797686]

[38]   Fernández Y, Forarada L, García-Aranda N, Mancilla S, Suárez-López L, Céspedes MV, *et al.* Bioluminescent Imaging of Animal Models for Human Colorectal Cancer Tumor Growth and Metastatic Dissemination to Clinically Significant Sites. J Mol Biol Mol Imaging 2015; 2(2): 1019-33.

[39]   Li Y, Du Y, Liu X, *et al.* Monitoring Tumor Targeting and Treatment Effects of IRDye 800CW and GX1-Conjugated Polylactic Acid Nanoparticles Encapsulating Endostar on Glioma by Optical Molecular Imaging. Mol Imaging 2015; 14: 356-65.
[PMID: 26162457]

CHAPTER 8

How does "Protein Corona" Affect the *In vivo* Efficiency of Polymeric Nanoparticles? State of Art

F. Pederzoli, M. Galliani, F. Forni, M.A. Vandelli, D. Belletti, G. Tosi and **B. Ruozi**[*]

Te.Far.T.I., Department of Life Sciences, University of Modena and Reggio Emilia, Via Campi 103, 41124 Modena, Italy

Abstract: Nanomedicine is increasingly considered as one of the most promising ways to overcome the limits of traditional medicine and conventional pharmaceutical formulations. In particular, polymeric nanoparticles (NPs) represent one of the most important tools in the nanomedicine field due to their potential in a wide range of biomedical applications such as imaging, drug targeting and drug delivery. However, their application is strongly hampered by limited knowledge and control of their interactions with complex biological systems. In biological environments, NPs are enshrouded by a layer of biomolecules, predominantly proteins, which tend to associate with NPs, forming a new surface named 'protein corona' (PC). Thus, the resulting nano-structure is a new entity, defined as PC-NP complex, featured by new characteristics, different from the original features of the bare NPs. In this chapter, starting from the definition of PC, we critically discuss the physico-chemical properties of polymeric NPs (*e.g.*, size, shape, composition, surface functional groups, surface charge, hydrophilicity/hydrophobicity) and the environmental biological parameters (blood concentration, plasma gradient, temperature) affecting PC formation and composition. We further discuss how the new "entity" generated by the interactions between NPs and proteins *in vivo* mediates the ability of all the nanosystems to circulate, biodistribute and selectively release the drugs to the target site. We conclude by highlighting the gaps in the knowledge of the PC in relation to polymeric NPs and by discussing the main issues to be addressed and investigated in order to speed up the translatability of NPs into clinical protocols.

Keywords: *In vivo* outcome, Protein corona (PC), Polymeric nanoparticles (NPs), PC-NP complex, Protein –NPs interaction.

INTRODUCTION

Recent years witnessed a progressive growing interest in developing innovative formulations for drug delivery to solve the limits of conventional pharmaceutical

[*] **Corresponding author Barbara Ruozi:** Department of Life Sciences, University of Modena and Reggio Emilia, Via Campi 103, 41124 Modena, Italy; Tel: +39-059-2058562; E-mail: barbara.ruozi@unimore.it

Giovanni Tosi (Ed)

formulations. Innovation in pharmaceutical technology means to stabilize and selective deliver drugs to the site of action without affecting healthy organs, and achieve drug dose maintenance in the organism without the need of repeated administrations [1]. These needs have led to the development of nanomedicine, defined as the science of studying nanoscale-sized structures for diagnostics, therapeutics and specific drug-delivery [2].

One of the primary challenges of nanomedicine is to deliver a drug to the target site, avoiding side effects to non-targeted organs [3]. During the last several decades, numerous lipidic and polymeric nanosized drug delivery systems were proposed. Examples of lipidic carriers are liposomes and solid lipid nanoparticles, while polymeric carriers are mainly represented by nanoparticles (NPs), dendrimers and micelles [4].

These nano-systems can be directed to a specific target site by means of different strategies, *i.e.* passive targeting and active targeting. Passive targeting exploits differences in patho-physiological features of diseased tissues, enabling delivery of drugs to the target, as diseased tissues are often altered in terms of facilitated accumulation or permability aspects. As an example, within tumoral tissues, nano-sized systems can escape from nonspecific trapping by the reticuloendothelial system and accumulate in target tissues after circulating in the blood by passive targeting exploiting the enhanced permeability and retention (EPR) effect (high interstitial pressure, enhanced vascular permeability, and the lack of functional lymphatic drainage). On the other hand, active targeting is complicated regarding the design of surface engineered nanocarriers; frequently, these systems are stabilized by polyethylene glycol (PEG) moieties and conjugated with ligands, specifically able to recognize structures or specific environments on or inside the target. This kind of targeting also includes carriers sensitive to physical stimuli such as temperature, pH or magnetism [5].

In this context, NPs were extensively studied in the last couple of decades, due to their potential for a wide range of biomedical applications. These systems represent a versatile tool in drug delivery, able to load a large variety of drugs with different chemico-physical features.

Additionally, chemical moieties present on the surface of NPs can be suitable for functionalization with different ligands directed to a specific target [6 - 10]. This surface modification approach allowed to reach significant results with pre-clinical application considering both polymeric and inorganic NPs, which highlighted their role in medicine fields [11 - 14]. These evidences, even if considered "promising" *in vivo* proof-of concept and results, were considered as "enough for now" over a long time [15].

Nowadays, aiming to speed up the translatability of nanomedicine and NPs into clinical protocols, another concept and relatively un-explored field are turning on around NPs: understanding their fate *in vivo*.

The destiny of NPs after their administration and their interaction with biological fluids is an interesting but complex field of research. The first observation was that, as happens with any foreign materials, NPs are immediately covered by proteins from the blood stream, leading to the formation of what is called "protein corona" (PC) [16].

Several studies pointed out that the PC plays an important role in the NPs behavior *in vivo* and may impact on biodistribution, drug targeting, intracellular uptake and toxicity of NPs [17]. Thus, the characterization of NPs is not sufficient anymore, but the relative PC must be also characterized and possibly controlled, in order to completely predict the real fate and efficiency of these drug delivery systems *in vivo*.

In literature, the studies on PC mainly involve inorganic NPs; moreover, in this review, we tried to combine and to critically comment the outputs relating to the behavior of polymeric NPs. By understanding the PC, we focalized on the parameter affecting the formation of the new biochemical entity, namely NP-PC and then on the evaluation of the complex interaction NP-PC/body.

Protein Corona: Composition and Structure

Nowadays it is almost clear that any foreign material that enters in contact with a biological fluid interacts with its components, particularly its "resident" proteins. This event happens also to NPs when injected into the bloodstream; their surface is immediately covered by circulating proteins [18], leading to the formation of a complex and variable structure, called PC [16, 19, 20]. There is no "universal" corona for all the nanomaterials: the PC composition strongly depends on the synthetic identity of the NPs. In addition, the relative densities of the adsorbed proteins generally do not necessarily correlate with their relative abundances in plasma [20]. Walkey and Chan identified a subset of plasma proteins detected on at least one nanomaterial surface, and called it "adsorbome". According to their results, in general, the plasma PC consists of 2-6 proteins adsorbed with high abundance and many other adsorbed with low abundance. In particular, they pointed out that the most abundant identified protein generally represents the 29% of the total adsorbed proteins, while the top 3 most abundant proteins represent 56% of the total amount [21].

The structure of PC consists of two components, known as *hard* and *soft* corona (Fig. **1**) .

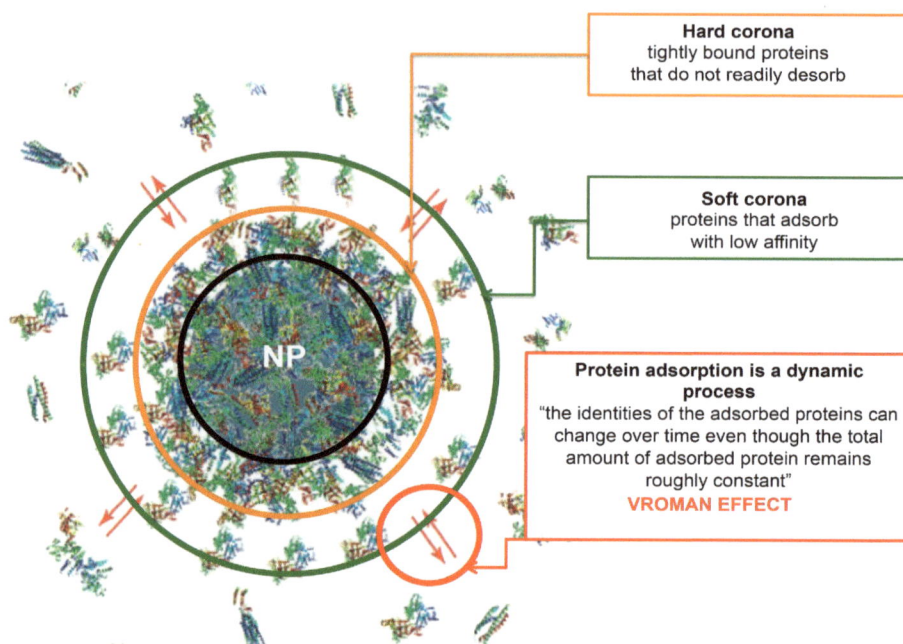

Fig. (1). Schematic representation of soft and hard protein corona and of the concept of dynamism in protein adsorption and desorption processes. This dynamic process determines the exchange time and lifetime of proteins in the PC and is generally ascribable to the Vroman effect.

The hard corona is formed by proteins which adsorb onto NPs surface with high affinity, featured by long typical exchange times, while the soft corona consists of loosely bound proteins with low affinity for the NPs surface and short exchange rates. Simberg *et al.* proposed a model for the PC which consists of 'primary binders' that directly recognize the NPs surface, and 'secondary binders' that interact with the primary binders *via* protein-protein interactions. These secondary binders can alter the activity of the primary binders or 'mask' them, therefore preventing their interactions with the surrounding medium [21, 22].

Dynamics and Mechanism of PC Formation

The PC is not a solid and fixed layer, since circulating proteins have a very wide range of affinities for the particle surface and a huge range of equilibrium constants that represent the different binding mechanisms present in the game. Therefore, PC is an extremely dynamic layer, in which the composition changes over the time. In fact, the composition of the PC at any given time will be determined by the concentrations of the several thousands of proteins in plasma and their respective equilibrium binding constants for the particular NP [23].

As the NPs come in contact with plasma, the most abundant proteins cover the NPs surface, but over-time these proteins will be replaced by other proteins with lower concentrations and higher affinities for the NPs [24].

As evident, one of the most critical parameters which influences the fate of NPs once in the biological fluids consists on the association and dissociation rates of different proteins from NPs. As example: a tightly associated protein may follow the NPs during endocytosis, while an intracellular protein will replace a protein with fast exchange during or after this transfer [25].

Several confusing conclusion were reported by the mechanism of protein interaction with a NP surface [21, 26, 27].

After NPs entrance into the bloodstream, a widely accepted hypothesis described the initial migration of the circulating free proteins occuring onto the NPs surface by diffusion or by traveling down a potential energy gradient. Then, thermodynamically favorable events lead to adsorption of the proteins.

The stability of the NP-PC complex depends on the net binding energy of the adsorption event: proteins which adsorb with large energy have a low probability of desorption and tend to stay associated with the particles, while proteins that adsorb with small energy tend to desorb easily and return to solution [21]. During adsorption on the surface of the NPs, proteins may undergo structural rearrangements called "conformational changes". These changes are thermodynamically favorable if they allow a hydrophobic or charged sequence of a protein to interact with a hydrophobic or charged NP surface, respectively [20, 21].

Bekale *et al.* studied the interactions between chitosan NPs and two model proteins, namely bovine (BSA) and human serum albumins (HSA). Results revealed that the chitosan NPs interact with BSA to form chitosan-BSA complexes, mainly through hydrophobic contacts while HSA-chitosan complexation is mainly *via* electrostatic interactions. Furthermore, the association between polymer and protein were shown to cause a partial protein conformational change leading to a major reduction of α-helix from 63% (free BSA) to 57% in chitosan-BSA and from 57% (free HSA) to 51% in chitosan-HAS [28].

Further, Pan and co-workers provided evidence of adsorption-induced unfolding of a model protein (GB1) on latex NPs asserting that this process can be controlled by external factors such as pH and ionic strength [29]. Another study proposed by Norde and Giacomelli demonstrated that structural changes of bovine serum albumin (BSA) on polystyrene NPs were irreversible and that the irreversibility is mainly influenced by the hydrophobicity of the sorbent surface [30].

Generally, many studies relating to the change in protein conformation of protein and its temporal stability refer to inorganic NPs, which are more stable and reproducible when compared to polymeric NPs; nevertheless the results are often confusing.

Moreover, the adsorption of proteins does not necessarily require direct interaction with the NP surface, but may occur through protein-protein interactions. These interactions can be either specific or non-specific.

Specific interactions occur if two or more proteins interact through domains (independent units formed the proteins) that contain complementary aminoacid sequences. To this end, Andersson and co-workers observed that adsorption of the complement factor C3 onto a polystyrene surface may induce the exposure of epitopes of this protein that specifically binds with other complement factors, leading to the assembly of protein complexes and activation of the alternative complement pathway [31]. On the other hand, non-specific interactions result from the exposure of charged or hydrophobic domains in a protein that interact with other proteins. Generally nonspecific protein–protein interactions are unusual in *natural* biochemistry, and are interpreted as a 'danger signal' by the body, triggering an immune response [32]. As an example Deng *et al.* showed that the polymeric coating with poly (acrylic acid) in gold NPs bound fibrinogen from blood plasma and induced its unfolding, which in turn activated the receptor Mac-1 on THP-1 cells, causing a release of inflammatory cytokines [33].

Vroman Effect and Kinetics of Protein Adsorption

Proteins with both high concentrations in plasma and high association rates will initially form the surface of the NPs. If these proteins have short residence times, they will soon be replaced by less abundant proteins featured by slower association rates, but feature by higher affinity and longer residence times for the NP surface. This phenomenon is known as *'Vroman effect'* [16, 21, 34, 35].

Although the identities of the adsorbed proteins can change over time, the total amount of adsorbed proteins on a surface remains roughly constant [20, 36]. The Vroman effect can be described in two different stages, namely "early" and "late" stages.

The *early stage* consists of the rapid adsorption of proteins abundant in the bloodstream such as albumin, fibrinogen and IgG; during the *late stage*, proteins with moderate affinities for the NP surface are usually replaced by those with very high affinities [20]. That way, Allémann and co-workers investigated the kinetics of blood protein adsorption onto poly(D,L-lactic acid) (PLA) NPs, sizing about 300 nm, incubated over variable periods of time in human serum and citrated plasma. They observed that albumin was detected at high amounts for short incubation periods, but it decreased over time, while an increase in the level of apolipoproteins occurred as a function of the incubation time period [37].

Starting from the previously described results, also confirmed by a broad amount of literature in the field [38, 38], the studies on the protein corona were mainly focused on the rates of association and dissociation of a protein to the surface, described by the parameters K_{on} and K_{off}, respectively.

The value of K_{on} is related to the frequency of contact between the protein and the NPs. Large values of K_{on} are typical for proteins in high concentrations, which diffuse rapidly and which can interact with the surface of the NPs.

On the contrary, the values of K_{off} depend on the binding energy of the NPs-protein complex: if the binding energy is high, K_{off} will have a lower value. Thus, the affinity of a protein to the surface of the NP is determined by the balance between these two constants and defined by the association constant K_d [21].

The parameters impacting the rates of association/dissociation are mainly dependent on the *nature* and *properties* of both the protein and the NPs.

As example, Cedervall and co-workers studied human serum albumin (HSA) association and dissociation rates from NPs, sizing 200 nm and featured by a different rate of hydrophobicity/hydrophilicity, as made of isopropylacrylamide/ N-*tert*- butylacrylamide NPs (NIPAM-BAM NPs) at different polymer ratios. The study shows that HSA displays a longer residence time on the more hydrophilic particles [25].

Finally, proteins adsorbing to and desorbing from NPs, depending on their own K_{on} and K_{off} values can be divided into "fast" and "slow" components, corresponding to the hard and soft coronas, respectively. From a timing point of view, the fast component is formed in seconds, while the slow component is formed on a timescale of minutes to hours. Fast and slow components switch roles when considering the K_{off}, as the hard corona proteins desorb less rapidly than proteins forming the soft corona [21, 25, 40].

PARAMETERS AFFECTING PROTEIN CORONA

Understanding the dynamics that occur during PC formation and then the parameters (of NPs and environment) affecting the formation of PC is required to assess their "*in vivo*" behavior and consequently their efficacy to modulate drug release and to target the specific cells.

Although this aspect seems to be intuitive, the "nanomedicine application" demands a rigorous understanding of the physical chemistry of the complex as a entirely new biological entity.

Focusing the attention on polymeric nanoparticles, the biological environment greatly influences the stability and the biodegradability of NPs, changing the chemico-physical parameters of the NPs, making the study of plasma protein/NPs subject to the evaluation of a number of variables (Table **1**).

Chemico-physical Parameters of NPs

NP Size

NPs are frequently referred to as nano-systems with a diameter in the size ranging from 1 to 200 nm. As a consequence, NPs possess extremely high surface to volume ratios meaning that the importance of size *vs* surface area cannot be overemphasized when considering NPs. For example, a nanoparticle with a diameter of 70 nm at a concentration of 0.01 mg/ml provides a total surface area of 0.8 m^2/l, whereas a particle with a diameter of 200 nm at the same concentration (0.01 mg/ml) has a total surface area of 0.3 m^2/l, illustrating the dramatic increase in surface area [23]. This aspect strongly impacts on the serum protein absorption and suggests that the experiments should be compared only when NPs demonstrate monomodality and monodispersity with high reproducibility in terms of the dimensional population.

Table 1. Chemico-physical parameters affecting PC interaction with polymeric NPs.

	Polymeric NPs		**Experimental Method(s)**	**Incubation medium**	**Results**	**Reference**
	Composition	**Size (nm)**				
		Surface charge (mV)				
NPs Size	NIPAM-BAM in different ratio (85:15, 65:35, and 50:50)	range: 70-700	SDS-PAGE LC-MS/MS	Plasma	Amount of bound proteins changes with size, but the identity of the protein pattern is the same for all NPs and apolipoprotein AI is always the most abundant protein recovered	[26]
		/				
	Mixed (poly(acrylic acid)/gold) NPs*	range: 7-22	Surface plasmon resonance	Protein solution (fibrinogen 2,5 mg/ml)	Size influences number of bound protein molecules and affect the formation of aggregates	[72]
		-25/-50				

(Table 1) contd.....

NPs Shape	Mixed (phospholipid PEG/gold) NPs*	28	SDS-PAGE LC-MS/MS	Plasma	Binding of specific size range proteins to the different surface roughness	[50]
		-35				
NP Surface properties/polymeric chains	Functionalized polystirene latex NPs	50 and 100	Mass spectrometry	Plasma	Size and surface properties play a very significant role in determining the NP coronas composition	[43]
		a)-COOH modified NPs, -40; b)NH₂-modified NPs, +30				
	a) PLGA NPs stabilized with DDAB with and without PEG b) HSA NPs with and without PEG	a) range: 70-100 b) range: 130-150	SDS-PAGE LC-MS/MS	FBS	Influence of the core material, surface charge, and surface modification on the amount and nature of the adsorbed proteins	[41]
		a) 46/+14 b) -54/-53				
	Mixed (PEG/gold) NPs*	range:15-90	SDS- PAGE	10% v/v diluted human serum	High density PEG coating leads a minor amount of protein adsorbed on NPs surface than low density PEG-NPs	[55]
	Mixed (Dextran/PLA) NPs*	a) range:155-160 b) range: 165-174	Lowry-Peterson assay	BSA solution (50ug/ml)	Dextran coating reduces BSA adsorption on NPs surface	[57]
		a) -9 b) -7				
	Functionalized Polystyrene NPs	a)range 23-224; b)-COOH modified NPs; range 27-224; c)NH₂ modified NPs; range 57-284	CD and Intrinsic Tryptophan Fluorescence Spectroscopy	HSA, apo AI, rHDL-apo AI,chicken lysozyme in PBS or EDTA buffer	Different functionalizations on NP surface affect the secondary structure of adsorbed proteins in a specific way	[64]
		/				
NPs Surface Charge	Functionalized latex NPs	range: 82- 109	BCA	Human serum	Difference in protein adsorption pattern depends on surface charge	[60]
		a)acidic groups,-5/ -10; b)basic groups, +2/ +8				
	Mixed (PLGA/surfactants¹) NPs	~200	DLS BCA SDS-PAGE	Plasma	Positively charge NPs display the maximum size variation and protein adsorption	[61]
		a)PLGA/PEI NPs, +30/+49 b)PLGA/PVA NPs, neutral c)PLGA/casein NPs, -21/-31				
	Functionalized polystyrene NPs	~200	ZP	MEM with 10% FBS	Both groups of NPs (cationic and anionic) become negative after incubation, with ZP of -20/-25 mV.	[62], [63]
		a)cationic groups, +20; b)anionic groups, -30				

(Table 1) contd.....

	NIPAM-BAM in different ratio (85:15, 65:35, and 50:50)	70 and 200	ITC	HSA	Number of protein molecules bounded increases with the particle hydrophobicity.	[26]
		/				
NPs Hydrophility/Hidrophobocity	NIPAM-BAM in different ratio (85:15, 65:35, and 50:50)	70- 120- 200- 400- 700	ITC	HSA in Hepes/NaOH buffer	hydrophobic NPs bind high amount of HSA,(except at high curvature)	[65] [73]
		/				
	Latex NPs	91-89- 87- 120 340	2D-electrophoresis	Plasma	Decreasing hydrophobicity lead to a reduced protein adsorption and to a qualitatively different PC on NPs surface	[66]
		/				

Abbreviations: NIPAM-BAM = N-isopropylacrylamide-co-N-tert-butylacrylamide copolymer; PEG= polyethylene glycol; PLGA= poly-lactic-co-glycolic acid; DDAB= Dimethyldioctadecylammonium bromide; HSA= human serum albumin; PLA= poly-lactic acid; SDS-PAGE= Sodium Dodecyl Sulphate - PolyAcrylamide Gel Electrophoresis; LC-MS/MS= Liquid chromatography-tandem mass spectrometry; CD= Circular dichroism; BCA= bicinchoninic acid assay; DLS= dynamic light scattering; ZP= zeta potential measurement; ITC= Isothermal Titration Calorimetry; FBS= fetal bovine serum; BSA= bovine serum albumin; apo AI= apolipoprotein AI; rHDL= reconstituted high density lipoprotein; PBS= Phosphate-buffered saline buffer; EDTA= Ethylenediaminetetraacetic acid; MEM= Minimum Essential Medium.
mixed NPs*= polymeric coating of organic/inorganic NPs
surfactants ^{1}= PVA (polyvinyl alcohol), PEI (polyethyleneimine) and casein

Gossmann and collegues tried to overcome the difficulties of the specific surface area of different sized NPs. They normalized the amount of NPs used for protein adsorption experiments on the basis of their total surface area [41]. This alternative seems more problematic to be applied mainly due to the difficulty of reliably assessing the surface area of the nanosystems and to also fit the data with the NP curvatures.

The change in NP curvature observed when varying the dimensions of NPs really affects the amount and type of proteins adsorbed. Generally, low curvature leads to increased protein adsorption [41]. In addiction, the experiments of Klein and co-workers indicated that very small NPs with highly curved surfaces arrive to suppress protein adsorption to the point where it no longer occurs. Moreover, this effect is likely to be selective to larger proteins, offering a route to differentially control protein adsoption [42].

Cedervall and colleagues investigated the role of size and surface curvature using 50:50 N-isopropylacrylamide (NIPAM) and N-tert-butylacrylamide (BAM) particles with diameters ranging from 70 to 700 nm. Results showed that the amount of bound proteins after incubation with plasma varied with size and scaled with the amount of available surface area, but the protein pattern remained almost constant independent on the NP sizes. Therefore size and surface curvature influence the amount, but not the identities of bound proteins [26].

On the other hand, Lundqvist and co-workers, using six different polystyrene NPs with three different surface chemistries and two size of each (50 and 100 nm) incubated in plasma, showed that NP size could plays a significant role in the composition of the PC. In this case, NPs with the same surface characteristics, but different sizes showed differences in the composition of the corona: 100 nm negatively charged NPs adsorbed a greater amount of immunoglobulins than 50 nm NPs with the same surface characteristics. On the contrary, 50 nm positively charged NPs were surrounded by apolipoproteins with a larger extent than 100 nm NPs with an equivalent surface charge [43]. Similarly, Pedersen *et al.* found differences in the protein pattern composition on dextran NPs with different sizes. In particular they noted that classical IgM-dependent complement activation is most efficient on dextran particles in the optimal size range, of ~ 250 nm, whereas larger particles do not attract as much IgM, and therefore, do not activate to the same extent [44].

Thus, size and NP curvature represent fundamental parameters in determining the PC of a specific type of NPs. Also noted, NPs dispersed in protein-free media often tend to agglomerate and this event completely changes the interactions and conformation of plasma proteins adsorbed. Agglomerates describe a new behavior compared to with separated NPs, for the variation both in size and homogeneity and in other chemico- physical parameters such as shape and surface characteristics [45].

Moreover, several studies seem to indirectly confirm the possible close correlation between size of NPs and apolipoprotein. These proteins are known to be involved in lipoprotein complexes, which themselves display dimensions within the nanoscale range, namely from 100 nm (chylomicron) to ~10 nm (high density lipoproteins). Besides, apolipoprotein displays a remarkable high affinity to adsorb onto hydrophobic surfaces. The evidences is that apolipoproteins are among the most representative adsorbed protein onto polymeric NPs, during the late stage of 'Vroman effect' and therefore possible specific size-dependent interactions that drives the binding of apolipoproteins to NPs, must not be excluded [26].

NP Shape

The shape of NPs can substantially affect the PC: different morphologies of NPs are able to influence protein adsorption both at the composition and at the structural levels. This feature was principally investigated by means of inorganic metal NPs due to their higher versatility in governing shape-formations. Our opinion is that these results cannot be transposed on polymeric NPs. In fact, not only the shape, but also the parameters correlated with shape (type and number of atoms coordinated to NP surface, surface energies) play a key role in defining the curvature effects and-protein binding [46].

In this context, Roach and colleagues observed that a decreased surface curvature of spherical NPs corresponds to an enhanced loss of secondary structure of the model protein human serum albumin (HSA) [47]. Similarly, Gagner and co-workers, using model proteins (lysozyme and a-chymotryps) found higher levels of perturbation on the protein secondary structure upon nanorods structured, compared to the nanosphere morphology [48]. Thus, some interesting experiments can provide for valuable data on inorganic NPs that can be of "stimulus" and "suggestions" for experiments on polymeric NPs.

Surely, there is a strong need to investigate the role of NP morphology/shape on nanoparticle-protein interactions: this investigation would ultimately lead to the ability to plan and adapt the shape of nanosystems (and then the procedure of NPs formulation) to govern protein absorption and the PC-NP stability, which is actually essential for the successful *in vivo*-application of NPs.

As note, we mentioned the interesting experiments carried out by Ashkarran and co-workers that correlated the results relating the adsorption of protein on inorganic NPs with different structures (cube, sphere, wire and triangle). Maintaining the same experimental conditions, they achieved clear indications that *protein/NP interactions are a morphology-dependent event*. In particular, both high molecular weight proteins (>90 kDa) and low molecular weight proteins (<65 kDa) tend to absorb on cubic and triangle surfaces to a higher exent than spherical and wire NPs [49]. In a similar way, Mahomundi and Serpooshan demonstrated the preferential binding of proteins with specific size range to the different surface roughness in magnetic-polymeric NPs. Especially, 310–120 kDa and 70–30 kDa proteins were shown to be more affine to the smooth surfaces, while 120-70 kDa and 30-10 kDa proteins to the jagged surfaces [50].

Surface Properties: Composition and Polymeric Chain on NP Surface

Composition of PC (hard and soft) is not a universal rule to be translated for granted for all types of NPs and, as obviously, it greater depends on the composition of NPs. Several articles have described different PC compositions for different polymeric and inorganic NPs. In recent years, Karmali and Simbergs proposed a classification of plasma proteins adhering to different nanoparticles [51], while Walkey and Chen compiled a list, named "adsorbome", of the identified plasma proteins and their relative abundances in the PC for 63 different nanomaterials [21].

Thus, even if there is not a universal rule about the relationship between PC and NPs, analyzing these studies, a trend of protein adsorption onto different nanomaterials could be established.

As examples, polymeric NPs (*e.g.* poly-lactic, poly-lactic co-glycolic, cyanoacrylate, polystirene and caprolacton NPs), differently from inorganic NPs, show high affinity for apolipoproteins [24, 52 - 54]. On the other hand, the most abundant proteins in plasma (albumin and fibrinogen) were found onto many types of NPs both polymeric and metallic ones [21]. Other proteins like transferrin, haptoglobin, fetuin A (alpha-2-HS-glycoprotein), kininogen, histidine-rich glycoprotein represent a sub-set of adsorbed protein for different nanomaterials, since these proteins can be attracted by polymer NPs and NPs with hydrophobic surface components or hydrophilic inorganic NPs [20].

Besides material composition, the polymeric chain on the NP surface also influences the PC. Walkey and colleagues found that over 70 different serum proteins are heterogeneously adsorbed to the surface of gold NPs grafted with polyethylene glycol (PEG) at different densities. The relative abundance of each of these adsorbed proteins was found to depend not only on nanoparticle size, but also on the poly(ethylene glycol) grafting density. In general, high density PEG-NPs adsorbed less protein than low density PEG-NPs [55]. Accordingly with this finding, Perry *et al.* reported that brush PEG conformations adsorbed less protein than mushroom conformations [56]. Nevertheless when increasing the PEG content in the nanoparticles over 5 wt %, no further reduction in protein adsorption was assessed [12].

On the other hand, in parallel to PEG, the development of NPs coated with polysaccharides should be considered. Literature data demonstrated that the coating of polycaprolactone NPs with phenoxy dextran (DEX) is able to induce a decrease in immunoglobuline adsorption and, at the same time, an increase of apolipoproteins on NP surface [57].

Finally it should be mentioned that other coatings and ligands on NP surface could affect PC formation. However the available data are not fully developed and deeper studies are still required to be able to govern the composition of the PC by planning and tailoring a proper NP surface engineering.

Surface Charge

Regarding the role of surface charge in PC composition, several confusing studies are reported in literature. More than for other parameters, the protein adsorption onto NPs is complicated by surface charge due to the composition of NPs (polymer, presence of residual surfactant, residual active agent adsorbed), but also due to the possible ligand (PEG, antibody etc) and serum protein presentation. In summary, neutral particles display a slower opsonization rate than charged particles, demonstrating a direct correlation between surface charge and protein binding [58, 16]. Gessner *et al.* investigated the effect of surface charge density increasing on protein adsorption patterns by using polymeric NPs with a constant diameter and surface hydrophobicity. The surface charge density of their NPs varied in the range of -3.7 to -8.2 $\mu C/cm^2$. The results showed an increase in plasma protein adsorption correlated to the increase of the surface charge density, but no differences were assessed in the detected protein types [59].

In contrast, studies on latex NPs with either basic or acidic functional groups showed differences in the protein adsorption patterns depending on the surface charge. In fact, positively charged particles (bearing basic functional groups) showed a preferential binding of proteins with isoelectric points less than 5.5, such as albumin, while negatively charged particles (presenting acidic functional groups) tended to bind proteins with isoelectric points greater than 5.5, such as IgG. The same study also pointed out a preference for distinct proteins by different functional groups. In fact, albumin and IgG preferentially bind particles functionalized with strong basic or weak acidic groups, respectively [60].

Lundqvist *et al.* reported that positively charged NPs adsorbed more apolipoproteins than neutral and negatively charged particles, while neutral NPs adsorbed immunoglobulins and complement factors with a larger extent than charged particles [43]. More recently, Gopikrishna *et al.* described the behavior of negative, near neutral and positively charged poly-lactic-co-glycolic acid (PLGA) NPs in the presence of plasma proteins. NPs, with similar size range, were prepared by using casein, poly-vinyl-alcohol and poly(ethylene imine) respectively as surface stabilizers. A significant transient variation in hydrodynamic diameter of PLGA NPs was observed in the presence of plasma proteins, which correlated to the amount of proteins adsorbed to each surface. Positively charged particles displayed the maximum size variation and protein adsorption [61]. Similarly, a comparative study between positively charged NPs based on PLGA-didodecyldimethylammonium bromide (DMAB) and negatively charged nanoparticles based on human serum albumin (HSA) was performed by Gossmann and co-workers. The results demonstrated that the amount and diversity of adsorbed proteins on positively charged PLGA –DMAB NPs were much higher than for negatively charged HSA nanoparticles. On the surface of PLGA NPs about fivefold more proteins were identified. Moreover, differences in the chemical and physical nature of the adsorbed proteins were found due to the different surface charges of the two particle systems: proteins with lower isoelectric point (pI) were preferentially bound on the surface of PLGA NPs; in contrast, proteins with higher pI adsorbed onto HSA NPs [41].

However, since most plasma proteins are negatively charged, it is also possible that the formation of a PC lead to a change in NP surface charge, specifically to anionic patterns [17]. Experiments performed on polystyrene NPs (200 nm) functionalized with either cationic or anionic groups pointed out that, after the corona layer was established, the two kinds of NPs (anionic/cationic) are indistinguishable, as they both become anionic [62, 63].

Finally, not only may the surface charge affect the identity and amount, but also the structure of adsorbed proteins. This effect can be investigated using circular dichroism, fluorescence spectroscopy, and limited proteolysis experiments. The same protein adsorbed on polystyrene nanoparticles with different surface charges were shown different secondary and tertiary structures. Plain and negatively charged NPs showed to induce a helical structure in apolipoprotein AI (negative net charge), whereas positively charged NPs were able to reduce the amount of helical structure. Moreover, plain and negatively charged particles were demonstrated to induce a small blue shift in the tryptophan fluorescence spectrum, which was not noticed with positively charged particles [64].

Hydrophilicity/Hydrophobicity

Surface hydrophobicity of NPs is another parameter that clearly influences protein binding. Cedervall and collaborators demonstrated that the hydrophobicity of NP surface influences both the amount and the identity of the proteins bound to the NPs. In fact, they observed that 50:50 NIPAM-BAM copolymer NPs (with a size in the range 70-700 nm) bound several proteins, such as apolipoproteins AI, AII, AIC and E, as well as HSA and fibrinogen. On the contrary, the more hydrophilic 85:15 NIPAM-BAM particles of the same size only bound small amounts of HSA [26]. Thus, it seems that more hydrophobic is the NPs surface, more abundant is the amount of some adsorbed proteins.

To complete the description of the interaction of these kind of NPs and the more adsorbed proteins (*i.e.* HSA and fibrinogen), results clearly indicated that the hydrophobicity also influences the exchange rates of bound proteins. In fact, HSA and fibrinogen display a faster dissociation rate from the more hydrophobic particles [25].

Lindman and co-workers also studied the adsorption of HSA to polymeric (NIPAM-BAM) NPs with different hydrophobicity and the generated curvature; these studies confirmed that the most hydrophobic particles bound the higher amounts of albumin [65]. Otherwise, the binding of apolipoproteins were found to be a major part of the formation of PC on hydrophobic latex NPs. Gessner and collaborators synthesized different latex NPs with decreasing surface hydrophobicity as colloial carrier models. They found that the decreasing surface hydrophobicity leads to quantitative decrease in the amount of adsorbed proteins and to qualitative change in protein adsorption patterns. In particular, more hydrophobic NPs preferentially adsorbed apoliproteins while hydrophilic NPs typically adsorbed IgG, fibrinogen, and albumin [66].

Biological Parameters (Blood Concentration, Plasma gradient, Temperature)

The plasma protein concentration and the protein gradient notably influence the formation and the stability of the PC.

During their "*in vivo* journey", in accordance with the route of administration (*e.g.*, subcutaneous, intradermal, intramuscular, intravenous, intraosseous, intralumbar, and inhalation), NPs would be exposed to a variety of biological fluids, which contain several types of proteins at different concentrations. Obviously, the composition and amount of PC on NP surface correlate also with the gradient complex media that should be evaluated in order to optimize the NP pathways in the human body [67]. Recently, Monopoli and co-workers, by using both silica and polystyrene NPs and by comparing the results from several analytical techniques, demonstrated that an increase concentration of plasma protein corresponds to an increase of the thickness, particularly of the hard PC [24].

The composition of PC seems to evolve when NP-PC complexes are moved from one biological environment into another, for example from blood plasma into cytosolic fluid. Some proteins from the "*original*" corona are quickly replaced by proteins from the new biological fluid, while other proteins are still preserved in the protein layer notwithstanding the change in the environments of NPs. Thus, the corona seems to change in composition, but retains a sort of *fingerprint* of its birth and initial evolution [68].

In this case, several results related to *in vitro* studies on different nanosystems strengthen the influence of the *in vivo* observation. For example, Maiorano and colleagues investigated the composition and dynamics of the PC around NPs incubated in two widely used cellular media that differ in the concentration of glucose and of single amino acids. They demonstrated a different composition, dynamics of NP-protein interactions and an increase in size [69].

Remarkably, the source of the incubation medium also affects the formation of the PC; studies on liposomes (size range 200-250 nm), incubated either in human plasma or in mouse plasma, showed that in the latter serum type, the corona was more enriched in proteins than in the first one [70].

Temperature is a very important factor for the *in vivo* applications of NPs. The mean body temperature for different individuals varies from 35.8 to 37.2°C and it also varies for different parts of the body. Furthermore, the body temperature of females is slightly higher than males and can be influenced by the hormonal cycle. Reasonably, the temperature could influence the structure and the stability of proteins and then the formation of the PC. The physiological changes described above and, most importantly, the phatological alteration of temperature (as in inflamed or cancers tissues) should be considered as critical parameter affecting the fate of NPs in the organism [20]. We only cited the study carried out by Mahmoudi and co-workers on superparamegnetic/inorganic NPs as it can be considered significantly indicative for further evaluation of temperature in NP-PC interaction studies. They demonstrated that the composition and the dynamics of the PC formation on NPs synthesized from different materials with different surface chemistries (negative, positive and neutral, with size of 33, 79 and 33 nm respectively) changed by varying the temperature from 37 to 41°C. Forcing the temperature conditions, thickness of the PC around NPs increased with protein concentration at 23°C, while this increase was less pronounced at 43°C. The authors hypothesize that this divergence may lie on the fact that, at higher temperatures, polymer shell is characterized by enhanced permeability and not only do proteins adsorb onto the NPs surface, but also partially penetrate in the polymer coating, thus leading to a minor increase in diameter. Additionally, at 43°C, binding affinity of proteins was higher than at 23°C. Lastly, the authors also observed an increase in cellular uptake of NPs corresponding to the increase in temperature [71].

PC-NPs *vs* ADMINISTRATION

The relation between the original NPs and the PC is still complicate and it remains very elusive.

As a common point, the formation of a PC leads to an increase in the diameter of the NPs (reported to be in the range of 20-30%); as previously described, all the other surface/structure properties of NPs change dramatically in function of the experimental conditions (composition of NPs, surface charge, presence of surfactant, ligands on NP surface, medium etc.). Therefore it is almost impossible to create a model of NP-PC interaction. Notwithstanding this evidence, a common feeling and stated knowledge to declare that the new "entity" generated by the interaction between NPs and proteins *in vivo* will mediate the ability of all the nanosystems to circulate, biodistribute and selectively release the drug to the target site. The properties of NPs, but also the nature, the dynamicism and the conformation of the PC play a key role in biological events (*e.g.* opsonization, cellular uptake, trafficking, toxicity *etc*) describing the applicability of these nanocarriers (fig. **2**). All these variables strongly hamper a possible prediction of the *in vivo* behavior of NPs.

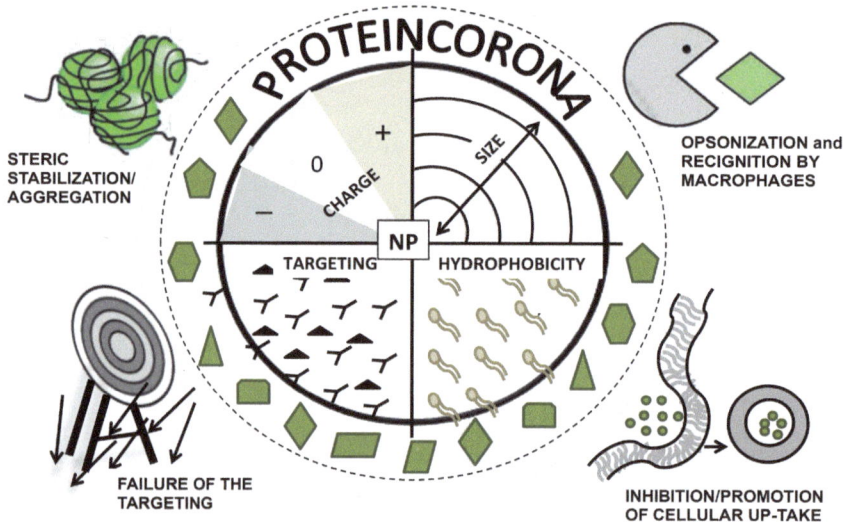

Fig. (2). Schematic representation of the biological fate of a polymeric NP in physiological media. Pre-existing or initial NP characteristics (*e.g.* charge, size, surface functionalization, hydrophobicity/hydrophility) contribute to the formation of a specific pattern of adsorbed protein on NP surface (the green geometric forms). These adsorbed proteins can influence the in-vivo behaviour of the NP. The four external pictures illustrate different scenarios of some potential effects following protein adsorption

Influence on Stability

Aggregation of NPs

Under physiological conditions, NP stabilization could decrease due to high electrolyte concentrations and also due to interaction with biomolecules (that neutralize the surface charge for example) (Table **2**). Several experiments demonstrated that aggregates present a lower surface area with unpredictable diffusion properties and possibility of sedimentation and consequently failure of therapy [74]. Over the destabilization due to the possible aggregation of NPs after the interaction with plasma protein, the timing of these events is also a critical parameter to define the availability of NP-distribution to the target site. An interesting series of trials carried out by Gebauer and co-workers are based on the introduction of a new statistical model to describe a binding curve and to quantify a binding affinity. This approach also considers the time of agglomeration, suggesting that the stability of NPs under physiological conditions can be almost entirely independent of the particle formulation. Particularly, the authors demonstrated how steric effects could replace the charge stabilization and alter the stability of NPs. However, many colloids are well known to be unstable in serum suggesting that other parameters presumably contribute to the stability of NPs. For example, the selective bridging effects between individual proteins could also contribute to a possible destabilization [75].

Table 2. PC/polymeric NPs *vs* administration. *In vitro* and *in vivo* experiments to evaluate stability, immunological response, biodistribution, drug targeting, uptake and toxicity evidences.

PC/Polymeric NPs		Type of experiment	Objective	Results	References
PC identity	NPs composition, Size (nm), ZP (mV)				
7-8 nm of PC thickness after incubation in fibrinogen solution for NPs with 22 nm of HD	Mixed (Poly-acrylic acid/gold) NPs*, size range 7-22, ZP range -25/-50	*In vitro* (fluorescence spectroscopy, DLS)	Stability of NPs in protein solution (fibrinogen 2,5 mg/ml)	In excess of NPs, fibrinogen induced aggregation of NPs with size > 12 nm, that could adsorb more than one protein molecule.	[72]
/	Mixed (dextran/iron oxide) NPs*, different size (50, 100, 250, 600)	*In vitro* (hemolytic assay)	Relationship between complement system and NPs with different size	IgMs preferentially adsorb on NPs having a diameter around 250 nm inducing the activation of complement system	[44]
NPs with 160 nm of HD tend to aggregate while other NPs not shown altered HD after the incubation in BSA solution	Mixed (dextran/PIBCA) NPs *, size range 90-230	*In vitro* (serum incubation and 2D immunoelectrophoresis)	Immunological response in relation with NP coating	Surface density of dextran coating reduce complement activation	[85]
Variable amount of the adsorbed proteins on the basis of NP surface functionalization: NH_2>COOH>OCH_3	Lipid/PLGA/PEG[1] NPs, size a)107, b)88 c)110, ZP a) -25 b)-10 c) +10 [1]PEGunits a)COOH-PEG; b)OCH₃-PEG; c)NH₂-PEG,	*In vitro* (ELISA PAGE LC-MS/MS)	Immunological response in relation with NP surface functional group	Hybrid NPs with methoxyl surface groups induced the lowest complement activation, probably due minor amount of adsorbed proteins their high affinity with complement deactivating protein (fH)	[88]

(Table 2) contd.....

PC/Polymeric NPs		Type of experiment	Objective	Results	References
PC identity	**NPs composition, Size (nm), ZP (mV)**				
Thin layer of PC with average thickness of 0.8 nm after NP incubation in the protein solution with highest concentration	Mixed (Dextran/PIBCA) NPs*, size 25	*In vitro* (Radial immunodiffusion Neutron scattering)	Proof of concept of the NP opsonization process	Upon adsorption onto NPs, BSA undergoes a structural change that possibly triggers complement system activation.	[84]
/	Polystirene NPs, size 50	*In vitro* (SDS-PAGE, LC-MS/MS)	Evolving of PC in different cellular compartment	PC evolves on NP surface as a result of transfer of the same NPs from one biological fluid into another.	[68]
/	Polystyrene NPs, size 50, ZP - 21	*In vitro* (Western blot Fluorescence spectroscopy)	Influence of specific protein adsorption in biodistribution process	Fetuin on NP surface mediated the recognition and internalization of NPs by Kupffer cells *via* scavenger receptors	[94]
/	PLGA NPs, size 227, ZP -18	*In vitro* (LC-MS/MS)	Proof of concept of the possible influence of PC in biodistribution process	PLGA NPs show to adsorb ApoE on their surface that could represent a possible strategy for the brain targeting	[54]
Protein amount after serum incubation ~ 350 ng/ µg of NPs	Polystyrene NPs with and without BSA coating, size 50	*In vivo* experiments (rat model)	Biodistribution experiments	BSA coated NPs show longer blood-circulating property	[97]
More concentrated plasma leads to a thicker PC (up to 38% of HD increase) and to a lower ZP (less negative)	Polystyrene NPs, size range 50- 200, ZP-40	*In vitro* (DLS ZP, TEM, SDS-PAGE)	Influence of plasma concentration in PC formation	Different plasma concentrations correlate with different thickness and qualitative profiles of PC	[73]
Reduction (8%) of HD after incubation in highest concentrated protein medium indicating loss of TfR binding when a protein corona is formed on the NP surface	Functionalized[2] silica NPs, size 50 [2]*transferrin ligand*	*In vitro* (SDS-PAGE, flow cytometry)	Influence of PC on drug targeting activity of NPs	After incubation in serum, transferrin loses its specificity for transferrin receptor, suggesting that proteins are able to shield this ligand	[100]

(Table 2) contd.....

PC/Polymeric NPs		Type of experiment	Objective	Results	References
PC identity	**NPs composition, Size (nm), ZP (mV)**				
/	PBCA NPs[3], size 300 [3]loaded with loperamide	*In vivo* (mouse model)	Influence of different apolipoproteins in brain targeting delivery	ApoE coating enhanced BBB crossing of NPs.	[95]
Adsorbed protein amount is higher for bare NPs compared with PEG-NP (56.2 *vs* 35.1 μg/mg of NPs)	PHDCA with/without PEG coating and PHDCA-PEG precoated with different apolipopreteins All NP size ~ 140, ZP~ -20	*In vitro* (rat brain endothelial cells uptake experiment)	Study of the mechanism by which NPs could penetrate into rat brain endothelial cells	ApoE and ApoB-100 adsorption onto NP surface leads to a more efficient NPs uptake by the endothelial cells	[108]
Thicker PC (up to 31% of HD increase) after NP incubation in more concentrated plasma medium	Lipid(DOTAP)/DNA NPs, size 220 nm	*In vitro* (Flow –cytometry, colocalization-assay)	Study of PC influence in drug targeting	NPs incubated in plasma acquire a selective targeting for cancer cell overexpressing integrin receptor due to their PC enriched *in vitro*nectin	[131]
Increase in HD (up to 72%) and moderately negative ZP (~-20/-30 mV) after serum incubation.	Functionalized polystyrene NPs[4], size ~ 100 nm, ZP range from +8 to -61 mV [4] COOH, NH3, SO3, PO3, PO3-Lut, PO3-SDS	*In vitro* (Flow cytometry Confocal microscopy)	Correlation study between PC and NP uptake	Only few specific protein composing PC are able to influences cellular uptake	[125]
Increase in HD (up to 15%) and moderately negative ZP (~-20/-30 mV) after serum incubation	Functionalized polystyrene NPs[5], size ~ 200 nm, ZP range -4 /+8 mV[5] AA, VBPA, SEMA, AEMH	*In vitro* (Flow cytometry)	Influence of serum protein in cellular uptake	Cellular uptake is lower in experiments with FCS supplemented medium than in the FCS free medium	[115]
ZP of ~ -20 mV for both NPs after serum proteins incubation	Functionalized polystyrene NPs[6], size ~ 200 nm, ZP a) cationic groups +20 mV b)anionic groups-30 mV [6]COOH, NH2	*In vitro* (Uptake experiment by monkey kidney epithelial cells)	Influence of NP properties on PC formation and in cellular uptake	Uptake was enhanced only for PC/NPs complexes formed from cationic NPs, although the complexes were both anionic.	[132]
Increasing concentrations of BSA reduce the effective surface charge of the NP with a shift to negative values (up to -20 mV)	Latex NPs, size ~ 90 nm, ZP +43	*In vitro* (Single particle tracking experiment)	Track the cellular internalization of NPs exposed to extracellular serum proteins	BSA is present during uptake by the cell suggesting a role of PC during cellular internalization of NPs	[119]

(Table 2) contd.....

PC/Polymeric NPs		Type of experiment	Objective	Results	References
PC identity	NPs composition, Size (nm), ZP (mV)				
Increase (~20%) of HD and altered ZP (-20 mV) after NPs incubation in protein medium	Poly-methacrylic acid NPs, size 500, ZP-39	*In vitro* (Fluorescence Deconvolution Microscopy Flow cytometry)	Influence of PC in NP uptake by differentiated and un-differentiated monocytic cells	BSA adsorbed on NPs surface undergoes to conformational changes. This protein unfolding differently influences NPs uptake on the basis of the cellular type	[121]
ZP of ~ -20 mV for all types of NPs	Functionalized polystyrene NPs, size range 40- 200, ZP range+20/ -31	*In vitro* (Fluorescence microscopy Single particle tracking)	Study of PC/NPs and cell interaction	PC with same composition adsorbed on surface of different NPs binds different cellular receptors determining different uptake process	[63]
Increase (~25%) of HD and altered ZP (~-10) as a consequence of charge compensation after incubation in cDMEM	Polystyrene NPs, size ~50, ZP~ +40	*In vitro* (Flow cytometry)	Influence of PC in cell damage induced by cationic NPs	PC around NPs is retained during cellular uptake and cell apoptosis occurs later than after uptake of bare NPs	[133]

Abbreviations: HD= hydrodynamic diameter; TfR= transferring receptor; PIBCA= poly-isobutylcyanoacrylate; PBCA=poly-butylcyanoacrylate; PEG= polyethylene glycol; PLGA= poly-lactic--o-glycolic acid; PHDCA= poly- hexadecyl cyanoacrylate; DOTAP= N-[1-(2,3-Dioleoyloxy)propyl]-N-N,N-trimethylammonium methyl-sulfate; DNA= deoxyribonucleic acid; Lut= Lutensol AT50; SDS= sodium dodecyl sulfate; AA= acrylic acid; VBPA= vinyl(benzylphosphonicacid); SEMA= 2-sulfoethyl methacrylate; AEMH= 2-aminoethyl methacrylate hydrochloride; DLS= dynamic light scattering; ZP= zeta potential measurement; LC-MS/MS= liquid chromatography-tandem mass spectrometry; ELISA= enzyme-linked immunosorbent assay; PAGE= PolyAcrylamide Gel Electrophoresis; TEM= transmission electron microscopy; ApoE= Apolipoprotein E; ApoB100= apolipoprotein B100; BSA= bovine serum albumin; FCS= fetal calf serum; cDMEM= complete Dulbecco's Modified Eagle's Medium.
mixed NPs*= polymeric coating of organic/inorganic NPs.

Actually, PC could produce several effects, negative or positive on serum stability. Destabilization was principally referred to the action of same protein forming PC that may trigger NP aggregation through protein bridges, resulting in the formation of larger clusters [17, 72]. On the contrary, the prevention of aggregation due to the molecular layer of other protein adsorbed around the particles may prevent the surfaces of the NPs from coming in contact with one another [40, 76, 77].

RES Uptake and Immunological Response

Protein binding can cause a change in the size and surface charge of NPs influencing the RES uptake (Table **2**). As a general rule, the opsonization of hydrophobic particles, as compared to hydrophilic particles, occurs more quickly due the enhanced adsorbability of blood serum proteins on these surfaces [78]. By the same modality, larger NPs will be more efficiently captured by the RES and may cause embolization in the liver and lung [79 - 81].

Focusing on the changes in biodistribution caused by the type of protein bound or un-bound on the NPs, it is almost clear that immunological recognition by the RES cells and the consequent elimination of these systems from blood may be triggered by immunostimolatory proteins named 'opsonins' adsorbed onto the NPs [78, 82]. The most commonly described opsonins are fibrinogen, immuno-globulins, albumin and complement components.

Immunoglobulins are the most present opsonins and are involved in both in the recognition and uptake of NPs by macrophages; they bind to the Fc-receptor on the surface of several immune cells, such as neutrophils, eosinophils, monocytes, some macrophages (including Kupffer cells), mast cells and some dendritic cells, immunoglobulins augmenting the capture of particles and reduce their bioavailability [17].

Albumin, immunoglobulins and fibrinogen bind to nanoparticles due to their large abundance in the blood, but their binding affinities may be weak. These proteins could initially dominate the surface of the NPs but could be displaced by proteins of lower abundance, that have higher affinity, and slower kinetics. This displacement might lead to other NP-bound proteins, such as apolipoproteins that could govern their fate and organ distribution [83].

Immunoglobulins can also stimulate phagocytosis by activating the complement system, a network of proteins able to recognize and eliminate foreign nano and microparticles. Polymeric nanomaterials have especially been found to trigger complement activation. Once the complement system is activated, NPs are eliminated by the immune system. Certain complement proteins (*e.g.*, C3, C5) are known to play a key role in complement activation and NP clearance *via* the reticuloendothelial system (RES) [51]. Recognition and elimination of foreign particles by macrophages may occur through adsorption of the complement C3 large protein. After binding to a particle, C3 changes its conformation and activates the complement cascade, which leads to the elimination of the particles by macrophages.

With this mind, recently, Vauthier and co-workers have shown that the C3 protein might not bind to the particle directly; by investigating the complement system activation triggered by adsorption of BSA onto NPs (25 nm) composed of dextran-poly(isobutylcyanoacrylate) (PICBA) copolymer, researchers demonstrated that albumin undergoes a conformational alteration during adsorption inducing binding of the complement protein C3 and complement system activation [84, 85]. Similarly, fibrinogen is also capable of acting as an opsonin as a result of conformational changes in its γ-chain. This conformational change occurs during fibrinogen adsorption and seems to be mainly due to highly curved NPs [33]. Overall, variable relationships exist between material properties and adverse immunological responses. Different structural determinants makes it difficult to precisely map-out 'structure-complement activation' relationships for better design [86].

In order to prevent macrophage recognition, one of the most common strategy consist of decorating the NP surface with repelling hydrophilic polymers. As a matter of fact, the presence of elements that generate steric constraints (*e.g.*, projected surface polymers) to particle–macrophage interaction can induce resistance in spite of opsonization [86]. In particular, PEG functionalization improves circulation properties and decreases macrophage recognition of many types of NPs [87]. Salvador-Morales and collaborators demonstrated a method to control the levels of complement activation of PLGA–lipid PEG NPs by controlling surface chemical modification with methoxyl, carboxyl, and amine groups on NPs. In this study, hybrid polymeric structures of NPs with methoxyl surface groups bound complement deactivating proteins (fH) and, thus, induce the lowest complement activation [88]. Several papers highlighted the importance of the use PEGylated NPs in order to improve their stability and steric hindrance, inhibition of proteins (opsonins) adsorption, reducing the uptake of the PEGylated particles by the mononuclear phagocytic system of the RES.

As referred in the previous paragraph, the density and shape of PEG chains on a liposomal surface governs the degree of protein interactions. The brush conformation is more effective at protecting the NPs from protein binding. Nevertheless, upon repeated administration of pegylated NPs the body may develop antibodies against PEG [89]. Moreover, we should underline that generally pegylation produce a conflicting effect at target site. Notwithstanding, a long retention times in the body that are highly desirable to obtain longer circulation of the NPs in the blood vessels, thus offering a higher chance to arrive at target site, a reduced NP uptake by target cells is often observed [90 - 92].

Influence on Biodistribution

Obviously, the protein binding governs the biodistribution of NPs in the body, not only to RES, as plasma proteins exhibit distinct affinities for different tissues. Both the absence and the presence of adsorbed plasma proteins on the surfaces of NPs influence the biodistribution (Table **2**). Changes in NPs size due to protein adsorbed on NPs surface strongly impact in body biodistribution. As known, NPs smaller than 5 nm are excreted by the kidneys [81], while NPs larger than 100 nm tend to accumulate in the liver; at the same time, certain diseases characterized by hypovascularity and hypopermeability require treatment with particles smaller than 50 nm [93].

The amount, conformation and type of plasma proteins can direct lead to distribution of the nanoparticle to a given organ or tissue, inducing the researchers to formulate NPs already functionalized with proteins on the surface or likely to bound/adsorb such proteins once administered. As example, a study involving polystyrene nanospheres showed the formation of a complex with fibronectin, allowing these particles to be primarily taken up by liver macrophages (Kupffer cells) [94].

On the other hand, covalent attachment of various apolipoproteins (ApoE, ApoA-I, and ApoB-100 mainly) to HSA or poly(butylcyanoacrylate) NPs represents an approach for the drug transport into the brain, taking advantage of the interaction of these protein-engineered NPs with brain endothelial cells [95, 96]. On the basis of these observation, Sempf and co-workers speculate that PLGA NPs only adsorbing apolipoprotein E on the surface could represent a valid candidate for a brain targeted delivery system [54]. In other experiments Owara and co-workers demonstrated that albumin adsorbed on NP surface can reduce liver deposition and prolong blood circulation times [97]. Tumors also exhibit increased uptake of albumin, due to the enhanced permeability and retention effect (EPR), and receptor-mediated endothelial transcytosis of the protein [[93] and this phenomenon was exploited for the development of albumin-bound therapeutics for cancer therapy [93].

At the same time, researchers changed the surface of NPs by conjugating hydrophilic polymers as poloxamine [87], polaxamers [98], dextrans [52], pluronics [99] in order to prevent protein binding, or at least decrease the amount, and/or change the overall profile of proteins bound changing the biodistribution particularly to the the reticuloendothelial system (RES).

Influence on Drug Targeting

Surface functionalization of NPs with ligands for specific cell receptors is a widespread strategy to target drug delivery systems to selected tissues or cell populations (Table **2**). In particular, considering CNS drug-targeting, it represents one of the most interesting strategies to by-pass BBB with a non-invasive approach. However, the formation of a protein coating around the particle can mask these specific ligands and hamper their targeting activity. This event was observed in different targeted silica NPs (*i.e.* transferrin-conjugated silica NPs [100] or bicyclononyne -functionalized silica NPs [101], but also in lipidic and polymeric nanoparticles coated with tumor-specific antibodies that does not enhance tumor accumulation *in vivo* [102, 103]). Simply, the most obvious consideration is that the PC surrounding the NP hinders interactions between the NPs ligands and their targets on the cell surface.

Despite this possibility, in some cases, bound proteins are able to direct or target the NPs to a particular area of the body, since plasma proteins exhibit distinct affinities for different tissues. As already stated, opsonins induce liver and spleen accumulation, leading to a rapid clearance of the NPs from the blood circulation.

Knowing the correlation of "surface properties–adsorption pattern–organ/tissue distribution", one can theoretically design nanocarriers with surface properties, which lead automatically to a preferential adsorption of proteins that mediate the adherence to the desired target area [104]. For example, Caracciolo and co-workers created NPs that acquire a selective targeting capability with a PC adsorbed on the surface. In these experiments, particles made of a lipid 1,2-dioleoyl-3-trimethylammonium propane (DOTAP) and DNA (used as polymer in this case), upon interaction with human plasma components, spontaneously become coated with vitronectin that promotes efficient uptake in cancer cells expressing high levels of the vitronectin $\alpha v\beta 3$ integrin receptor

On the other hand, the most prominent examples are referred to CNS drug-delivery. In particular, it has been shown that Tween 80-coated polymeric nanoparticles loaded with darlagin [105] preferentially adsorb apolipoprotein E leading to uptake and penetration across the BBB.

Even if apolipoprotein-E is also involved in hepatic uptake of nanomaterials, therefore promoting the clearance from the blood circulation [106], apolipoproteins (A, B and E) were shown to assist in transport of several other types of NPs, as (poly(butyl)cyanoacrylate (PBCA), poly(ethylene glycol-c--hexadecyl)cyanoacrylate (PEG-PHDCA) NPs) able to cross the blood brain barrier [95, 107, 108]. Receptor mediated endocytosis was speculated to be the main pathway involved in NPs uptake. In particular, a recent paper describes ApoA-I crossing the BBB by a saturable transport mechanism. Mahmoudi *et al.* demonstrate that this ability persists when ApoA-I is coated on the surface of

SPION NPs, inducing their NPs to enter in endothelial barrier cells. On these bases, the authors speculate that NPs with a higher association of ApoA-I in their corona compositions would be better able to reach the CNS by transporters at BBB level [109]. Alternatively, a sort of pre-formed PC could be applied on NP surface by functionalization with specific protein in order to achieve specific drug-targeting. Lalani *et al.* [110] compared transferrin- and lactoferrin-functionalized PLGA NPs in order to obtain a preferential release of lamotrigine (sodium and calcium channel blocker) at brain level. Their biodistribution and pharmacodynamic studies proved that lactoferrin is more efficacious than transferrin in brain targeting after intravenous administration. The authors suggested that the enhanced uptake of lactoferrin-conjugated NPs was due to the low endogenous concentration and unidirectional transport of lactoferrin. However, PC formation and protein-protein interaction after intravenous injection was not evaluated in this study. This lack is particularly evident if we consider that PC has been widely demonstrated to alter the biodistribution of NPs [100]. Therefore, a novel rule in NPs targeting evalutation could be proposed by deeply considering both the surface ligand potential and the proteins composing hard and soft corona related to NPs.

Influence on Cell Uptake

The interaction between adsorbed proteins, NPs, and cells is a complex issue, as the adsorbed proteins can control the interaction of NPs with cells and viceversa as NPs themselves can change the structure of the absorbed proteins [11] (Table **2**). Thus, the impact of PC-NPs-complex on the cellular uptake should be evaluated case by case. Moreover, the surface characteristics of NPs discussed in section 2 (size, shape, surface charge, surface functional groups, hydrophilicity) are certanly more important than the bulk/core characteristics to define the PC and thereby the possible biological impacts.

Some important considerations related to the effect of the PC on cellular uptake could be pointed out.

-un-modified nanoparticles

a) Several *in vitro* studies demonstrated the influence of serum proteins on the cellular uptake of NPs [112– 115]. The PC controls the cellular interaction of the NPs by defining the cell surface receptor to which the protein NP complex binds [116 - 118]. The cell surface receptor then directs the internalization and intracellular transport of the NPs [119 - 122].

In vitro studies can be considered only as informative of the behavior of NP-uptake; typically, cellular studies use low amounts (10% dilution or less, depending on cell type) of animal derived serum instead of full plasma, which is present in *in vivo* studies, and thus, NPs coronas are formed at a very different protein-to-NPs ratios under *in vitro* and *in vivo* conditions. Then, frequently, there is not chance to have comparable information between *in vitro* and *in vivo* outputs, increasing the distance between vitro and vivo approaches and thus generating confusion.

b) Depending on the chemical nature of the protein–NP interaction, adsorption forces exerted on proteins may be sufficiently strong to change the delicate architecture of these weakly stabilized, flexible macromolecules and then their conformation at physiological temperatures, varing the capability of NP-PC to bind the cell receptors [123]. Then, it is not surprising that similar NPs with identical PC compositions are able to bind different cellular receptors; as demonstrated by the experiments of Fleischer and co-workers, the difference in the structure of the adsorbed protein may be responsible for the alteration in cellular binding of the NP-PC complexes. Starting from NPs functionalized with either amine or carboxylate groups to provide a cationic or anionic surface, respectively, researchers [126] demonstrated the influence on cellular uptake of the structure of proteins adsorbed onto NPs surface. In this study, serum proteins were adsorbed onto the surface of both cationic and anionic NPs, forming a net anionic protein/NP complex. Although these NP-PC complexes have similar diameters and effective surface charges, they showed the exact opposite behavior in terms of cellular binding. In the presence of bovine serum albumin (BSA), the cellular binding of BSA-NP complexes formed from cationic NPs is enhanced, whereas the cellular binding of BSA-NPs complexes formed by anionic NPs was inhibited. These trends are independent from NPs diameter or cell type. Another study investigated cellular binding of polystyrene NPs (200 nm) either positively or negatively charged after exposure to fetal bovine serum (FBS). Both kinds of NPs formed negatively charged NP-PC complexes, but showed an opposite behavior in terms of cellular binding. NP-PC complexes formed from cationic NPs were taken up more readily than PC/NPs complexes formed from anionic NPs. This behavior was further investigated using only BSA: competition assays showed that BSA/NPs complexes formed from anionic NPs were taken up by albumin cellular receptors, while BSA/NPs complexes deriving from cationic NPs bound to scavenger receptors, that are cellular surface receptors that bind misfolded albumin. Thus, authors suggested that the divergence in NPs cellular uptake was due to a difference in protein conformation after adsorption onto the NP surface, that caused binding to different receptor [63].

c) Literature is lacking of data related the persistence of the PC on NPs after their cellular uptake. The adsorbed proteins seem to be internalized by the cells together with the NPs and, therefore, may enter cellular compartments that they would not normally reach [42, 124]. Understanding the destiny of the PC on NPs during the internalization is, therefore, critically important for the elucidation of the cellular NP uptake.

-modified nanoparticles

a) Modification with PEG reduces cellular uptake, likely because PEG reduces protein adsorption, but it cannot completely prevent PC formation. As a result, the adsorbed proteins, rather than the NPs themselves, determine the cellular receptors used for binding, internalization mechanism, intracellular transport pathway, and immune response [12, 125].

b) The PC surronding the NPs, modified with selective ligands, as well described, can mask the targeting capabilities and the cellular uptake of the NPs.

Influence on NP Toxicity

The interactions between plasma components and NPs may have implications for toxicology. In fact, the PC layer around the NPs can influence positively (decrease) or negatively (increase) the toxicity of NPs (Table **2**).

In particular, after i.v. administration, the adsorption of serum proteins on the NP surface can lead to a modification of endogenous proteins that disrupts their normal function and potentially causes complications, since misfolded proteins are associated with various diseases. For instance, the unfolding of fibrinogen occurs after adsorption promotes interaction with the integrin receptor, Mac-1, and activates the inflammatory signaling pathway [33].

On the other hand, a protective activity of the PC against cytotoxicity is also possible. In these cases, the presence of a protein coat can decrease cell damage during the NP cell interaction. These studies are primarily referred on inorganic NPs [126 - 128] with an ambiguous toxicity correlating with their composition and small size. Studies on carbon NPs demonstrated that serum protein adsorption attenuated the inherent cytotoxicity of NPs and resulted in decreased cytotoxicity by increasing the amount of serum proteins adsorbed on the particle surface. It was hypothesized that the possible mechanism governing this behavior was connected to the presence of serum proteins in the medium which significantly reduced the NPs intracellular uptake. As such, serum proteins adsorbed on NPs can inhibit their toxicity by shielding impurities of metal catalyzers and suppressing the competitive adsorption of other proteins in the medium [129]. Translating the evidence on polymeric NPs, an interestingly experiment was carried out on cytotoxic polycyanoacrylate NPs; PEGylation decreased both protein binding and cytotoxicity, and increased blood circulation time [130].

CONCLUSION AND FUTURE DIRECTIONS

The biodistribution as the efficiency in drug delivery and targeting of polymeric nanoparticles are strongly affected by the biological environment. As a result of wide literature in this area (Table **2**), the principal player in nanodelivery is not the "naked" drug delivery system, but the "entity" generated by the interaction of nanoparticles with the components of the biological system, particularly plasma proteins. Frequently, it is referred to the plasma protein/nanoparticles complexes (PC/NPs). The understanding of the mechanisms that drive the formation, stability, composition of the PC, especially in relation to the administration of polymeric nanoparticles, represents an essential step for the optimization of these systems to be successfully applied. To date, this topic is still very ambiguous.

In general, there are only few amount and poor reproducible experiments on polymeric NPs in respect to inorganic NPs. These un-complete and discontinuos data generate a relevant gap in knowledge that slows the translatability and clinical application of these drug delivery systems.

Regarding the future directions, it is almost clear that the optimization of analytical technologies and procedures able to efficienly define the delicate and dynamic equilibrium proteins/NPs will help the comprehension of the interaction with cells and biodistribution processe, by governing the *in vivo* efficiency of these nanocarries. In our opinion, it is almost impossible to obtain universal "standard models" for polymeric NPs due their compositive complexity and different features of the polymers used in the formulation. In this view, by planning "simple" NPs, that are not extremely complex in material composition or by multiple surface engineering, the understanding of their fate and, generally, these "novel entities" will be faster and more reliable.

CONFLICT OF INTEREST

The authors confirm that they have no conflict of interest to declare for this publication.

ACKNOWLEDGMENTS

Declared none.

REFERENCES

[1]   Parveen S, Misra R, Sahoo SK. Nanoparticles: a boon to drug delivery, therapeutics, diagnostics and imaging. Nanomedicine (Lond) 2012; 8(2): 147-66.
[PMID: 21703993]

[2]   Elsabahy M, Wooley KL. Design of polymeric nanoparticles for biomedical delivery applications. Chem Soc Rev 2012; 41(7): 2545-61.

[http://dx.doi.org/10.1039/c2cs15327k] [PMID: 22334259]

[3]   Coelho JF, Ferreira PC, Alves P, *et al.* Drug delivery systems: Advanced technologies potentially applicable in personalized treatments. EPMA J 2010; 1(1): 164-209.
        [http://dx.doi.org/10.1007/s13167-010-0001-x] [PMID: 23199049]

[4]   Estanqueiro M, Amaral MH, Conceição J, Sousa Lobo JM. Nanotechnological carriers for cancer chemotherapy: the state of the art. Colloids Surf B Biointerfaces 2015; 126: 631-48.
        [http://dx.doi.org/10.1016/j.colsurfb.2014.12.041] [PMID: 25591851]

[5]   Wilczewska AZ, Niemirowicz K, Markiewicz KH, Car H. Nanoparticles as drug delivery systems. Pharmacol Rep 2012; 64(5): 1020-37.
        [http://dx.doi.org/10.1016/S1734-1140(12)70901-5] [PMID: 23238461]

[6]   Sperling RA, Parak WJ. Surface modification, functionalization and bioconjugation of colloidal inorganic nanoparticles. Philos Trans A Math Phys. Eng Sci 2010; 368: 1333-83.

[7]   Thanh NT, Green LW. Functionalisation of nanoparticles for biomedical applications. Nano Today 2010; 5: 213-30.
        [http://dx.doi.org/10.1016/j.nantod.2010.05.003]

[8]   Chou LY, Ming K, Chan WC. Strategies for the intracellular delivery of nanoparticles. Chem Soc Rev 2011; 40(1): 233-45.
        [http://dx.doi.org/10.1039/C0CS00003E] [PMID: 20886124]

[9]   Alexander-Bryant AA, Vanden Berg-Foels WS, Wen X. Bioengineering strategies for designing targeted cancer therapies. Adv Cancer Res 2013; 118: 1-59.
        [http://dx.doi.org/10.1016/B978-0-12-407173-5.00002-9] [PMID: 23768509]

[10]   Conde J, Dias JT, Grazú V, Moros M, Baptista PV, de la Fuente JM. Revisiting 30 years of biofunctionalization and surface chemistry of inorganic nanoparticles for nanomedicine. Front Chem 2014; 2: 48.
        [http://dx.doi.org/10.3389/fchem.2014.00048] [PMID: 25077142]

[11]   Liong M, Lu J, Kovochich M, *et al.* Multifunctional inorganic nanoparticles for imaging, targeting, and drug delivery. ACS Nano 2008; 2(5): 889-96.
        [http://dx.doi.org/10.1021/nn800072t] [PMID: 19206485]

[12]   Gref R, Lück M, Quellec P, *et al.* Stealth corona-core nanoparticles surface modified by polyethylene glycol (PEG): influences of the corona (PEG chain length and surface density) and of the core composition on phagocytic uptake and plasma protein adsorption. Colloids Surf B Biointerfaces 2000; 18(3-4): 301-13.
        [http://dx.doi.org/10.1016/S0927-7765(99)00156-3] [PMID: 10915952]

[13]   Tosi G, Costantino L, Rivasi F, *et al.* Targeting the central nervous system: *in vivo* experiments with peptide-derivatized nanoparticles loaded with Loperamide and Rhodamine-123. J Control Release 2007; 122(1): 1-9.
        [http://dx.doi.org/10.1016/j.jconrel.2007.05.022] [PMID: 17651855]

[14]   Monsalve Y, Tosi G, Ruozi B, *et al.* PEG-g-chitosan nanoparticles functionalized with the monoclonal antibody OX26 for brain drug targeting. Nanomedicine (Lond) 2015; 10(11): 1735-50.
        [http://dx.doi.org/10.2217/nnm.15.29] [PMID: 26080696]

[15]   Valenza M, Chen JY, Di Paolo E, *et al.* Cholesterol-loaded nanoparticles ameliorate synaptic and cognitive function in Huntingtons disease mice. EMBO Mol Med 2015; 7(12): 1547-64.
        [http://dx.doi.org/10.15252/emmm.201505413] [PMID: 26589247]

[16]   Aggarwal P, Hall JB, McLeland CB, Dobrovolskaia MA, McNeil SE. Nanoparticle interaction with plasma proteins as it relates to particle biodistribution, biocompatibility and therapeutic efficacy. Adv Drug Deliv Rev 2009; 61(6): 428-37.
        [http://dx.doi.org/10.1016/j.addr.2009.03.009] [PMID: 19376175]

[17]   Wolfram J, Yang Y, Shen J, *et al.* The nano-plasma interface: Implications of the protein corona. Colloids Surf B Biointerfaces 2014; 124(124): 17-24.
[http://dx.doi.org/10.1016/j.colsurfb.2014.02.035] [PMID: 24656615]

[18]   Hellstrand E, Lynch I, Andersson A, *et al.* Complete high-density lipoproteins in nanoparticle corona. FEBS J 2009; 276(12): 3372-81.
[http://dx.doi.org/10.1111/j.1742-4658.2009.07062.x] [PMID: 19438706]

[19]   Lynch I, Dawson KA. Protein-nanoparticle interactions. Nano Today 2008; 3: 40-7.
[http://dx.doi.org/10.1016/S1748-0132(08)70014-8]

[20]   Rahman M, Laurent S, Tawil N, Yahia L, Mahmoudi M. Protein-Nanoparticle Interactions. The Bio-nano interface. Springer Ser Biophys 2013; •••: 15. [Ed Springer- Verlag, Berlin, Heidelberg.].

[21]   Walkey CD, Chan WC. Understanding and controlling the interaction of nanomaterials with proteins in a physiological environment. Chem Soc Rev 2012; 41(7): 2780-99.
[http://dx.doi.org/10.1039/C1CS15233E] [PMID: 22086677]

[22]   Simberg D, Park JH, Karmali PP, *et al.* Differential proteomics analysis of the surface heterogeneity of dextran iron oxide nanoparticles and the implications for their *in vivo* clearance. Biomaterials 2009; 30(23-24): 3926-33.
[http://dx.doi.org/10.1016/j.biomaterials.2009.03.056] [PMID: 19394687]

[23]   Lynch I, Cedervall T, Lundqvist M, Cabaleiro-Lago C, Linse S, Dawson KA. The nanoparticle-protein complex as a biological entity; a complex fluids and surface science challenge for the 21st century. Adv Colloid Interface Sci 2007; 134-135: 167-74.
[http://dx.doi.org/10.1016/j.cis.2007.04.021] [PMID: 17574200]

[24]   Monopoli MP, Walczyk D, Campbell A, *et al.* Physical-chemical aspects of protein corona: relevance to *in vitro* and *in vivo* biological impacts of nanoparticles. J Am Chem Soc 2011; 133(8): 2525-34.
[http://dx.doi.org/10.1021/ja107583h] [PMID: 21288025]

[25]   Cedervall T, Lynch I, Lindman S, *et al.* Understanding the nanoparticle-protein corona using methods to quantify exchange rates and affinities of proteins for nanoparticles. Proc Natl Acad Sci USA 2007; 104(7): 2050-5.
[http://dx.doi.org/10.1073/pnas.0608582104] [PMID: 17267609]

[26]   Cedervall T, Lynch I, Foy M, *et al.* Detailed identification of plasma proteins adsorbed on copolymer nanoparticles. Angew Chem Int Ed Engl 2007; 46(30): 5754-6.
[http://dx.doi.org/10.1002/anie.200700465] [PMID: 17591736]

[27]   Walczyk D, Bombelli FB, Monopoli MP, Lynch I, Dawson KA. What the cell sees in bionanoscience. J Am Chem Soc 2010; 132(16): 5761-8.
[http://dx.doi.org/10.1021/ja910675v] [PMID: 20356039]

[28]   Bekale L, Agudelo D, Tajmir-Riahi HA. Effect of polymer molecular weight on chitosan-protein interaction. Colloids Surf B Biointerfaces 2015; 125: 309-17.
[http://dx.doi.org/10.1016/j.colsurfb.2014.11.037] [PMID: 25524222]

[29]   Pan H, Qin M, Meng W, Cao Y, Wang W. How do proteins unfold upon adsorption on nanoparticle surfaces? Langmuir 2012; 28(35): 12779-87.
[http://dx.doi.org/10.1021/la302258k] [PMID: 22913793]

[30]   Norde W, Giacomelli CE. BSA structural changes during homomolecular exchange between the adsorbed and the dissolved states. J Biotechnol 2000; 79(3): 259-68.
[http://dx.doi.org/10.1016/S0168-1656(00)00242-X] [PMID: 10867186]

[31]   Anderson NL, Anderson NG. The human plasma proteome: history, character, and diagnostic prospects. Mol Cell Proteomics 2002; 1(11): 845-67.
[http://dx.doi.org/10.1074/mcp.R200007-MCP200] [PMID: 12488461]

[32]   Seong SY, Matzinger P. Hydrophobicity: an ancient damage-associated molecular pattern that initiates innate immune responses. Nat Rev Immunol 2004; 4(6): 469-78.
[http://dx.doi.org/10.1038/nri1372] [PMID: 15173835]

[33]   Deng ZJ, Liang M, Monteiro M, Toth I, Minchin RF. Nanoparticle-induced unfolding of fibrinogen promotes Mac-1 receptor activation and inflammation. Nat Nanotechnol 2011; 6(1): 39-44.
[http://dx.doi.org/10.1038/nnano.2010.250] [PMID: 21170037]

[34]   Mahmoudi M, Lynch I, Ejtehadi MR, Monopoli MP, Bombelli FB, Laurent S. Protein-nanoparticle interactions: opportunities and challenges. Chem Rev 2011; 111(9): 5610-37.
[http://dx.doi.org/10.1021/cr100440g] [PMID: 21688848]

[35]   Jansch M, Stumpf P, Graf C, Rühl E, Müller RH. Adsorption kinetics of plasma proteins on ultrasmall superparamagnetic iron oxide (USPIO) nanoparticles. Int J Pharm 2012; 428(1-2): 125-33.
[http://dx.doi.org/10.1016/j.ijpharm.2012.01.060] [PMID: 22342465]

[36]   Vroman L, Adams AL, Fischer GC, Munoz PC. Interaction of high molecular weight kininogen, factor XII, and fibrinogen in plasma at interfaces. Blood 1980; 55(1): 156-9.
[PMID: 7350935]

[37]   Allémann E, Gravel P, Leroux JC, Balant L, Gurny R. Kinetics of blood component adsorption on poly(D,L-lactic acid) nanoparticles: evidence of complement C3 component involvement. J Biomed Mater Res 1997; 37(2): 229-34.
[http://dx.doi.org/10.1002/(SICI)1097-4636(199711)37:2<229::AID-JBM12>3.0.CO;2-9]   [PMID: 9358316]

[38]   DellOrco D, Lundqvist M, Oslakovic C, Cedervall T, Linse S. Modeling the time evolution of the nanoparticle-protein corona in a body fluid. PLoS One 2010; 5(6): e10949.
[http://dx.doi.org/10.1371/journal.pone.0010949] [PMID: 20532175]

[39]   Darabi Sahneh F, Scoglio C, Riviere J. Dynamics of nanoparticle-protein corona complex formation: analytical results from population balance equations. PLoS One 2013; 8(5): e64690.
[http://dx.doi.org/10.1371/journal.pone.0064690] [PMID: 23741371]

[40]   Casals E, Pfaller T, Duschl A, Oostingh GJ, Puntes V. Time evolution of the nanoparticle protein corona. ACS Nano 2010; 4(7): 3623-32.
[http://dx.doi.org/10.1021/nn901372t] [PMID: 20553005]

[41]   Gossmann R, Fahrländer E, Hummel M, Mulac D, Brockmeyer J, Langer K. Comparative examination of adsorption of serum proteins on HSA- and PLGA-based nanoparticles using SDS-PAGE and LC-MS. Eur J Pharm Biopharm 2015; 93: 80-7.
[http://dx.doi.org/10.1016/j.ejpb.2015.03.021] [PMID: 25813886]

[42]   Klein J. Probing the interactions of proteins and nanoparticles. Proc Natl Acad Sci USA 2007; 104(7): 2029-30.
[http://dx.doi.org/10.1073/pnas.0611610104] [PMID: 17284585]

[43]   Lundqvist M, Stigler J, Elia G, Lynch I, Cedervall T, Dawson KA. Nanoparticle size and surface properties determine the protein corona with possible implications for biological impacts. Proc Natl Acad Sci USA 2008; 105(38): 14265-70.
[http://dx.doi.org/10.1073/pnas.0805135105] [PMID: 18809927]

[44]   Pedersen MB, Zhou X, Larsen EK, *et al.* Curvature of synthetic and natural surfaces is an important target feature in classical pathway complement activation. J Immunol 2010; 184(4): 1931-45.
[http://dx.doi.org/10.4049/jimmunol.0902214] [PMID: 20053940]

[45]   Bihari P, Vippola M, Schultes S, *et al.* Optimized dispersion of nanoparticles for biological *in vitro* and *in vivo* studies. Part Fibre Toxicol 2008; 5: 14.
[http://dx.doi.org/10.1186/1743-8977-5-14] [PMID: 18990217]

[46] Durán N, Silveira CP, Durán M, Martinez DS. Silver nanoparticle protein corona and toxicity: a mini-review. J Nanobiotechnology 2015; 13: 55.
[http://dx.doi.org/10.1186/s12951-015-0114-4] [PMID: 26337542]

[47] Roach P, Farrar D, Perry CC. Surface tailoring for controlled protein adsorption: effect of topography at the nanometer scale and chemistry. J Am Chem Soc 2006; 128(12): 3939-45.
[http://dx.doi.org/10.1021/ja056278e] [PMID: 16551101]

[48] Gagner JE, Lopez MD, Dordick JS, Siegel RW. Effect of gold nanoparticle morphology on adsorbed protein structure and function. Biomaterials 2011; 32(29): 7241-52.
[http://dx.doi.org/10.1016/j.biomaterials.2011.05.091] [PMID: 21705074]

[49] Ashkarran AA, Ghavami M, Aghaverdi H, Stroeve P, Mahmoudi M. Bacterial effects and protein corona evaluations: crucial ignored factors in the prediction of bio-efficacy of various forms of silver nanoparticles. Chem Res Toxicol 2012; 25(6): 1231-42.
[http://dx.doi.org/10.1021/tx300083s] [PMID: 22551528]

[50] Mahmoudi M, Serpooshan V. Large protein absorptions from small changes on the surface of nanoparticles. J Phys Chem C 2011; 115: 18275-83.
[http://dx.doi.org/10.1021/jp2056255]

[51] Karmali PP, Simberg D. Interactions of nanoparticles with plasma proteins: implication on clearance and toxicity of drug delivery systems. Expert Opin Drug Deliv 2011; 8(3): 343-57.
[http://dx.doi.org/10.1517/17425247.2011.554818] [PMID: 21291354]

[52] Lemarchand C, Gref R, Passirani C, et al. Influence of polysaccharide coating on the interactions of nanoparticles with biological systems. Biomaterials 2006; 27(1): 108-18.
[http://dx.doi.org/10.1016/j.biomaterials.2005.04.041] [PMID: 16118015]

[53] Kim YK, Kim VN. Processing of intronic microRNAs. EMBO J 2007; 26(3): 775-83.
[http://dx.doi.org/10.1038/sj.emboj.7601512] [PMID: 17255951]

[54] Sempf K, Arrey T, Gelperina S, et al. Adsorption of plasma proteins on uncoated PLGA nanoparticles. Eur J Pharm Biopharm 2013; 85(1): 53-60.
[http://dx.doi.org/10.1016/j.ejpb.2012.11.030] [PMID: 23395970]

[55] Walkey CD, Olsen JB, Guo H, Emili A, Chan WC. Nanoparticle size and surface chemistry determine serum protein adsorption and macrophage uptake. J Am Chem Soc 2012; 134(4): 2139-47.
[http://dx.doi.org/10.1021/ja2084338] [PMID: 22191645]

[56] Perry JL, Reuter KG, Kai MP, et al. PEGylated PRINT nanoparticles: the impact of PEG density on protein binding, macrophage association, biodistribution, and pharmacokinetics. Nano Lett 2012; 12(10): 5304-10.
[http://dx.doi.org/10.1021/nl302638g] [PMID: 22920324]

[57] Rouzes C, Gref R, Leonard M, De Sousa Delgado A, Dellacherie E. Surface modification of poly(lactic acid) nanospheres using hydrophobically modified dextrans as stabilizers in an o/w emulsion/evaporation technique. J Biomed Mater Res 2000; 50(4): 557-65.
[http://dx.doi.org/10.1002/(SICI)1097-4636(20000615)50:4<557::AID-JBM11>3.0.CO;2-R] [PMID: 10756314]

[58] Roser M, Fischer D, Kissel T. Surface-modified biodegradable albumin nano- and microspheres. II: effect of surface charges on in vitro phagocytosis and biodistribution in rats. Eur J Pharm Biopharm 1998; 46(3): 255-63.
[http://dx.doi.org/10.1016/S0939-6411(98)00038-1] [PMID: 9885296]

[59] Gessner A, Lieske A, Paulke B, Müller R. Influence of surface charge density on protein adsorption on polymeric nanoparticles: analysis by two-dimensional electrophoresis. Eur J Pharm Biopharm 2002; 54(2): 165-70.
[http://dx.doi.org/10.1016/S0939-6411(02)00081-4] [PMID: 12191688]

[60]   Gessner A, Lieske A, Paulke BR, Müller RH. Functional groups on polystyrene model nanoparticles: influence on protein adsorption. J Biomed Mater Res A 2003; 65(3): 319-26.
[http://dx.doi.org/10.1002/jbm.a.10371] [PMID: 12746878]

[61]   Pillai GJ, Greeshma MM, Menon D. Impact of poly(lactic-co-glycolic acid) nanoparticle surface charge on protein, cellular and haematological interactions. Colloids Surf B Biointerfaces 2015; 136: 1058-66.
[http://dx.doi.org/10.1016/j.colsurfb.2015.10.047] [PMID: 26590899]

[62]   Fleischer CC, Payne CK. Nanoparticle-cell interactions: molecular structure of the protein corona and cellular outcomes. Acc Chem Res 2014; 47(8): 2651-9.
[http://dx.doi.org/10.1021/ar500190q] [PMID: 25014679]

[63]   Fleischer CC, Payne CK. Secondary structure of corona proteins determines the cell surface receptors used by nanoparticles. J Phys Chem B 2014; 118(49): 14017-26.
[http://dx.doi.org/10.1021/jp502624n] [PMID: 24779411]

[64]   Cukalevski R, Lundqvist M, Oslakovic C, Dahlbäck B, Linse S, Cedervall T. Structural changes in apolipoproteins bound to nanoparticles. Langmuir 2011; 27(23): 14360-9.
[http://dx.doi.org/10.1021/la203290a] [PMID: 21978381]

[65]   Lindman S, Lynch I, Thulin E, *et al.* Systematic investigation of the thermodynamics of HSA adsorption to N-tert-Butylacrylamide copolymer nanoparticles. Effects of particle size and hydrophobicity. Nano Lett 2007; 7: 914-20.
[http://dx.doi.org/10.1021/nl062743+] [PMID: 17335269]

[66]   Gessner A, Waicz R, Lieske A, Paulke B, Mäder K, Müller RH. Nanoparticles with decreasing surface hydrophobicities: influence on plasma protein adsorption. Int J Pharm 2000; 196(2): 245-9.
[http://dx.doi.org/10.1016/S0378-5173(99)00432-9] [PMID: 10699728]

[67]   Ghavami M, Saffar S, Emamy BA, *et al.* Plasma concentration gradient influences the protein corona decoration on nanoparticles. RSC Advances 2013; 3: 1119-26.
[http://dx.doi.org/10.1039/C2RA22093H]

[68]   Lundqvist M, Stigler J, Cedervall T, *et al.* The evolution of the protein corona around nanoparticles: a test study. ACS Nano 2011; 5(9): 7503-9.
[http://dx.doi.org/10.1021/nn202458g] [PMID: 21861491]

[69]   Maiorano G, Sabella S, Sorce B, *et al.* Effects of cell culture media on the dynamic formation of protein-nanoparticle complexes and influence on the cellular response. ACS Nano 2010; 4(12): 7481-91.
[http://dx.doi.org/10.1021/nn101557e] [PMID: 21082814]

[70]   Pozzi D, Colapicchioni V, Caracciolo G, *et al.* Effect of polyethyleneglycol (PEG) chain length on the bio-nano-interactions between PEGylated lipid nanoparticles and biological fluids: from nanostructure to uptake in cancer cells. Nanoscale 2014; 6(5): 2782-92.
[http://dx.doi.org/10.1039/c3nr05559k] [PMID: 24463404]

[71]   Deng ZJ, Liang M, Toth I, Monteiro MJ, Minchin RF. Molecular interaction of poly(acrylic acid) gold nanoparticles with human fibrinogen. ACS Nano 2012; 6(10): 8962-9.
[http://dx.doi.org/10.1021/nn3029953] [PMID: 22998416]

[72]   Mahmoudi M, Abdelmonem AM, Behzadi S, *et al.* Temperature: the ignored factor at the NanoBio interface. ACS Nano 2013; 7(8): 6555-62.
[http://dx.doi.org/10.1021/nn305337c] [PMID: 23808533]

[73]   Monopoli MP, Walczyk D, Campbell A, *et al.* Physical-chemical aspects of protein corona: relevance to *in vitro* and *in vivo* biological impacts of nanoparticles. J Am Chem Soc 2011; 133(8): 2525-34.
[http://dx.doi.org/10.1021/ja107583h] [PMID: 21288025]

[74] Cho EC, Zhang Q, Xia Y. The effect of sedimentation and diffusion on cellular uptake of gold nanoparticles. Nat Nanotechnol 2011; 6(6): 385-91.
[http://dx.doi.org/10.1038/nnano.2011.58] [PMID: 21516092]

[75] Gebauer JS, Malissek M, Simon S, *et al.* Impact of the nanoparticle-protein corona on colloidal stability and protein structure. Langmuir 2012; 28(25): 9673-9.
[http://dx.doi.org/10.1021/la301104a] [PMID: 22524519]

[76] Dominguez-Medina S, Blankenburg J, Olson J, Landes CF, Link S. Adsorption of a protein monolayer *via* hydrophobic interactions revents nanoparticle aggregation under harsh environmental conditions. ACS Sustain Chem& Eng 2013; 1(7): 833-42.
[http://dx.doi.org/10.1021/sc400042h] [PMID: 23914342]

[77] Wells MA, Abid A, Kennedy IM, Barakat AI. Serum proteins prevent aggregation of Fe2O3 and ZnO nanoparticles. Nanotoxicology 2012; 6: 837-46.
[http://dx.doi.org/10.3109/17435390.2011.625131] [PMID: 22149273]

[78] Owens DE III, Peppas NA. Opsonization, biodistribution, and pharmacokinetics of polymeric nanoparticles. Int J Pharm 2006; 307(1): 93-102.
[http://dx.doi.org/10.1016/j.ijpharm.2005.10.010] [PMID: 16303268]

[79] Choi HS, Ashitate Y, Lee JH, *et al.* Rapid translocation of nanoparticles from the lung airspaces to the body. Nat Biotechnol 2010; 28(12): 1300-3.
[http://dx.doi.org/10.1038/nbt.1696] [PMID: 21057497]

[80] Choi HS, Liu W, Liu F, *et al.* Design considerations for tumour-targeted nanoparticles. Nat Nanotechnol 2010; 5(1): 42-7.
[http://dx.doi.org/10.1038/nnano.2009.314] [PMID: 19893516]

[81] Choi HS, Liu W, Misra P, *et al.* Renal clearance of quantum dots. Nat Biotechnol 2007; 25(10): 1165-70.
[http://dx.doi.org/10.1038/nbt1340] [PMID: 17891134]

[82] Lunov O, Syrovets T, Loos C, *et al.* Differential uptake of functionalized polystyrene nanoparticles by human macrophages and a monocytic cell line. ACS Nano 2011; 5(3): 1657-69.
[http://dx.doi.org/10.1021/nn2000756] [PMID: 21344890]

[83] Dufort S, Sancey L, Coll JL. Physico-chemical parameters that govern nanoparticles fate also dictate rules for their molecular evolution. Adv Drug Deliv Rev 2012; 64(2): 179-89.
[http://dx.doi.org/10.1016/j.addr.2011.09.009] [PMID: 21983079]

[84] Vauthier C, Lindner P, Cabane B. Configuration of bovine serum albumin adsorbed on polymer particles with grafted dextran corona. Colloids Surf B Biointerfaces 2009; 69(2): 207-15.
[http://dx.doi.org/10.1016/j.colsurfb.2008.11.017] [PMID: 19135340]

[85] Vauthier C, Persson B, Lindner P, Cabane B. Protein adsorption and complement activation for di-block copolymer nanoparticles. Biomaterials 2011; 32(6): 1646-56.
[http://dx.doi.org/10.1016/j.biomaterials.2010.10.026] [PMID: 21093043]

[86] Moghimi SM, Andersen AJ, Ahmadvand D, Wibroe PP, Andresen TL, Hunter AC. Material properties in complement activation. Adv Drug Deliv Rev 2011; 63(12): 1000-7.
[http://dx.doi.org/10.1016/j.addr.2011.06.002] [PMID: 21689701]

[87] Moghimi SM, Muir IS, Illum L, Davis SS, Kolb-Bachofen V. Coating particles with a block co-polymer (poloxamine-908) suppresses opsonization but permits the activity of dysopsonins in the serum. Biochim Biophys Acta 1993; 1179(2): 157-65.
[http://dx.doi.org/10.1016/0167-4889(93)90137-E] [PMID: 8218358]

[88] Salvador-Morales C, Zhang L, Langer R, Farokhzad OC. Immunocompatibility properties of lipid-polymer hybrid nanoparticles with heterogeneous surface functional groups. Biomaterials 2009; 30(12): 2231-40.
[http://dx.doi.org/10.1016/j.biomaterials.2009.01.005] [PMID: 19167749]

[89]   Knop K, Hoogenboom R, Fischer D, Schubert US. Poly(ethylene glycol) in drug delivery: pros and cons as well as potential alternatives. Angew Chem Int Ed Engl 2010; 49(36): 6288-308. [http://dx.doi.org/10.1002/anie.200902672] [PMID: 20648499]

[90]   Pelaz B, del Pino P, Maffre P, *et al.* Surface functionalization of nanoparticles with polyethylene glycol: effects on protein adsorption and cellular uptake. ACS Nano 2015; 9(7): 6996-7008. [http://dx.doi.org/10.1021/acsnano.5b01326] [PMID: 26079146]

[91]   Brandenberger C, Mühlfeld C, Ali Z, *et al.* Quantitative evaluation of cellular uptake and trafficking of plain and polyethylene glycol-coated gold nanoparticles. Small 2010; 6(15): 1669-78. [http://dx.doi.org/10.1002/smll.201000528] [PMID: 20602428]

[92]   Van Hoecke K, De Schamphelaere KA, Ali Z, *et al.* Ecotoxicity and uptake of polymer coated gold nanoparticles. Nanotoxicology 2013; 7(1): 37-47. [http://dx.doi.org/10.3109/17435390.2011.626566] [PMID: 22023156]

[93]   Cabral H, Matsumoto Y, Mizuno K, *et al.* Accumulation of sub-100 nm polymeric micelles in poorly permeable tumours depends on size. Nat Nanotechnol 2011; 6(12): 815-23. [http://dx.doi.org/10.1038/nnano.2011.166] [PMID: 22020122]

[94]   Nagayama S, Ogawara K, Minato K, *et al.* Fetuin mediates hepatic uptake of negatively charged nanoparticles *via* scavenger receptor. Int J Pharm 2007; 329(1-2): 192-8. [http://dx.doi.org/10.1016/j.ijpharm.2006.08.025] [PMID: 17005341]

[95]   Kreuter J, Shamenkov D, Petrov V, *et al.* Apolipoprotein-mediated transport of nanoparticle-bound drugs across the blood-brain barrier. J Drug Target 2002; 10(4): 317-25. [http://dx.doi.org/10.1080/10611860290031877] [PMID: 12164380]

[96]   Kreuter J, Hekmatara T, Dreis S, Vogel T, Gelperina S, Langer K. Covalent attachment of apolipoprotein A-I and apolipoprotein B-100 to albumin nanoparticles enables drug transport into the brain. J Control Release 2007; 118(1): 54-8. [http://dx.doi.org/10.1016/j.jconrel.2006.12.012] [PMID: 17250920]

[97]   Ogawara K, Furumoto K, Nagayama S, *et al.* Pre-coating with serum albumin reduces receptor-mediated hepatic disposition of polystyrene nanosphere: implications for rational design of nanoparticles. J Control Release 2004; 100(3): 451-5. [http://dx.doi.org/10.1016/j.jconrel.2004.07.028] [PMID: 15567509]

[98]   Stolnik S, Daudali B, Arien A, *et al.* The effect of surface coverage and conformation of poly(ethylene oxide) (PEO) chains of poloxamer 407 on the biological fate of model colloidal drug carriers. Biochim Biophys Acta 2001; 1514(2): 261-79. [http://dx.doi.org/10.1016/S0005-2736(01)00376-5] [PMID: 11557026]

[99]   Dutta D, Sundaram SK, Teeguarden JG, *et al.* Adsorbed proteins influence the biological activity and molecular targeting of nanomaterials. Toxicol Sci 2007; 100(1): 303-15. [http://dx.doi.org/10.1093/toxsci/kfm217] [PMID: 17709331]

[100]   Salvati A, Pitek AS, Monopoli MP, *et al.* Transferrin-functionalized nanoparticles lose their targeting capabilities when a biomolecule corona adsorbs on the surface. Nat Nanotechnol 2013; 8(2): 137-43. [http://dx.doi.org/10.1038/nnano.2012.237] [PMID: 23334168]

[101]   Mirshafiee V, Mahmoudi M, Lou K, Cheng J, Kraft ML. Protein corona significantly reduces active targeting yield. Chem Commun (Camb) 2013; 49(25): 2557-9. [http://dx.doi.org/10.1039/c3cc37307j] [PMID: 23423192]

[102]   Kirpotin DB, Drummond DC, Shao Y, *et al.* Antibody targeting of long-circulating lipidic nanoparticles does not increase tumor localization but does increase internalization in animal models. Cancer Res 2006; 66(13): 6732-40. [http://dx.doi.org/10.1158/0008-5472.CAN-05-4199] [PMID: 16818648]

[103] Cirstoiu-Hapca A, Buchegger F, Lange N, Bossy L, Gurny R, Delie F. Benefit of anti-HER2-coated paclitaxel-loaded immuno-nanoparticles in the treatment of disseminated ovarian cancer: Therapeutic efficacy and biodistribution in mice. J Control Release 2010; 144(3): 324-31.
[http://dx.doi.org/10.1016/j.jconrel.2010.02.026] [PMID: 20219607]

[104] Müller RH, Heinemann S. Surface modelling of microparticles as parenteral systems with high tissue affinity. Ed. Bioadhesion- Possible and future trends; Stuttgart, 1989.

[105] Kreuter J, Petrov VE, Kharkevich DA, Alyautdin RN. Influence of the type of surfactant on the analgesic effects induced by the peptide dalargin after its delivery across the blood–brain barrier using surfactant-coated nanoparticles. J Control Release 1997; 49: 81-7.
[http://dx.doi.org/10.1016/S0168-3659(97)00061-8]

[106] Yan X, Kuipers F, Havekes LM, et al. The role of apolipoprotein E in the elimination of liposomes from blood by hepatocytes in the mouse. Biochem Biophys Res Commun 2005; 328(1): 57-62.
[http://dx.doi.org/10.1016/j.bbrc.2004.12.137] [PMID: 15670750]

[107] Michaelis K, Hoffmann MM, Dreis S, et al. Covalent linkage of apolipoprotein e to albumin nanoparticles strongly enhances drug transport into the brain. J Pharmacol Exp Ther 2006; 317(3): 1246-53.
[http://dx.doi.org/10.1124/jpet.105.097139] [PMID: 16554356]

[108] Kim HR, Andrieux K, Gil S, et al. Translocation of poly(ethylene glycol-co-hexadecyl)cyanoacrylate nanoparticles into rat brain endothelial cells: role of apolipoproteins in receptor-mediated endocytosis. Biomacromolecules 2007; 8(3): 793-9.
[http://dx.doi.org/10.1021/bm060711a] [PMID: 17309294]

[109] Mahmoudi M, Sheibani S, Milani AS, et al. Crucial role of the protein corona for the specific targeting of nanoparticles. Nanomedicine (Lond) 2015; 10(2): 215-26.
[http://dx.doi.org/10.2217/nnm.14.69] [PMID: 25600967]

[110] Lalani J, Patil S, Kolate A, Lalani R, Misra A. Protein-functionalized PLGA nanoparticles of lamotrigine for neuropathic pain management. AAPS PharmSciTech 2015; 16(2): 413-27.
[http://dx.doi.org/10.1208/s12249-014-0235-3] [PMID: 25354788]

[111] Caracciolo G, Cardarelli F, Pozzi D, et al. Selective targeting capability acquired with a protein corona adsorbed on the surface of 1,2-dioleoyl-3-trimethylammonium propane/DNA nanoparticles. ACS Appl Mater Interfaces 2013; 5(24): 13171-9.
[http://dx.doi.org/10.1021/am404171h] [PMID: 24245615]

[112] Saptarshi SR, Duschl A, Lopata AL. Interaction of nanoparticles with proteins: relation to bio-reactivity of the nanoparticle. J Nanobiotechnology 2013; 11: 26.
[http://dx.doi.org/10.1186/1477-3155-11-26] [PMID: 23870291]

[113] Kettler K, Veltman K, van de Meent D, van Wezel A, Hendriks AJ. Cellular uptake of nanoparticles as determined by particle properties, experimental conditions, and cell type. Environ Toxicol Chem 2014; 33(3): 481-92.
[http://dx.doi.org/10.1002/etc.2470] [PMID: 24273100]

[114] Cartiera MS, Johnson KM, Rajendran V, Caplan MJ, Saltzman WM. The uptake and intracellular fate of PLGA nanoparticles in epithelial cells. Biomaterials 2009; 30(14): 2790-8.
[http://dx.doi.org/10.1016/j.biomaterials.2009.01.057] [PMID: 19232712]

[115] Ehrenberg MS, Friedman AE, Finkelstein JN, Oberdörster G, McGrath JL. The influence of protein adsorption on nanoparticle association with cultured endothelial cells. Biomaterials 2009; 30(4): 603-10.
[http://dx.doi.org/10.1016/j.biomaterials.2008.09.050] [PMID: 19012960]

[116] Baier G, Costa C, Zeller A, *et al.* BSA adsorption on differently charged polystyrene nanoparticles using isothermal titration calorimetry and the influence on cellular uptake. Macromol Biosci 2011; 11(5): 628-38.
[http://dx.doi.org/10.1002/mabi.201000395] [PMID: 21384550]

[117] Fleischer CC, Payne CK. Nanoparticle surface charge mediates the cellular receptors used by protein-nanoparticle complexes. J Phys Chem B 2012; 116(30): 8901-7.
[http://dx.doi.org/10.1021/jp304630q] [PMID: 22774860]

[118] Fleischer CC, Kumar U, Payne CK. Cellular binding of anionic nanoparticles is inhibited by serum proteins independent of nanoparticle composition. Biomater Sci 2013; 1(9): 975-82.
[http://dx.doi.org/10.1039/c3bm60121h] [PMID: 23956836]

[119] Mao Z, Zhou X, Gao C. Influence of structure and properties of colloidal biomaterials on cellular uptake and cell functions. Biomater Sci 2013; 1: 896.
[http://dx.doi.org/10.1039/c3bm00137g]

[120] Doorley GW, Payne CK. Nanoparticles act as protein carriers during cellular internalization. Chem Commun (Camb) 2012; 48(24): 2961-3.
[http://dx.doi.org/10.1039/c2cc16937a] [PMID: 22328990]

[121] Patel PC, Giljohann DA, Daniel WL, Zheng D, Prigodich AE, Mirkin CA. Scavenger receptors mediate cellular uptake of polyvalent oligonucleotide-functionalized gold nanoparticles. Bioconjug Chem 2010; 21(12): 2250-6.
[http://dx.doi.org/10.1021/bc1002423] [PMID: 21070003]

[122] Yan Y, Gause KT, Kamphuis MM, *et al.* Differential roles of the protein corona in the cellular uptake of nanoporous polymer particles by monocyte and macrophage cell lines. ACS Nano 2013; 7(12): 10960-70.
[http://dx.doi.org/10.1021/nn404481f] [PMID: 24256422]

[123] Tenzer S, Docter D, Kuharev J, *et al.* Rapid formation of plasma protein corona critically affects nanoparticle pathophysiology. Nat Nanotechnol 2013; 8(10): 772-81.
[http://dx.doi.org/10.1038/nnano.2013.181] [PMID: 24056901]

[124] Nienhaus GU, Heinzl J, Huenges E, Parak F. Protein crystal dynamics studied by time-resolved analysis of X-ray duffuse scattering. Lett to Nat 1989; 338: 665-6.
[http://dx.doi.org/10.1038/338665a0]

[125] Jiang X, Weise S, Hafner M, *et al.* Quantitative analysis of the protein corona on FePt nanoparticles formed by transferrin binding. J R Soc Interface 2010; 7 (Suppl. 1): S5-S13.
[http://dx.doi.org/10.1098/rsif.2009.0272.focus] [PMID: 19776149]

[126] Ritz S, Schöttler S, Kotman N, *et al.* Protein corona of nanoparticles: distinct proteins regulate the cellular uptake. Biomacromolecules 2015; 16(4): 1311-21.
[http://dx.doi.org/10.1021/acs.biomac.5b00108] [PMID: 25794196]

[127] Ge C, Du J, Zhao L, *et al.* Binding of blood proteins to carbon nanotubes reduces cytotoxicity. Proc Natl Acad Sci USA 2011; 108(41): 16968-73.
[http://dx.doi.org/10.1073/pnas.1105270108] [PMID: 21969544]

[128] Hu W, Peng C, Lv M, *et al.* Protein corona-mediated mitigation of cytotoxicity of graphene oxide. ACS Nano 2011; 5(5): 3693-700.
[http://dx.doi.org/10.1021/nn200021j] [PMID: 21500856]

[129] Wang L, Li J, Pan J, *et al.* Revealing the binding structure of the protein corona on gold nanorods using synchrotron radiation-based techniques: understanding the reduced damage in cell membranes. J Am Chem Soc 2013; 135(46): 17359-68.
[http://dx.doi.org/10.1021/ja406924v] [PMID: 24215358]

[130] Zhu Y, Li W, Li Q, *et al.* Effects of serum proteins on intracellular uptake and cytotoxicity of carbon nanoparticles. Carbon NY 2009; 47: 1351-8.
[http://dx.doi.org/10.1016/j.carbon.2009.01.026]

[131] Peracchia MT, Fattal E, Desmaële D, *et al.* Stealth PEGylated polycyanoacrylate nanoparticles for intravenous administration and splenic targeting. J Control Release 1999; 60(1): 121-8.
[http://dx.doi.org/10.1016/S0168-3659(99)00063-2] [PMID: 10370176]

[132] Wang F, Yu L, Monopoli MP, *et al.* The biomolecular corona is retained during nanoparticle uptake and protects the cells from the damage induced by cationic nanoparticles until degraded in the lysosomes. Nanomedicine (Lond) 2013; 9(8): 1159-68.
[PMID: 23660460]

[133] Fleischer CC, Payne CK. Secondary structure of corona proteins determines the cell surface receptors used by nanoparticles. J Phys Chem B 2014; 118(49): 14017-26.
[http://dx.doi.org/10.1021/jp502624n] [PMID: 24779411]

CHAPTER 9

Safety of Nanomedicine: Neuroendocrine Disrupting Potential of Nanoparticles and Neurodegeneration

Eva Rollerova[1,*], Alzbeta Bujnakova Mlynarcikova[2], Jana Tulinska[1], Jevgenij Kovriznych[1], Alexander Kiss[2] and Sona Scsukova[2]

[1] *Slovak Medical University, Faculty of Public Health, Department of Toxicology and Faculty of Medicine, Laboratory of Immunotoxicology, Bratislava, Slovak Republic*

[2] *Biomedical Research Center, Slovak Academy of Sciences, Institute of Experimental Endocrinology, Bratislava, Slovak Republic*

Abstract: The development of nanomaterials (NMs) for applications in biomedicine inclusive of drug delivery as well as medical imaging is currently undergoing an enormous expansion. NMs may have many different forms and characteristics, depending on their size, chemical composition, manufacturing method, and surface modification. The use of NMs in the field of neurodegenerative diseases diagnosis and treatment implies the ability of NMs to cross the blood-brain barrier (BBB) and enter the central nervous system (CNS) in dependence on their physico-chemical properties, composition, and functionalization. The same properties that make the NMs beneficial for their applications may also affect their interactions with biological systems and have unintended consequences on human health. Several *in vivo* and *in vitro* studies have demonstrated that intentional exposure to NMs with potential use for diagnostic and therapeutic purposes might induce neurotoxic effects resulting in neuro-degeneration in different CNS regions. Recent evidence has indicated that neuro-endocrine disrupting effects by the action of NMs in dopaminergic, serotoninergic, and gonadotropic systems might be relevant to neuropathogenesis and neurodegeneration. In line with developmental origin of adult diseases, it is forewarning the evidence that pre- and post-natal exposure to different risk factors including NMs may lead to phenotypic heterogeneity and susceptibility to neurodegenerative diseases in later stages of the life. In the light of the above mentioned events, relevant test models are required to assess: i) the role of NMs in the development and progression of neurodegenerative disease; ii) the effects of NMs on neurodevelopment upon *in utero* exposure of foetuses or neonatal exposure of pups; or iii) the neuroendocrine disrupting effects during critical period being crucial for the development of neurodegenerative diseases. Early identification of potential negative features of NMs using interdisciplinary research approaches (biological, toxicological, clinical, engineering) could minimize the risk of newly designed/developed nanomedicines.

* Corresponding author Eva Rollerova: Slovak Medical University Bratislava, Faculty of Public Health, Department of Toxicology, Limbova 14, 833 01 Bratislava 37, Slovak Republic; E-mail: eva.rollerova@szu.sk

Giovanni Tosi (Ed)

Keywords: Endocrine disruption, Gonadotropins, Nanosafety, Nanomedicine, Nanotoxicology, Nanoparticles, Neuroendocrinology, Neurodegeneration.

NANOTOXICOLOGY AND SAFETY OF NANOMATERIALS

Nanomaterials (NMs)/**nanoparticles** (NPs) cover a heterogeneous group of materials, including inorganic metal and metal oxide NMs, polymeric particulate materials and carbon-based NMs in a wide range of shapes. NMs possess unique physico-chemical properties, such as ultra small size (1-100 nm), large surface area to mass ratio, and high reactivity, which considerably distinguish from the bulk microscale material of the same composition. A wide range of NMs is already accessible on the market, and NMs for future applications like, novel robotic devices, targeted drug delivery systems, molecule-by-molecule design, and self-assembly structures are in the course of development. According to the European Commission (EC), the global quantity of NMs may achieve around 11.5 million tones with a market value of circa 20 bn € per year [1].

Nanotoxicology is a newly-formed discipline which focuses on the understanding of the properties of engineered NMs and their interactions with biological systems emphasized to elucidate the relationship between the physico-chemical properties of NMs and induction of toxic biological responses [2, 3].

Several leading scientists [4] have suggested five grand challenges that need to be achieved in line with safety and sustainability of the developed **nanotechnologies** (NTs). These require to develop: 1) instruments to monitor NM exposure in water and air; 2) validation of methods for the evaluation of the toxicity of NMs; 3) models predicting the impact of NMs on the human health and environment; 4) robust systems for evaluation of NMs impact on health and environment over entire life cycle; and 5) strategic programs intent on relevant risk-focused research. These challenges have been chosen to initiate strategic research aimed at the safety of NT.

Many questions should be opened before NPs would be widely implemented in the marketplace. These are concerning the medicine and environment and say: will NPs induce nano-specific qualitatively distinct and novel toxic effects; how will be measured and predicted nano-specific effects; what will be the relationship between the shape, size, and surface chemistry of NPs on the one hand and their *in vivo* behavior on the other hand; how will be the NMs degraded or metabolized, will be the NMs and/or their degradation products effectively excreted from the body?

At the European level, the discussion about NMs at legislative and scientific level has been ongoing for several years. To date, the current regulatory guidelines are summarized in Table **1**.

Table 1. Legislative activities in the European Union on regulation of nanomaterials.

Date	Regulatory subject	Action	Conclusions	Ref.
May 2004	EC	Communication "Towards a European strategy for nanotechnology"	- proposed actions to promote a strong role of Europe in nanoscience and nanotechnology - the need to address potential risks for health and environment	a
June 2005	EC	Action plan "Nanosciences and nanotechnologies" for 2005-2009		
2008	EC	First Regulatory Review of EU legislation with respect to nanomaterials	"Current legislation covers in principle the potential health, safety and environmental risks in relation to NMs. The protection of health, safety and environment needs mostly to be enhanced by improving implementation of current legislation."	
2009	EP	EP resolution of April 2009 on regulatory aspects of nanomaterials	1. Call for a regulatory and policy framework that explicitly addresses NMs. 2. Call on the Commission to review all relevant legislation. 3. Call for an inventory and product labeling. 4. Call on the Commission to evaluate the need to review REACH concerning: a) simplified registration for NMs manufactured or imported below 1 tonne; b) consideration of all NMs as new substances; c) a chemical safety report with exposure assessment for all registered NMs; d) notification requirements for all NMs placed on the market on their own, in preparations or in articles	b
October 2012	EC	Second Regulatory Review on Nanomaterials	- the REACH registration and proof of safety use for NMs should be based on a case by case approach, and each type of NM should be clearly described	c
February 2013	EC	REACH Review	- revision of annexes	d

(Table 1) contd.....

Date	Regulatory subject	Action	Conclusions	Ref.
May 2013	EC	Public consultation on the modification of the REACH annexes	- public consultation on how the annexes of REACH could be amended to ensure that NMs are registered more clearly under REACH and that the safe use of NMs is adequately demonstrated within the registration dossiers	e
2014	EC	13th Meeting of Competent Authorities for REACH and CLP (CARACAL): document CA/36/2013	-final results of the public consultation	f

EC - European Commission; EP - European Parliament; NMs - nanomaterials
a) http://ec.europa.eu/nanotechnology/pdf/nano_com_en_new.pdf
b) http://www.europarl.europa.eu/sides/getDoc.do?type=TA&reference=P6-TA-2009-0328&language=EN
c) http://eur-lex.europa.eu/LexUriServ/LexUriServ.do?uri=COM:2012:0572:FIN:en:PDF
d) http://eur-lex.europa.eu/LexUriServ/LexUriServ.do?uri=COM:2013:0049:FIN:EN:PDF
e) http://ec.europa.eu/enterprise/sectors/chemicals/reach/nanomaterials/index_en.htm
f) http://circabc.europa.eu/sd/d/1e829ba0-9450-4a08-8b8a-64695a2f6e70/04%20-20CA_36_2013_Nano% 201A_v2.doc

The **REACH (Registration, Evaluation, Authorisation and Restriction of Chemicals)** generally concerning with the chemical substances has no special provisions that explicitly refer to NMs. Actually NMs are principally covered by the REACH and their regulation under REACH has to perform specific requirements. The EC has published the Second Regulatory Review on Nanomaterials in October 2012 and the REACH Review in February 2013. Both papers have addressed questions regarding the regulation of NMs in the REACH. The EC has proposed to improve the forthcoming situation by adaptation of the REACH Regulation and has planned to revise the annexes only. Many experts have shared the same opinion that the test strategies, testing requirements, and test methods under the REACH are in principle practicable to the substances at nanoscale. However, some adaptations to the specificities of NMs are needed [5, 6].

European Nanosafety Cluster Compendium (www.nanosafetycluster.eu) is forum for Sixth Framework Programme (FP6) and Seventh Framework Programme (FP7) of the EC projects on the nanosafety and related national projects in the EU member states. Forum aims to maximize the cooperations between the existing projects by addressing the risk assessment, toxicology, ecotoxicology, exposure assessment, mechanisms of interaction, and standardization issues. **NanoFUTURES** is a European multi-sectorial, integrating platform with the objective to connect and establish the collaboration and representation of all relevant technology platforms that utilize NTs in their industrial sector and products (www.nanofutures.eu). **The European Academies**

Science Advisory Council (EASAC) and the **Joint Research Centre of the EC (JRC)** have published a report entitled "Impact of engineered NMs on Health: Considerations for Benefit-Risk Assessment" that endeavors to identify gaps in the current knowledge and offers a collection of recommendations for the future research; the authors highlight the significance of "safety-by-design" as requirement for the successful implementation of the emerging NTs (www.easac.eu and www.jrc.europa.eu) (reviewed in [7]).

NANOBIOTECHNOLOGIES USED IN NANOMEDICINE

During the past 50 years, progress in material science and physics has led to a broad use of NMs in the field of life science and healthcare with various applications in medicine and biotechnologies [8, 9]. Nanobiotechnologies (NTs) have found a large potential in the medical areas such as **nanodiagnostics, nanodevices, and nanopharmacology**. The use of NTs in molecular diagnostics allowed to produce faster and cheaper diagnostic tools: *nanobiosensors* for point-of care testing of patients, *nanochips* and *nanoarrays* for trace detection of proteins in biological liquids not determined by conventional immunoassays [10]. Inorganic fluorophores quantum dots (QDs) or semiconductor nanocrystals have a wide range of applications, comprising *in vivo* imaging, cell labeling, and diagnostics, mainly diagnosis of cancer [11]. Magnetic NPs, such as superparamagnetic iron oxide nanoparticles (SPIONs), have found ubiquitous appliance in magnetic resonance imaging (MRI) [12, 13] as well as hyperthermic destruction of tumor tissue [14].

Nanodevices *for diagnosis* such as nanobiosensors - nanoradiotransmitters, and nanoacoustical devicies, and CT scanners are able to identify diseases at the cellular level in very early stage. *Nanodevices used for drug delivery* serve as *microcontainers* for medications inserted directly at the site of injury or illness. *Nanovalves* can trap or release the molecules at will. *Nanorobots/nanobots* are developed to perform precise intracellular surgery or *gutbots* for efficient, accurate, and less invasive diagnosis in the digestive tract. *Nanoscale laser surgery/femtolasers* used both for ablation and repair of tissues without the cells injury and hitting the other structures has already been introduced in the corneal surgery [15 - 18].

In **nanopharmacology**, new NTs have been applied for design of progressive drug formulations and improving their delivering to the target cells. NTs in form of NPs, nanoemulsions, and NP aerosol vaccines particularly have refined of *biological therapies* including vaccines, cell therapy, gene therapy, antisense therapy, and RNA interference [17]. NTs also play a substantial role in the development of *stem cell-based therapies*, such as SPIONs, which are emerging as an ideal probe for the noninvasive cell tracking *in vivo* [19]. Nanomaterial constructs provide number of possibilities to transfect, label, visualize, and monitor cells/tissues used in *transplantation*. Nanoliposomes, gelatine NPs, calcium phosphate NPs, dendrimers, and apatite NPs can be used for *nonviral gene delivery* [20].

Besides the applications of nanoparticulate systems in diagnostic processes, nanomedicine has been focused to increase the efficacy of drugs with known dose-limiting toxicity, poor bioavailability and biological stability to improve the selectivity of drug delivery and drug uptake by tissues [21].

To fulfill all the requirements for the construction of **innovative nanodrug formulations,** several approaches have been applied. *Reduction of particle size* to nanometer scale increased the surface area and thereby the rate of dissolution in water for insoluble or poorly soluble drugs. *Improvement of absorption of insoluble compounds and macromolecules* increased the bioavailability and release rates, potentially increasing safety through reduced dose required and decrease of side effects. *Improvement of physical stability* extended the half-life and prevented the formation of secondary aggregates more susceptible to clearance by the mononuclear phagocytic system. Sustained-release profiles up to 24 h provided *improvement of patient compliance with drug regimens*. Conjunction of NPs with relevant biorecognitive ligands *increased the targeted drug delivery*. Moreover, combination of imaging and therapy for diagnosis and thermal ablation of lesions have evolved the so-called *theranostics* [9, 17, 22].

The physicochemical properties of NMs and their potential changes during absorption, distribution, and metabolism affect the fate (and toxicity) of NMs in animal and human body. In addition, the mode of administration and sites of deposition are determining factor for severity of their toxicity. Actually, certain basic questions are emerging regarding the fate of NMs in human body: are they able to penetrate cells and tissues; how and where are they distributed; how are they biodegradated and excreted [9, 23]?

The physicochemical characteristics of NMs for drug delivery influence drug loading, drug release, system stability, and cellular uptake. The chemical and surface properties of NPs are important variables for the interaction of intravenously injected nanodrugs/nanocarriers with plasma proteins and other soluble molecules, forming biomolecular corona/bioshell, which determines another interaction with the cells [8]. Further, the choice of NMs for therapeutic applications depends on the request for biodegradation of the carriers, achievement of increased efficacy of the therapy, and avoidance of tissue reaction to the NPs [24]. In the case that safety-by-design of NPs cannot be guaranteed, potentially advantageous properties of therapeutic NMs/NPs may produce toxicological problems [25, 26].

SAFETY OF NANOMEDICINE

Nanomedicines are undoubtedly in concurrency with other pharmaceutical entities as the safety of every new drug or drug carrier (NMs) always has to be thoroughly evaluated, within pre-clinical and clinical trials, prior to its approval by the relevant regulatory agencies. Nevertheless, there are some very specific issues in regard to the application of NMs in the clinical settings. Nyström *et al.* [27] have suggested the key challenges that need to be addressed in next years:

1. Constitute standardized, validated *in vitro* assays for nanosafety testing incorporating also a set of reference materials;
2. Introduce *ex vivo* models relevant for the specific routes of administration of nanomedicines;
3. Establish *in silico* modeling approaches to predict the biological and toxicological responses of nanomedicines;
4. Achieve an understanding of the absorption, distribution, metabolism, and excretion (ADME) of NMs *in vivo*;
5. Develop a paradigm for the understanding of factors (*e.g.* biocorona) that determine NM interactions with living systems;
6. Organize interdisciplinary training and set up interdisciplinary research teams for the development of nanomedicines from the clinical, biological, engineering, and toxicological point of view.

In summary, to design safe nanomedicines, it may be advisable integrate safety assessment ("toxicology") by design of nanoscale drugs or drug carriers that should be: a) biocompatible (see review in [28]); b) deliverable; c) biodegradable (or excretable); d) traceable; and e) scalable [27].

The Nanotechnology Characterization Laboratory has published a set of protocols relevant for safety assessment of nanomedicines (www.ncl.cancer.gov). These assays present a good basis in the sense of the minimum requirements for physico-chemical characterization. Demand for implementation of well-characterized standards as reference materials is underlined. One such reference standard is the citrate-stabilized gold NPs from the **National Institute of Standards and Technology** (NIST, www.nist.gov) [27].

Safety assessment of NMs engineered for the use in nanomedicine are further discussed and reviewed by El-Ansary and Al-Daihan [29], Chowdhury(regulatory guidance) [30], Fadeel and Garcia-Bennett (inorganic NPs for biomedical applications) [31], and Keck and Muller (classification of NMs) [32].

In the light of the above-mentioned issue, widespread use of different types of NTs/NMs in various biomedical fields suggested their capacity to revolutionize medical diagnostics, imaging, and therapeutics. On the other hand, it concerns their increasing access to many organs in animal and human body, and consequently the potential toxic effects [8] that may be involved in the onset and progression of neurodegeneration [33].

NANOMATERIALS/NANOPARTICLES AND NEURODEGENERATION - BENEFITS AND NEGATIVES

The **central nervous system** (**CNS**) plays a key role in the regulation of neuroendocrine processes. On the other hand, the releasing of neurohormones modulates the activity/stability of different brain regions. **Neurodegenerative diseases** affecting specific neuronal populations may cause neuroendocrine dysfunctions and in turn, these modifications may influence the neurodegenerative process progression [34]. As neuronal changes along with the decline in cognitive abilities may closely correlate with endocrine functions, the implication of endocrine feedback on the CNS has crucial role in neurodegeneration research. It has been demonstrated that reproductive steroids and peptide hormones directly affect synaptic plasticity, neuronal wellbeing, and cognition. Further, it has been shown that metabolic peptide hormones are critical for energy homeostasis, alteration of synaptic strength and intensification of learning and memory [35].

Genesis of neurodegenerative diseases has not been elucidated yet, although, several *genetic factors* have been identified which contribute to their onset and development. *Gender and age-associated endocrine system dysfunction* are thought to be major factors in the onset and progression of neurodegenerative diseases. In addition, an attention has been focused on the other risk factors, including NMs/NPs, which may potentially damage the developing nervous system through common as well as epigenetic mechanisms, resulting in neurodegenerative diseases later in life [36].

NMs/NPs are highly reactive entities, which are able to interact with cells and subcellular structures due to their large surface area compared to their overall mass and small size in highly efficient but insufficiently characterized ways [8]. The use of NMs in the diagnosis and treatment of neurodegenerative diseases implies the capability of NMs to cross the blood-brain barrier (BBB) and enter the CNS in dependence on their physico-chemical properties, composition, and functionalization [37, 38]. The important issues are also the affinity between the drug and the nanocarrier and after all, subsequent removal of the nanodevices/nanocarriers from the brain [33].

A wide variety of NPs has already been described to cross BBB, including TiO_2 NPs, Ag-NPs, Cu-NPs, polymers, lipids, and others [39], that could be potential nanocarriers applied in the treatment of CNS diseases. In recent years, some NP systems have already been designed particularly for treatment of neurodegenerative diseases, *i.e.* chitosan NPs, loaded with tacrine or rivastigmine for the treatment of the Alzheimer`s disease (AD) [40, 41], lactoferrin conjugated PEG-PLGA NPs loaded with urocortin for treatment of Parkinson`s disease [42] or polysorbate-80 coated poly(butyl cyanoacrylate) nanocarriers loaded with nerve growth factor (NGF) for reduction of symptoms of Parkinsonism [43].

Despite exhibiting great potential benefits, several *in vivo* and *in vitro* studies have demonstrated that intentional exposure to NMs/NPs with potential use for diagnostic and therapeutic purposes might induce neurotoxicity effects resulting in neurodegeneration in different CNS regions [44, 45, 42, 46, 47]. For example, Hu *et al.* [42] have demonstrated that mice chronically exposed to TiO_2-NPs *via* intragastric exposure exerted increased levels of caspase-3 and 9, Bax, cytochrome c, ROS production and decreased levels of Bcl-2 in hippocampus, promoting learning and memory deficits. Rats orally exposed to Ag-NPs showed several synaptic structures modifications and degeneration in hippocampus [46]. Cu-NPs administered intraperitoneally induced decreased number of neurons in CA1 (Cornu Ammonis 1) area, increased levels of ROS generation, and apoptosis in rat hippocampus promoting critical cognitive deficits [48]. *In vitro* study on Neuro-2a-cells has shown that treatment with Ag-NPs elicited the deposition of amyloid ß (Aß) plaques and increased expression of amyloid precursor protein (APP), suggesting induction of AD by altering the amyloidogenic pathway [49].

In summary, a number of adverse effects by intentional as well as incidental exposures to NMs/NPs that can contribute to neurodegeneration by inducing mitochondrial dysfunction, redox imbalance and apoptosis, autophagy and impaired lysosomal activity, cytoskeletal damage and vesicle trafficking perturbations, neuroinflammation and microglia activation, changes in neuronal morphology, and cell death, have been described [8, 50].

Furthermore, there is increasing evidence that induction of changes in the expression of genes involved in DNA methylation pathways and global changes in epigenetic marks such as DNA methylation, histone tail modifications, and RNA interference being important processes in embryonal development, cellular differentiation and aging, are also involved in neurodegeneration by the action of NMs/NPs [51 - 53]. Stoccoro and colleagues [52] have summarized the epigenetic effects of NMs/NPs and showed their potential to induce global DNA methylation changes as well as changes of gene-specific methylation patterns, including tumor suppressor genes (APC, p16, RASSF1A, p53), inflammatory genes (iNOS, IFNG, IL4), DNA repair genes (PARP-1), and impaired expression of genes involved in DNA methylation reactions (DNMT1, DNMT3A, DNMT3B, MBD2). In addition, they also have described the changes in acetylation and methylation of histone tails and global or gene-specific alteration of miRNA expression as a consequence of the exposure to NPs [52].

Finally, recent evidence has indicated that *neuroendocrine disrupting effects* by the action of NPs/NMs in dopaminergic, serotoninergic and gonadotropic systems might be relevant to neuropathogenesis and neurodegeneration (reviewed by [54, 55, 33]. In the connection, an interesting finding after prenatal exposure to TiO_2 NPs presented Takahashi and colleagues [56], who have described a deleterious effect on dopaminergic system of developing rat brain. Intranasally instilled TiO_2 NPs, subsequently translocated into murine brain, significantly increased norepinephrine and 5-hydroxytriptamine levels, while levels of dopamine, 3,4-dihydrophenylacetic acid, homovanilic and 5-hydroxyindole acetic acid were decreased [57]. Bai *et al.* [44] have observed that exposure to Cu-NPs has altered levels of dopamine, serotonine and norepinephrine in different regions of CNS. Disruption of gonadotropic system has been described in one-generation reproductive toxicity study, when exposure to Ni-NPs increased the levels of serum luteinizing (LH) and follicle stimulating (FSH) hormones and decreased estradiol (E_2) levels significantly in the female rats, meanwhile, the male rats serum FSH and testosterone (T) were significantly decreased [58] (summarized in Table **2**).

Table 2. Adverse effects of nanoparticles on CNS and reproductive functions.

NPs	Exposure mode	Dose	Test type	Tissue	Reference
TiO_2 NPs (anatase, 6.5 nm, 174.8 m²/g)	Intragastric administration (CD-1: ICR female mice)	5, 10, 50 mg/kg BW; every day for 60 days	↑ Caspase-3, and -9, Bax, cytochrome c, ROS ↓ Bcl-2	Hippocampus	[42]
TiO_2 NPs (anatase, 25-70 nm, 20-25 m²/g)	Subcutaneous administration (pregnant ICR mice)	100 µg; GD 6, 9, 12, 15, 18	↑ DA and their metabolites	Prefrontal area and striatum from 6-wee--old male pups	[56]
TiO_2 NPs (rutile: 80 nm, anatase: 155 nm)	Intranasal administration (CD-1: ICR female mice)	~ 500 g; every other day for 2, 10, 20 30 days	↑ GSH-Px, GST, GSH, SOD, MDA ↑ IL-1β, TNF-α	Brain	[57]

(Table 2) contd.....

NPs	Exposure mode	Dose	Test type	Tissue	Reference
TiO$_2$ NPs (anatase, 25-70 nm, 20-25 m^2/g)	Subcutaneous administration (pregnant Slc: ICR mice)	100 µg; 3, 7 10, 14 days postcoitum	↑ Caspase-3	Olfactory bulb from 6-wee--old male pups	[79]
TiO$_2$ NPs (anatase, 2570 nm, 2025 m^2/g)	Subcutaneous administration (pregnant ICR mice)	100 µg; GD 6, 9, 12, 15	changes in the expression of genes associated with brain development, cell death, response to oxidative stress, and mitochondria in the brain during the perinatal period, and those associated with inflammation and neurotransmitters in the later stage	Brains from male fetuses on ED16; Brains from male pups at PND2, 7, 14, 21	[80]
AgNPs (citrate-stabilized, 10 nm)	Oral administration (adult rats)	2-week period	Synaptic structures modification and degeneration ↓ Synapsin I, synaptophysin, PSD-50 protein	Hippocampus	[46]
AgNPs	ALT cells BV-2 cells N2a cells (murine)	5, 10, 12.5 µg/ml	↑ IL-1β, CXCL13, MARCO, GSS, APP, Aβ plaques ↓ NEP, LDLR	Astrocytes Microglial cells Neuron cells	[49]
nano-CuO (10-70 nm, 15.7 m^2/g)	Intraperitoneal administration (male Wistar rats)	0.5 mg/kg BW/day) for 14 days	↑ ROS, MDA; caspase-3, HNE ↓ SOD, GSH-Px activities	Hippocampus	[48]
CuNPs (23.5 nm, 2.95 m^2/g, 1.7x10^{10}/g)	Intranasal instillation (CD-1: ICR female mice)	1, 10, 40 mg/kg BW	Damages to nerve cells and astrocyte; Changes in DA, DOPAC, HVA; 5-HT, 5-HIAA	Hippocampus, cerebral cortex, cerebellum, striatum at 21 day	[44]
NiNPs	Administration by gavage (female Sprague-Dawley rats)	5, 15, 45 mg/kg BW/day	↑ serum LH, FSH and ↓ serum estradiol in females ↓ serum FSH, testosterone in males	-	[58]

(Table 2) contd.....

NPs	Exposure mode	Dose	Test type	Tissue	Reference
NR-DE (148.86 µg/m³, 1.83 × 10⁶ #/cm³) F-DE (3.10 µg/m³, 2.66 #/cm³)	Inhalation administration (pregnant Fischer rats)	5 h daily from GD1 to GD19	↑ serum LH, corticosterone, estradiol ↓ LHRH mRNA, progesterone	Ovary	[82]

CNS – central nervous system; BW – body weight; GD – gestation day; ED – embryonic day; PND – postnatal day; ROS – reactive oxygen species;GSH-Px – glutathione peroxidase; GST – glutathione--transferase; GSH – reduced glutathione; SOD – superoxidase dismutase, MDA – malondiadehyde; CXCL13 – C-X-C motif chemokine 13; MARCO – macrophage receptor with collagenous structure; GSS – glutathione synthetase; APP – amyloid precursor protein; NEP – neprilysin; LDLR – low-density lipoprotein receptor; HNE – 4-hydroxynonenal; DA – dopamine; DOPAC – 3,4-dihydroxyphenylaceticacid; HVA – homovanillic acid; 5-HT – 5-hydroxytryptamine; 5-HIAA – 5-hydroxyindoleacetic acid; LH – luteinizing hormone; FSH – follicle-stimulating hormone; NR-DE – nanoparticle-rich diesel exhaust; F-DE – filtered diesel exhaust

POTENTIAL NEURODEGENERATIVE MECHANISMS OF GONADO-TROPINS/DEVELOPMENTAL ORIGIN OF NEURODEGENERATION – IMPACT OF NANOPARTICLES

Gonadotropins and their receptors, LHR and FSHR [59], primarily found in the gonads have been shown to be expressed also in the hippocampus, hypothalamus, amygdala, cerebral cortex, cerebellum, brainstem, and spinal cord of rodents and humans as well as in the microglia and neuronal cells of the human parietal cortex [60 - 63]. LHR expression in brain has been shown to be developmentally regulated; in rats, it significantly increased between days 17 and 21 of gestation indicating for its *key role in the neural development* [64]. Additionally, extra-gonadal receptor dependent or independent processes studied with a transgenic LHR knockout having high LH serum suggested a *significant role of LHR in memory and neuroprotection* [65].

Recently, novel effects of gonadotropins have been described in the CNS implicating gonadotropin dysregulation in the development of neurodegenerative diseases [66, 67]. In the connection, potential mechanisms influenced by gonadotropins have been suggested to be one of the candidates in neurode-generation processes of AD such as amyloid ß (Aß) production, inflammation, and cholesterol homeostasis [66]. According to the "amyloid hypothesis" Aß peptide, a product of proteolytic processing of the amyloid precursor protein (APP), represents the toxic principle starting neurodegeneration cascades in AD [68]. Treatment of neuronal cells with gonadotropins at concentrations similar to those observed in menopause altered APP processing increasing Aß secretion [69] mediated probably through the LHR/cAMP signaling cascade [70, 64, 71]. In the connection, the results of *in vivo* experiments might implicate the gonadotropins in an exacerbation of AD neurodegeneration.

Gonadotropins, particularly LH, have been implicated in inflammatory processes, particularly in ovulation and regulation of immune responses during pregnancy [72]. Elevated levels of gonadotropins in postmenopause might overstimulate enzyme 5`lipoxygenase (5`LOX) inflammatory signaling pathway and increase susceptibility of areas expressing high LHR levels (hippocampus, cerebellum) to excitotoxic neurodegeneration and elevated inflammatory response [66].

The important role of gonadotropins in the regulation of gonadal cholesterol, as the substrate for sex hormone synthesis, speaks out for its involvement in the regulation of cholesterol in the CNS. Though the role of gonadotropins in the regulation and dysregulation of neurosteroidogenesis in the brain remain unresolved, it has been suggested that gonadotropin dysregulation may result in neurosteroid hormone imbalance, disrupted cholesterol homeostasis, and upregulation of Aß [73 - 75].

Actually, it has been demonstrated that variations in hormone synthesis throughout the life are crucial for the development of neurodegenerative diseases. Connection to AD due to hypothalamus-pituitary-gonads (HPG) axis dysfunction may be explained by a diminished ability of estrogens to inhibit the release of the hypothalamic gonadotropin releasing hormone (GnRH) [76] and significant secretion of gonadotropins in postmenopausal women [77] (Fig **1**).

In line with developmental origin of adult diseases, it is forewarning the evidence that pre- and post-natal exposure to different risk factors including NMs/NPs may lead to phenotypic diversity and susceptibility to neurodegenerative diseases in later life (reviewed by [36]). Within the context of predisposition to the developmental onset of disease in later life, it is worth noting that the potential toxicity of NMs/NPs in next generation has been evaluated in several studies which have been focused on the developmental toxicity using *in vivo* rodent models [78, 56, 79 - 81]. The obtained experimental data have shown that NPs may move across the placenta into foetus and induce adverse effects Fig. (**2**). Takahashi and colleagues [56] presented interesting finding by the action of prenatal exposure (subcutaneous administration) to TiO_2 NPs. They observed significantly increased levels of dopamine (DA) and DA metabolites in the strata and prefrontal area exposed rats compared to the control animals. The subcutaneous administration of nano-sized TiO_2 to pregnant mice affected genital and cranial nerve systems in male offspring, when TiO_2 NPs were found in the testes as well as brain of 6-week-old male mice [79]. The alteration of gene expression related to mice brain development and function has been observed in the study of Shimizu *et al.* [80] after subcutaneous injection of nano-sized anatase TiO_2 to pregnant mice. Moreover, in fetal as well as pups brains, the altered genes associated with response to oxidative stress, cell death, and mitochondrial activity have been observed [80]. Prenatal exposure to NP-rich diesel exhaust (NRDE-NPs) elicited disruption of gonadotropin system demonstrated by increased LH levels accompanied with decreased progesterone levels [82].

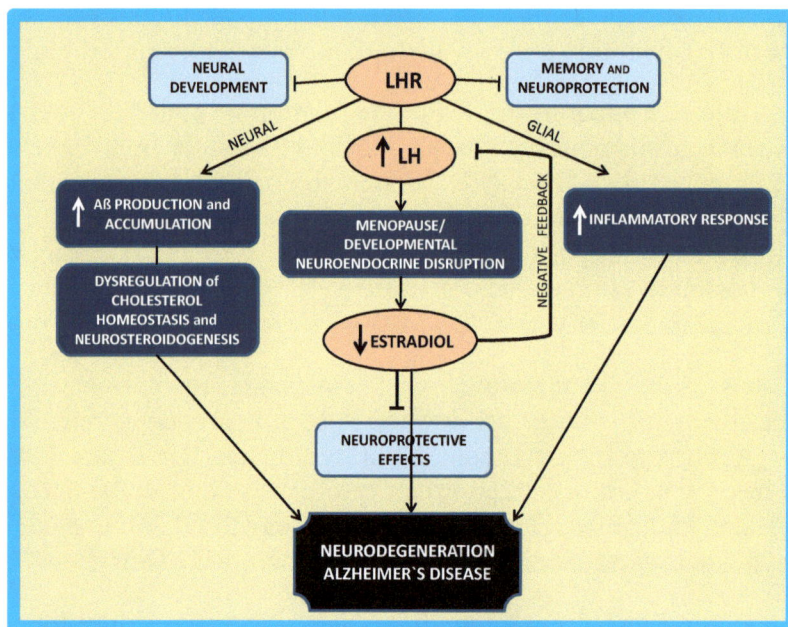

Fig. (1). Potential neurodegenerative mechanisms of luteinizing hormone (LH).
LH receptors (LHR) play key role in neural development as well as in memory and neuroprotection. Estrogens exert neuroprotective effects by prevention of amyloid ß (Aß) accumulation, reduction of Aß production and promoting Aß clearance. Postmenopausal elevated LH levels or LH levels elevated due to developmental neuroendocrine disruption could act through LHR to mediate dysregulation of cholesterol homeostasis and neurosteroidogenesis or stimulate production and accumulation of Aß in neural cells. Moreover, increased levels of LH might also stimulate inflammatory response in glial cells. During reproductive decline or as an outcome of endocrine disruption, gonadal failure to produce sex hormones abolishes negative feedback on the hypothalamus and pituitary and marked increases in LH levels occurs. Activation of putative pathways could converge in neurodegenerative changes typical of Alzheimer disease.

NANO-STRADEGIES IN CNS DRUG DELIVERY:

POLYMERIC NANOPARTICLES FOR THERAPEUTIC USE

Recently, several effective nano-strategies have been developed for the treatment of neurological disorders. In this sense, the polymeric NPs systems for the targeted drug delivery to the brain provide a promising approach due to the possibility to specifically engineer polymeric macromolecules and tune their chemical properties, thus controlling their biocompatibility, improving bioavailability and biodistribution, delivery kinetics and pharmacological activity

[83]. At present, two polymers, polylactide (PLA) and polylactide-co-glycolide (PLGA), are approved by the US Food and Drug Administration (US FDA) for the preparation of NPs, while clinical data evaluating polymer-based NPs as drug carriers are still lacking [84]. Among nanocarriers, polymeric micelles consisting of amphiphilic multi-block copolymers have attracted the interest because of their ability to form a core/shell structure and transport either hydrophilic or hydrophobic molecules [85]. Therefore, systemic administration of polymeric micelles for CNS drug delivery have shown lower efficiency due to their interaction with reticuloendothelial system (RES), particle surface coating with hydrophilic polymers was realized. This popular strategy has allowed preparation of nanoconstruct with minimal surface charge to reduce the opsonisation process, to avoid capture by RES, and to enhance the half-life in blood stream [86]. PEGylation, direct chemical linking of short chains of polyethylene glycol (PEG) to particle surface, is widely used to increase the circulation time in the host body [87]. In addition, improving delivery efficiency of NPs to specific targets of CNS as well as NPs accumulation in CNS (hippocampus, ventricles telencephalon, and raphe nuclei) after intravenous administration has been shown [88].

While number of NMs can induce neurotoxicity, neuroinflammation, and cognitive deficits, polymeric nanoparticles are biocompatible and biodegradable having important characteristics for nanosafety of CNS drug delivery systems [89]. On the other hand, several reports have highlighted the fact of complement activation by PEG and surfactants being cause of serious detrimental effects in the brain [90, 91]. Therefore, surface modifications of polymeric NPs for facilitating their entry into the brain should also take into consideration the safety aspects [8].

Recently, to broad safety profile of PEGylated PLA NPs (copolymer PEG-*b*-PLA [$CH_3O(CH_2CH_2O)_x(COCHCH_3O)_yH$], we have studied neuroendocrine and developmental toxicity by the action of PEG-*b*-PLA NPs after neonatal exposure of female Wistar rats [92]. Our study has demonstrated the adverse effects of short-time neonatal/developmental exposure to polymeric PEG-*b*-PLA NPs on the somatic and pubertal benchmarks and some endpoints of reproductive functions (estrous cyclicity) in the female rats. There are also indications that PEG-*b*-PLA NPs might interfere with the activation and function of the HPG axis and that hormonal effects might play an important role in the nanoreprotoxicity of PEG-*b*-PLA NPs at both central neuroendocrine and gonadal levels. Specifically, the alteration of pituitary LH release to luteinizing hormone releasing hormone (LHRH) stimulation (Table **3**), higher production of progesterone (P4) by the corpora lutea, and prevalence of the diestrus stage of the estrus cycle (reproductive dysfunction) might suggest possible disturbance of central control by a negative feedback mechanism and the pituitary as one of the possible target sites of PEG-*b*-PLA NPs action [92].

Titanium dioxide nanoparticles

Maternal red blood cells

Maternal blood

Mononucleartrophoblast
Syncytiotrophoblast I
Syncytiotrophoblast II
Basement membrane
Fetal endothelium

Placental barrier

Fetal nucleated red blood cells

Fetal blood

Tight junction
Fetal brain barrier endothelial cells
Basement membrane
Astrocyte
Neuron
Microglia

Fetal brain

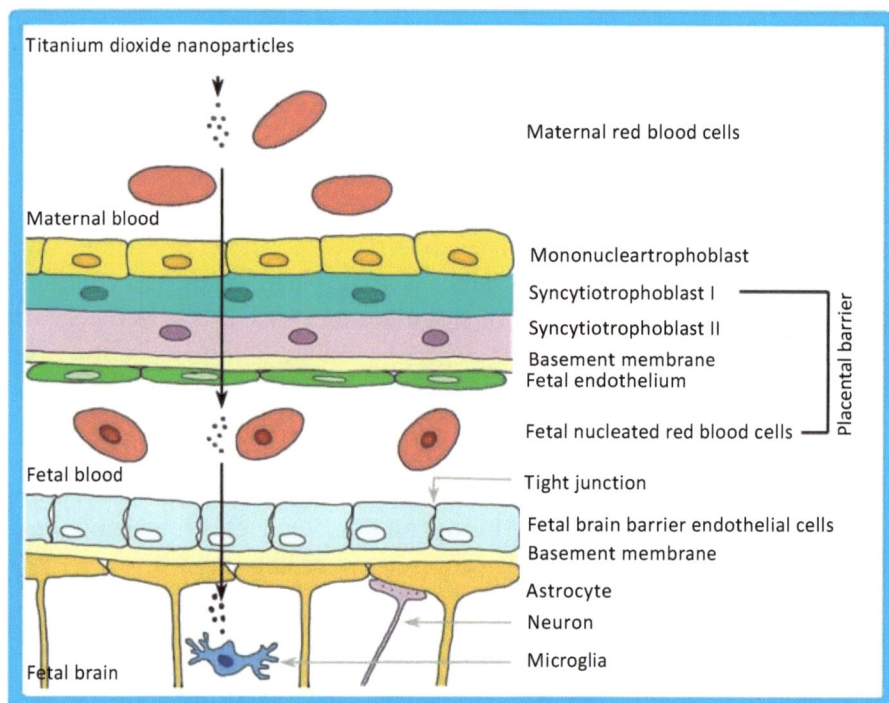

Fig. (2). Transition of TiO_2 across placental and fetal blood brain barrier after mother's exposure to ultrafine TiO_2 during pregnancy (with permission from Rollerova E, Tulinska J, Liskova A, Kuricova M, Kovriznych J, Mlynarcikova A, Kiss A, Scsukova S: Titanium dioxide nanoparticles: some aspects of toxicity/focus on the development. Endocrine Regul 2015; 49: 97-112).

Table 3. Effects of neonatal exposure to polymeric nanoparticle poly(ethylene glycol)-block-polylactide methyl ether (PEG-*b*-PLA) on *in vivo* and *in vitro* basal and LHRH-induced LH secretion.

In vivo	LH levels (ng/ml)				
	PND 15		PND 176		
LHRH (ng/kg b.w.)	0	100	0	100	100
Time (min)	0	15	0	15	30
CONTROL	2.36±0.33	41.23±3.17	1.30±0.17	4.63±0.45	3.15±0.42
PEG-*b*-PLA (20 mg/kg b.w.)	1.57±0.21	50.71±4.06*	1.42±0.16	5.14±0.62	4.79±0.29*

In vivo	LH levels (ng/ml)			
	PND 15		PND 176	
LHRH (M)	0	10^{-7}	0	10^{-7}
Time (min)	0	60	0	60

(Table 3) contd.....

In vivo	LH levels (ng/ml)			
	PND 15		PND 176	
CONTROL	12.03±0.98	23.59±3.58	4.21±0.30	4.91±0.36
PEG-*b*-PLA (20 mg/kg b.w.)	18.50±1.00***	38.38±1.97***	10.53±0.60***	11.53±0.26***

Data are presented as mean ± SEM. Data were analyzed by one-way analysis of variance (ANOVA). *p<0.05; ***p<0.001 compared to CONTROL

Moreover, our *in vitro* study has provided further evidence of the neuroendocrine disrupting effect of neonatal exposure to polymeric PEG-*b*-PLA NPs by examining *in vitro* LHRH-induced LH release from gonadotrophic cells isolated from female rats at two different stages, infantile (PND 15) and adult (PND 176) ones, of the postnatal development (Table **3**). The obtained data have indicated that neonatal exposure of female rats to polymeric PEG-*b*-PLA NPs may induce significant alterations in the basal as well as LHRH-induced responsiveness of pituitary gonadotrophic cells. Moreover, significant increase in the pituitary LH secretion by the action of PEG-*b*-PLA NPs has persisted from the infantile to adult life period [93].

CONCLUDING REMARKS

In conclusion, only few polymeric NPs have already found clinical application [94]. On the other hand, the safety profiles of many polymeric NPs are still unknown. In the connection, polymeric NP-induced neurotoxicity *in vivo*, immunogenicity and a safe route of entry into brain without infliction of any damage have to be thoroughly investigated before their pharmaceutical application [89]. In the light of the above issue, relevant test models should be required in order: i) to assess the role of NMs/NPs in the development and progression of neurodegenerative disease; ii) to investigate the effects of NPs on neurodevelopment upon *in utero* exposure of foetuses [8] or neonatal exposure of pups; or iii) to study neuroendocrine disrupting effects during critical period being crucial for the development of neurodegenerative diseases. With respect to the exposure during pregnancy and early postnatal life, the barrier development, such as placental as well as BBB, should be strictly considered because of the greater possibility for NPs to enter the fetal or neonatal tissues leading to a greater risk of maldevelopments [95].

CONFLICT OF INTEREST

The authors confirm that they have no conflict of interest to declare for this publication.

ACKNOWLEDGEMENT

Experimental work was supported by the Slovak Research and Development Agency under the contract No APVV-0404-11 and the project "Center of excellence of environmental health", ITMS No. 26240120033, based on the supporting operational Research and development program financed from the European Regional Development Fund.

REFERENCES

[1]   European Commission: Communication from the Commission to the European Parliament, the Council and the European Economic and Social Committee - second regulatory review on nanomaterials. 2012. http://eur-lex.europa.eu/LexUriServ/LexUriServ.do?uri=COM:2012:0572:-FIN:en:PDF.

[2]   Oberdörster G, Oberdörster E, Oberdörster J. Nanotoxicology: an emerging discipline evolving from studies of ultrafine particles. Environ Health Perspect 2005; 113(7): 823-39.
[http://dx.doi.org/10.1289/ehp.7339] [PMID: 16002369]

[3]   Oberdörster G, Stone V, Donaldson K. Toxicology of nanoparticles: A historical perspective. Nanotoxicology 2007; 1: 2-25.
[http://dx.doi.org/10.1080/17435390701314761]

[4]   Maynard AD, Aitken RJ, Butz T, *et al.* Safe handling of nanotechnology. Nature 2006; 444(7117): 267-9.
[http://dx.doi.org/10.1038/444267a] [PMID: 17108940]

[5]   Hankin SM, Peters SA, Poland CA, *et al.* Specific advise on fulfilling information requirements for nanomaterials under REACH (RIP-oN 2) - final project report. http://ec.europa.eu/environment/-chemicals/nanotech/pdf/reports_ripon2.pdf.

[6]   Organization for Economic Co-operation and Development:. Preliminary review of OECD test guidelines for their applicability to manufactured nanomaterials. 2009 Jul 10; http://search.oecd.org/officialdocuments/displaydocumentpdf/?doclanguage=en&cote=env/jm/mono%282009%2921.

[7]   Schwirn K, Tietjen L, Beer I. Why are nanomaterials different and how can they be appropriately regulated under REACH? Environmental Sciences Europe 2014; 26: 4.
[http://dx.doi.org/10.1186/2190-4715-26-4]

[8]   Cupaioli FA, Zucca FA, Boraschi D, Zecca L. Engineered nanoparticles. How brain friendly is this new guest? Prog Neurobiol 2014; 119-120: 20-38.
[http://dx.doi.org/10.1016/j.pneurobio.2014.05.002] [PMID: 24820405]

[9]   Juillerat-Jeanneret L, Dusinska M, Fjellsbø LM, Collins AR, Handy RD, Riediker M. Biological impact assessment of nanomaterial used in nanomedicine. introduction to the NanoTEST project. Nanotoxicology 2015; 9 (Suppl. 1): 5-12.
[http://dx.doi.org/10.3109/17435390.2013.826743] [PMID: 23875681]

[10] Jain KK. Applications of nanobiotechnology in clinical diagnostics. Clin Chem 2007; 53(11): 2002-9.
[http://dx.doi.org/10.1373/clinchem.2007.090795] [PMID: 17890442]

[11] Yong KT, Ding H, Roy I, *et al.* Imaging pancreatic cancer using bioconjugated InP quantum dots. ACS Nano 2009; 3(3): 502-10.
[http://dx.doi.org/10.1021/nn8008933] [PMID: 19243145]

[12] Lee JH, Smith MA, Liu W, *et al.* Enhanced stem cell tracking *via* electrostatically assembled fluorescent SPION-peptide complexes. Nanotechnology 2009; 20(35): 355102.
[http://dx.doi.org/10.1088/0957-4484/20/35/355102] [PMID: 19671960]

[13]   Ito A, Shinkai M, Honda H, Kobayashi T. Medical application of functionalized magnetic nanoparticles. J Biosci Bioeng 2005; 100(1): 1-11.
[http://dx.doi.org/10.1263/jbb.100.1] [PMID: 16233845]

[14]   Veiseh O, Gunn JW, Kievit FM, *et al.* Inhibition of tumor-cell invasion with chlorotoxin-bound superparamagnetic nanoparticles. Small 2009; 5(2): 256-64.
[http://dx.doi.org/10.1002/smll.200800646] [PMID: 19089837]

[15]   Gimi B, Leong T, Gu Z, *et al.* Self-assembled 3D radiofrequency-shielded (RS) containers for cell encapsulation. Biomed Microdevices 2005; 7: 341-5.
[http://dx.doi.org/10.1007/s10544-005-6076-9] [PMID: 16404512]

[16]   Jain KK. Role of nanobiotechnology in developing personalized medicine for cancer. Technol Cancer Res Treat 2005; 4(6): 645-50.
[http://dx.doi.org/10.1177/153303460500400608] [PMID: 16292884]

[17]   Jain KK. Nanomedicine: application of nanobiotechnology in medical practice. Med Princ Pract 2008; 17(2): 89-101.
[http://dx.doi.org/10.1159/000112961] [PMID: 18287791]

[18]   Nguyen TD, Tseng HR, Celestre PC, *et al.* A reversible molecular valve. Proc Natl Acad Sci USA 2005; 102(29): 10029-34.
[http://dx.doi.org/10.1073/pnas.0504109102] [PMID: 16006520]

[19]   Partlow KC, Chen J, Brant JA, *et al.* 19F magnetic resonance imaging for stem/progenitor cell tracking with multiple unique perfluorocarbon nanobeacons. FASEB J 2007; 21(8): 1647-54.
[http://dx.doi.org/10.1096/fj.06-6505com] [PMID: 17284484]

[20]   Chowdhury EH. pH-sensitive nano-crystals of carbonate apatite for smart and cell-specific transgene delivery. Expert Opin Drug Deliv 2007; 4(3): 193-6.
[http://dx.doi.org/10.1517/17425247.4.3.193] [PMID: 17489648]

[21]   Schütz CA, Juillerat-Jeanneret L, Mueller H, Lynch I, Riediker M. Therapeutic nanoparticles in clinics and under clinical evaluation. Nanomedicine (Lond) 2013; 8(3): 449-67.
[http://dx.doi.org/10.2217/nnm.13.8] [PMID: 23477336]

[22]   Gao X, Tao W, Lu W, *et al.* Lectin-conjugated PEG-PLA nanoparticles: preparation and brain delivery after intranasal administration. Biomaterials 2006; 27(18): 3482-90.
[http://dx.doi.org/10.1016/j.biomaterials.2006.01.038] [PMID: 16510178]

[23]   Yildirimer L, Thanh NT, Loizidou M, Seifalian AM. Toxicological considerations of clinically applicable nanoparitlces. Nano Today 2011; 6: 585-607.
[http://dx.doi.org/10.1016/j.nantod.2011.10.001] [PMID: 23293661]

[24]   Godin B, Driessen WH, Proneth B, *et al.* An integrated approach for the rational design of nanovectors for biomedical imaging and therapy. Adv Genet 2010; 69: 31-64.
[http://dx.doi.org/10.1016/S0065-2660(10)69009-8] [PMID: 20807601]

[25]   Drobne D. Nanotoxicology for safe and sustainable nanotechnology. Arh Hig Rada Toksikol 2007; 58(4): 471-8.
[http://dx.doi.org/10.2478/v10004-007-0040-4] [PMID: 18063532]

[26]   Teow Y, Asharani PV, Hande MP, Valiyaveettil S. Health impact and safety of engineered nanomaterials. Chem Commun (Camb) 2011; 47(25): 7025-38.
[http://dx.doi.org/10.1039/c0cc05271j] [PMID: 21479319]

[27]   Nyström AM, Fadeel B. Safety assessment of nanomaterials: implications for nanomedicine. J Control Release 2012; 161(2): 403-8.
[http://dx.doi.org/10.1016/j.jconrel.2012.01.027] [PMID: 22306428]

[28]   Kohane DS, Langer R. Biocompatibility and drug delivery systems. Chem Sci (Camb) 2010; 1: 441-6.
[http://dx.doi.org/10.1039/C0SC00203H]

[29] El-Ansary A, Al-Daihan S. On toxicity of the therapeutically used nanoparticles: An overview J Toxicol 2009; 2009: 754810.

[30] Chowdhury N. Regulation of nanomedicines in the EU: distilling lessons from the pediatric and the advanced therapy medicinal products approaches. Nanomedicine (Lond) 2010; 5(1): 135-42.
[http://dx.doi.org/10.2217/nnm.09.91] [PMID: 20025470]

[31] Fadeel B, Garcia-Bennett AE. Better safe than sorry: Understanding the toxicological properties of inorganic nanoparticles manufactured for biomedical applications. Adv Drug Deliv Rev 2010; 62(3): 362-74.
[http://dx.doi.org/10.1016/j.addr.2009.11.008] [PMID: 19900497]

[32] Keck CM, Müller RH. Nanotoxicological classification system (NCS) - a guide for the risk-benefit assessment of nanoparticulate drug delivery systems. Eur J Pharm Biopharm 2013; 84(3): 445-8.
[http://dx.doi.org/10.1016/j.ejpb.2013.01.001] [PMID: 23333302]

[33] Migliore L, Uboldi C, Di Bucchianico S, Coppedè F. Nanomaterials and neurodegeneration. Environ Mol Mutagen 2015; 56(2): 149-70.
[http://dx.doi.org/10.1002/em.21931] [PMID: 25627719]

[34] González De Aguilar JL, René F, Dupuis L, Loeffler JP. Neuroendocrinology of neurodegenerative diseases. Insights from transgenic mouse models. Neuroendocrinology 2003; 78(5): 244-52.
[http://dx.doi.org/10.1159/000074445] [PMID: 14657605]

[35] Palm R, Ayala-Fontanez N, Garcia Y, Lee HG, Smith MA, Casadesus G. Neuroendocrinology-based therapy for Alzheimers disease. Biofactors 2012; 38(2): 123-32.
[http://dx.doi.org/10.1002/biof.1011] [PMID: 22438197]

[36] Chin-Chan M, Navarro-Yepes J, Quintanilla-Vega B. Environmental pollutants as risk factors for neurodegenerative disorders: Alzheimer and Parkinson diseases. Front Cell Neurosci 2015; 9: 124.
[http://dx.doi.org/10.3389/fncel.2015.00124] [PMID: 25914621]

[37] Kreuter J. Influence of the surface properties on nanoparticle-mediated transport of drugs to the brain. J Nanosci Nanotechnol 2004; 4(5): 484-8.
[http://dx.doi.org/10.1166/jnn.2003.077] [PMID: 15503433]

[38] Tiwari SB, Amiji MM. A review of nanocarrier-based CNS delivery systems. Curr Drug Deliv 2006; 3(2): 219-32.
[http://dx.doi.org/10.2174/156720106776359230] [PMID: 16611008]

[39] Sharma HS, Sharma A. Neurotoxicity of engineered nanoparticles from metals. CNS Neurol Disord Drug Targets 2012; 11(1): 65-80.
[http://dx.doi.org/10.2174/187152712799960817] [PMID: 22229317]

[40] Wilson B, Samanta MK, Santhi K, Kumar KP, Ramasamy M, Suresh B. Chitosan nanoparticles as a new delivery system for the anti-Alzheimer drug tacrine. Nanomedicine (Lond) 2010; 6(1): 144-52.
[PMID: 19446656]

[41] Wilson B, Samanta MK, Muthu MS, Vinothapooshan G. Design and evaluation of chitosan nanoparticles as novel drug carrier for the delivery of rivastigmine to treat Alzheimers disease. Ther Deliv 2011; 2(5): 599-609.
[http://dx.doi.org/10.4155/tde.11.21] [PMID: 22833977]

[42] Hu R, Zheng L, Zhang T, *et al.* Molecular mechanism of hippocampal apoptosis of mice following exposure to titanium dioxide nanoparticles. J Hazard Mater 2011; 191(1-3): 32-40.
[http://dx.doi.org/10.1016/j.jhazmat.2011.04.027] [PMID: 21570177]

[43] Kurakhmaeva KB, Djindjikhashvili IA, Petrov VE, *et al.* Brain targeting of nerve growth factor using poly(butyl cyanoacrylate) nanoparticles. J Drug Target 2009; 17(8): 564-74.
[http://dx.doi.org/10.1080/10611860903112842] [PMID: 19694610]

[44]   Bai R, Zhang L, Liu Y, *et al.* Integrated analytical techniques with high sensitivity for studying brain translocation and potential impairment induced by intranasally instilled copper nanoparticles. Toxicol Lett 2014; 226(1): 70-80.
        [http://dx.doi.org/10.1016/j.toxlet.2014.01.041] [PMID: 24503010]

[45]   Boyes WK, Chen R, Chen C, Yokel RA. The neurotoxic potential of engineered nanomaterials. Neurotoxicology 2012; 33(4): 902-10.
        [http://dx.doi.org/10.1016/j.neuro.2011.12.013] [PMID: 22198707]

[46]   Skalska J, Frontczak-Baniewicz M, Strużyńska L. Synaptic degeneration in rat brain after prolonged oral exposure to silver nanoparticles. Neurotoxicology 2015; 46: 145-54.
        [http://dx.doi.org/10.1016/j.neuro.2014.11.002] [PMID: 25447321]

[47]   Xia H, Gao X, Gu G, *et al.* Penetratin-functionalized PEG-PLA nanoparticles for brain drug delivery. Int J Pharm 2012; 436(1-2): 840-50.
        [http://dx.doi.org/10.1016/j.ijpharm.2012.07.029] [PMID: 22841849]

[48]   An L, Liu S, Yang Z, Zhang T. Cognitive impairment in rats induced by nano-CuO and its possible mechanisms. Toxicol Lett 2012; 213(2): 220-7.
        [http://dx.doi.org/10.1016/j.toxlet.2012.07.007] [PMID: 22820425]

[49]   Huang CL, Hsiao IL, Lin HC, Wang CF, Huang YJ, Chuang CY. Silver nanoparticles affect on gene expression of inflammatory and neurodegenerative responses in mouse brain neural cells. Environ Res 2015; 136: 253-63.
        [http://dx.doi.org/10.1016/j.envres.2014.11.006] [PMID: 25460644]

[50]   Iqbal A, Ahmad I, Khalid MH, Nawaz MS, Gan SH, Kamal MA. Nanoneurotoxicity to nanoneuroprotection using biological and computational approaches. J Environ Sci Health C Environ Carcinog Ecotoxicol Rev 2013; 31(3): 256-84.
        [http://dx.doi.org/10.1080/10590501.2013.829706] [PMID: 24024521]

[51]   Marques SC, Oliveira CR, Pereira CM, Outeiro TF. Epigenetics in neurodegeneration: a new layer of complexity. Prog Neuropsychopharmacol Biol Psychiatry 2011; 35(2): 348-55.
        [http://dx.doi.org/10.1016/j.pnpbp.2010.08.008] [PMID: 20736041]

[52]   Stoccoro A, Karlsson HL, Coppedè F, Migliore L. Epigenetic effects of nano-sized materials. Toxicology 2013; 313(1): 3-14.
        [http://dx.doi.org/10.1016/j.tox.2012.12.002] [PMID: 23238276]

[53]   Smolkova B, El Yamani N, Collins AR, Gutleb AC, Dusinska M. Nanoparticles in food. Epigenetic changes induced by nanomaterials and possible impact on health. Food Chem Toxicol 2015; 77: 64-73.
        [http://dx.doi.org/10.1016/j.fct.2014.12.015] [PMID: 25554528]

[54]   Hu YL, Gao JQ. Potential neurotoxicity of nanoparticles. Int J Pharm 2010; 394(1-2): 115-21.
        [http://dx.doi.org/10.1016/j.ijpharm.2010.04.026] [PMID: 20433914]

[55]   Iavicoli I, Fontana L, Leso V, Bergamaschi A. The effects of nanomaterials as endocrine disruptors. Int J Mol Sci 2013; 14(8): 16732-801.
        [http://dx.doi.org/10.3390/ijms140816732] [PMID: 23949635]

[56]   Takahashi Y, Mizuo K, Shinkai Y, Oshio S, Takeda K. Prenatal exposure to titanium dioxide nanoparticles increases dopamine levels in the prefrontal cortex and neostriatum of mice. J Toxicol Sci 2010; 35(5): 749-56.
        [http://dx.doi.org/10.2131/jts.35.749] [PMID: 20930469]

[57]   Wang J, Liu Y, Jiao F, *et al.* Time-dependent translocation and potential impairment on central nervous system by intranasally instilled TiO(2) nanoparticles. Toxicology 2008; 254(1-2): 82-90.
        [http://dx.doi.org/10.1016/j.tox.2008.09.014] [PMID: 18929619]

[58]   Kong L, Tang M, Zhang T, *et al.* Nickel nanoparticles exposure and reproductive toxicity in healthy adult rats. Int J Mol Sci 2014; 15(11): 21253-69.

[http://dx.doi.org/10.3390/ijms151121253] [PMID: 25407529]

[59]   McFarland KC, Sprengel R, Phillips HS, *et al.* Lutropin-choriogonadotropin receptor: an unusual member of the G protein-coupled receptor family. Science 1989; 245(4917): 494-9.
[http://dx.doi.org/10.1126/science.2502842] [PMID: 2502842]

[60]   Bukovsky A, Indrapichate K, Fujiwara H, *et al.* Multiple luteinizing hormone receptor (LHR) protein variants, interspecies reactivity of anti-LHR mAb clone 3B5, subcellular localization of LHR in human placenta, pelvic floor and brain, and possible role for LHR in the development of abnormal pregnancy, pelvic floor disorders and Alzheimers disease. Reprod Biol Endocrinol 2003; 1: 46.
[http://dx.doi.org/10.1186/1477-7827-1-46] [PMID: 12816543]

[61]   Lei ZM, Rao CV, Kornyei JL, Licht P, Hiatt ES. Novel expression of human chorionic gonadotropin/luteinizing hormone receptor gene in brain. Endocrinology 1993; 132(5): 2262-70.
[PMID: 8477671]

[62]   Lukacs H, Hiatt ES, Lei ZM, Rao CV. Peripheral and intracerebroventricular administration of human chorionic gonadotropin alters several hippocampus-associated behaviors in cycling female rats. Horm Behav 1995; 29(1): 42-58.
[http://dx.doi.org/10.1006/hbeh.1995.1004] [PMID: 7782062]

[63]   Rao SC, Li X, Rao ChV, Magnuson DS. Human chorionic gonadotropin/luteinizing hormone receptor expression in the adult rat spinal cord. Neurosci Lett 2003; 336(3): 135-8.
[http://dx.doi.org/10.1016/S0304-3940(02)01157-6] [PMID: 12505611]

[64]   al-Hader AA, Tao YX, Lei ZM, Rao CV. Fetal rat brains contain luteinizing hormone/human chorionic gonadotropin receptors. Early Pregnancy 1997; 3(4): 323-9.
[PMID: 10086084]

[65]   Casadesus G, Milliken EL, Webber KM, *et al.* Increases in luteinizing hormone are associated with declines in cognitive performance. Mol Cell Endocrinol 2007; 269(1-2): 107-11.
[http://dx.doi.org/10.1016/j.mce.2006.06.013] [PMID: 17376589]

[66]   Barron AM, Verdile G, Martins RN. The role of gonadotropins in Alzheimers disesase. Endocrine 2006; 29: 257-69.
[http://dx.doi.org/10.1385/ENDO:29:2:257] [PMID: 16785601]

[67]   Palm R, Ayala-Fontanez N, Garcia Y, Lee H-G, Smith MA, Casadesus G. Neuroendocrinology-based therapy for Alzheimers disease. Biofactors 2012; 38(2): 123-32.
[http://dx.doi.org/10.1002/biof.1011] [PMID: 22438197]

[68]   Boyt AA, Suzuky T, Hone E, Gnejc A, Martins RM. The structure and multifaceted function of the amyloid precursor protein. Clin Biochem Rev 2000; 21: 22-41.

[69]   Bowen RL, Verdile G, Liu T, *et al.* Luteinizing hormone, a reproductive regulator that modulates the processing of amyloid-beta precursor protein and amyloid-beta deposition. J Biol Chem 2004; 279(19): 20539-45.
[http://dx.doi.org/10.1074/jbc.M311993200] [PMID: 14871891]

[70]   Su Y, Ryder J, Ni B. Inhibition of Abeta production and APP maturation by a specific PKA inhibitor. FEBS Lett 2003; 546(2-3): 407-10.
[http://dx.doi.org/10.1016/S0014-5793(03)00645-8] [PMID: 12832078]

[71]   Zhang W, Lei ZM, Rao CV. Immortalized hippocampal cells contain functional luteinizing hormone/human chorionic gonadotropin receptors. Life Sci 1999; 65(20): 2083-98.
[http://dx.doi.org/10.1016/S0024-3205(99)00474-9] [PMID: 10579462]

[72]   Shirai F, Kawaguchi M, Yutsudo M, Dohi Y. Human peripheral blood polymorphonuclear leukocytes at the ovulatory period are in an activated state. Mol Cell Endocrinol 2002; 196(1-2): 21-8.
[http://dx.doi.org/10.1016/S0303-7207(02)00228-9] [PMID: 12385822]

[73]   Frears ER, Stephens DJ, Walters CE, Davies H, Austen BM. The role of cholesterol in the biosynthesis of beta-amyloid. Neuroreport 1999; 10(8): 1699-705.
[http://dx.doi.org/10.1097/00001756-199906030-00014] [PMID: 10501560]

[74]   Paganini-Hill A, Dworsky R, Krauss RM. Hormone replacement therapy, hormone levels, and lipoprotein cholesterol concentrations in elderly women. Am J Obstet Gynecol 1996; 174(3): 897-902.
[http://dx.doi.org/10.1016/S0002-9378(96)70322-8] [PMID: 8633665]

[75]   Simons M, Keller P, De Strooper B, Beyreuther K, Dotti CG, Simons K. Cholesterol depletion inhibits the generation of beta-amyloid in hippocampal neurons. Proc Natl Acad Sci USA 1998; 95(11): 6460-4.
[http://dx.doi.org/10.1073/pnas.95.11.6460] [PMID: 9600988]

[76]   Blair JA, McGee H, Bhatta S, Palm R, Casadesus G. Hypothalamic-pituitary-gonadal axis involvement in learning and memory and Alzheimers disease: more than just estrogen. Front Endocrinol (Lausanne) 2015; 6: 45.
[http://dx.doi.org/10.3389/fendo.2015.00045] [PMID: 25859241]

[77]   Reyes FI, Winter JS, Faiman C. Pituitary-ovarian relationships preceding the menopause. I. A cross-sectional study of serum follicle-stimulating hormone, luteinizing hormone, prolactin, estradiol, and progesterone levels. Am J Obstet Gynecol 1977; 129(5): 557-64.
[PMID: 910845]

[78]   Hougaard KS, Jackson P, Jensen KA, *et al.* Effects of prenatal exposure to surface-coated nanosized titanium dioxide (UV-Titan). A study in mice. Part Fibre Toxicol 2010; 7: 16.
[http://dx.doi.org/10.1186/1743-8977-7-16] [PMID: 20546558]

[79]   Takeda K, Suzuki K, Ishihara A, *et al.* Nanoparticles transferred from pregnant mice to their offspring can damage the genital and cranial nerve systems. J Health Sci 2009; 55: 95-102.
[http://dx.doi.org/10.1248/jhs.55.95]

[80]   Shimizu M, Tainaka H, Oba T, Mizuo K, Umezawa M, Takeda K. Maternal exposure to nanoparticulate titanium dioxide during the prenatal period alters gene expression related to brain development in the mouse. Part Fibre Toxicol 2009; 6: 20.
[http://dx.doi.org/10.1186/1743-8977-6-20] [PMID: 19640265]

[81]   Yamashita K, Yoshioka Y, Higashisaka K, *et al.* Silica and titanium dioxide nanoparticles cause pregnancy complications in mice. Nat Nanotechnol 2011; 6(5): 321-8.
[http://dx.doi.org/10.1038/nnano.2011.41] [PMID: 21460826]

[82]   Li C, Li X, Suzuki AK, *et al.* Effects of exposure to nanoparticle-rich diesel exhaust on pregnancy in rats. J Reprod Dev 2013; 59(2): 145-50.
[http://dx.doi.org/10.1262/jrd.2012-145] [PMID: 23257834]

[83]   Gagliardi M, Bardi G, Bifone A. Polymeric nanocarriers for controlled and enhanced delivery of therapeutic agents to the CNS. Ther Deliv 2012; 3(7): 875-87.
[http://dx.doi.org/10.4155/tde.12.55] [PMID: 22900468]

[84]   Costantino L, Boraschi D. Is there a clinical future for polymeric nanoparticles as brain-targeting drug delivery agents? Drug Discov Today 2012; 17(7-8): 367-78.
[http://dx.doi.org/10.1016/j.drudis.2011.10.028] [PMID: 22094246]

[85]   Jones M, Leroux J. Polymeric micelles - a new generation of colloidal drug carriers. Eur J Pharm Biopharm 1999; 48(2): 101-11.
[http://dx.doi.org/10.1016/S0939-6411(99)00039-9] [PMID: 10469928]

[86]   Ricci M, Blasi P, Giovagnoli S, Rossi C. Delivering drugs to the central nervous system: a medicinal chemistry or a pharmaceutical technology issue? Curr Med Chem 2006; 13(15): 1757-75.
[http://dx.doi.org/10.2174/092986706777452461] [PMID: 16787219]

[87] Soppimath KS, Aminabhavi TM, Kulkarni AR, Rudzinski WE. Biodegradable polymeric nanoparticles as drug delivery devices. J Control Release 2001; 70(1-2): 1-20.
[http://dx.doi.org/10.1016/S0168-3659(00)00339-4] [PMID: 11166403]

[88] Xia H, Gao X, Gu G, *et al.* Penetratin-functionalized PEG-PLA nanoparticles for brain drug delivery. Int J Pharm 2012; 436(1-2): 840-50.
[http://dx.doi.org/10.1016/j.ijpharm.2012.07.029] [PMID: 22841849]

[89] Leite PE, Pereira MR, Granjeiro JM. Hazard effects of nanoparticles in central nervous system: Searching for biocompatible nanomaterials for drug delivery. Toxicol In Vitro 2015; 29(7): 1653-60.
[http://dx.doi.org/10.1016/j.tiv.2015.06.023] [PMID: 26116398]

[90] Andersen AJ, Windschiegl B, Ilbasmis-Tamer S, *et al.* Complement activation by PEG-functionalized multi-walled carbon nanotubes is independent of PEG molecular mass and surface density. Nanomedicine (Lond) 2013; 9(4): 469-73.
[PMID: 23434678]

[91] Moghimi SM, Andersen AJ, Hashemi SH, *et al.* Complement activation cascade triggered by PEG-PL engineered nanomedicines and carbon nanotubes: the challenges ahead. J Control Release 2010; 146(2): 175-81.
[http://dx.doi.org/10.1016/j.jconrel.2010.04.003] [PMID: 20388529]

[92] Rollerova E, Jurcovicova J, Mlynarcikova A, *et al.* Delayed adverse effects of neonatal exposure to polymeric nanoparticle poly(ethylene glycol)-block-polylactide methyl ether on hypothalamic-pituitary-ovarian axis development and function in Wistar rats. Reprod Toxicol 2015; 57: 165-75.
[http://dx.doi.org/10.1016/j.reprotox.2015.07.072] [PMID: 26193689]

[93] Scsukova S, Mlynarcikova A, Kiss A, Rollerova E. Effect of polymeric nanoparticle poly(ethylene glycol)-block-poly(lactic acid) (PEG-b-PLA) on *in vitro* luteinizing hormone release from anterior pituitary cells of infantile and adult female rats. Neuroendocrinol Lett 2015; 36 (Suppl. 1): 88-94.
[PMID: 26757115]

[94] Alexis F, Pridgen EM, Langer R, Farokhzad OC. Nanoparticle Technologies for cancer therapy. In: Schräter-Korting M, Ed. Drug Delivery. Springer-Verlag Berlin Heidelberg 2010; 197: pp. 55-86.Handbook of Experimental Pharmacology
[http://dx.doi.org/10.1007/978-3-642-00477-3_2]

[95] Kulvietis V, Zalgeviciene V, Didziapetriene J, Rotomskis R. Transport of nanoparticles through the placental barrier. Tohoku J Exp Med 2011; 225(4): 225-34.
[http://dx.doi.org/10.1620/tjem.225.225] [PMID: 22052087]

SUBJECT INDEX

www.ingramcontent.com/pod-product-compliance
Lightning Source LLC
Chambersburg PA
CBHW041725210326
41598CB00008B/784